# Fractional Differential Equations

# Fractional Differential Equations: Theory, Methods and Applications

Special Issue Editors

**Juan J. Nieto**

**Rosana Rodríguez-López**

MDPI • Basel • Beijing • Wuhan • Barcelona • Belgrade

*Special Issue Editors*
Juan J. Nieto
Universidade de Santiago de Compostela
Spain

Rosana Rodríguez López
Universidade de Santiago de Compostela
Spain

*Editorial Office*
MDPI
St. Alban-Anlage 66
4052 Basel, Switzerland

This is a reprint of articles from the Special Issue published online in the open access journal *Symmetry* (ISSN 2073-8994) from 2018 to 2019 (available at: https://www.mdpi.com/journal/symmetry/special_issues/Fractional_Differential_Equations_Theory_Methods_Applications).

For citation purposes, cite each article independently as indicated on the article page online and as indicated below:

LastName, A.A.; LastName, B.B.; LastName, C.C. Article Title. *Journal Name* **Year**, *Article Number*, Page Range.

**ISBN 978-3-03921-732-8 (Pbk)**
**ISBN 978-3-03921-733-5 (PDF)**

# Contents

# About the Special Issue Editors

**Juan J. Nieto** Full Professor of Mathematical Analysis at the University of Santiago de Compostela and coordinator of the research group on differential equations. He has been a Fulbright Scholar at the University of Texas. His main interests in research and divulgation are the study of dynamical systems, mathematical modeling, and applications to biomedical issues. He has been Editor of the journal *Nonlinear Analysis: Real World Applications* and is currently Editor, among others, of the *Journal of Mathematical Analysis and Applications, Entropy*, and the *International Journal of Biomathematics*, and a member of the editorial board of the *Journal of the Franklin Institute*. He appears in the latest lists, published by Clarivate Analytics, of Highly Cited Researchers.

**Rosana Rodríguez-López** Professor of Mathematical Analysis at the University of Santiago de Compostela and a member of the research group on nonlinear differential equations. She has been a beneficiary of a research staff training grant (FPI fellowship) at USC. She is presently the Vice-Dean of the Faculty of Mathematics. Her research interests are focused on the study of models based on differential equations and the properties of their solutions, in order to predict the behavior of processes and the evolution of real phenomena. She has also participated in activities about the divulgation of mathematics for high school students. She has been the Head of the Mathematical Analysis Department at USC. She is a member of the Editorial Board of the journal *Mathematical Problems in Engineering*, and appears in the 2014 and 2015 editions of the list, published by Thomson Reuters, and in 2016 and 2017, published by Clarivate Analytics, of Highly Cited Researchers.

# Preface to "Fractional Differential Equations: Theory, Methods and Applications"

A timely topic in mathematics and its applications is the theory of differential equations of fractional order. Many real-world problems lead to such mathematical models. The contributions in this book study some problems from different fractional calculus approaches ranging from the classical Riemann–Liouville and Caputo classical fractional calculus to the most recent ones such as q-difference calculus or Fabrizio–Caputo–Losada–Nieto fractional calculus. Some of the real-world problems considered are Burgers equation, thermostat model, Navier–Stokes equations, or Kirchhoff–Schrödinger-type equations. Therefore, different approaches and techniques have been proposed to model these types of problems. This book is a collection of the papers published in the journal Symmetry within one Special Issue: ''Fractional Differential Equations: Theory, Methods, and Applications''. The book consists of eleven contributions.

**Juan J. Nieto, Rosana Rodríguez-López**
*Special Issue Editors*

*symmetry*

MDPI

*Article*

# Global Mittag—Leffler Synchronization for Neural Networks Modeled by Impulsive Caputo Fractional Differential Equations with Distributed Delays

**Ravi Agarwal [1,2], Snezhana Hristova [3,*] and Donal O'Regan [4]**

1 Department of Mathematics, Texas A&M University-Kingsville, Kingsville, TX 78363, USA; agarwal@tamuk.edu
2 Distinguished University Professor of Mathematics, Florida Institute of Technology, Melbourne, FL 32901, USA
3 Department of Applied Mathematics and Modeling, University of Plovdiv, Tzar Asen 24, 4000 Plovdiv, Bulgaria
4 School of Mathematics, Statistics and Applied Mathematics, National University of Ireland, H91 CF50 Galway, Ireland; donal.oregan@nuigalway.ie
* Correspondence: snehri@gmail.com

Received: 17 September 2018; Accepted: 2 October 2018; Published: 10 October 2018

**Abstract:** The synchronization problem for impulsive fractional-order neural networks with both time-varying bounded and distributed delays is studied. We study the case when the neural networks and the fractional derivatives of all neurons depend significantly on the moments of impulses and we consider both the cases of state coupling controllers and output coupling controllers. The fractional generalization of the Razumikhin method and Lyapunov functions is applied. Initially, a brief overview of the basic fractional derivatives of Lyapunov functions used in the literature is given. Some sufficient conditions are derived to realize the global Mittag–Leffler synchronization of impulsive fractional-order neural networks. Our results are illustrated with examples.

**Keywords:** fractional-order neural networks; delays; distributed delays; impulses; Mittag–Leffler synchronization; Lyapunov functions; Razumikhin method

## 1. Introduction

Over the last few decades, fractional differential equations have gained considerable importance and attention due to their applications in science and engineering, i.e., in control, in stellar interiors, star clusters [1], in electrochemistry, in viscoelasticity [2] and in optics [3]. For example, the control of mechanical systems is currently one of the most active fields of research and the use of fractional order calculus increases the flexibility of controlling any system from a point to a space. Applications of fractional quantum mechanics cover dynamics of a free particle and a new representation for a free particle quantum mechanical kernel (see, for example, [4]).

The stability of fractional order systems is quite a recent topic (see, for example, Ref. [5] for the Ulam–Hyers–Mittag–Leffler stability of fractional-order delay differential equations, Ref. [6] for the Mittag–Leffler stability of impulsive fractional neural network, Ref. [7] for the Mittag–Leffler stability of fractional systems, Ref. [8] for the Mittag–Leffler stability for fractional nonlinear systems with delay, and Ref. [9] for the Mittag–Leffler stability of nonlinear fractional systems with impulses). One of the most useful approaches in studying stability for nonlinear fractional differential equations is the Lyapunov approach. Its application to fractional differential equations is connected with several difficulties. One of the main difficulties is connected with the appropriate definition of derivative of Lyapunov functions among thedifferential equations of fractional order. Impulsive differential equations arise in real world problems to describe the dynamics of processes in which sudden, discontinuous jumps occur.

Most research on the synchronization of delayed neural networks has been restricted to the case of discrete delays (see, for example, [10]) Since a neural network usually has a spatial nature due to the presence of an amount of parallel pathways of a variety of axis sizes and lengths, it is desirable to model them by introducing distributed delays. Note in [11] that both time-varying delays and distributed time delays are taken into account in studying fractional neural networks with impulses and constant strengths between two units. In all models of neural networks, one considers the case of constant rate with which the $i$-th neuron resets its potential to the resting state in isolation, and the constant synaptic connection strength of the $i$-th neuron to the $j$-th neuron (see, for example, [10]). In our paper, we consider the general case of time varying coefficients in the model that allows more appropriate modeling of the connections between the neurons. These more complicated mathematical equations lead to an application of new types of fractional derivatives of Lyapunov functions and new stability results.

In this paper, Caputo fractional delay differential equations with impulses and two types of delays-variable in time and distributed ones are studied. Some results for piecewise continuous Lyapunov functions based on the Razumikhin method are obtained. Appropriate derivatives of Lyapunov functions among the studied fractional equations are used. Our results are applied to study the synchronization of neural networks with Caputo fractional derivatives, variable delays, distributed delays, and impulses. We study the case when the lower limit of the fractional derivative is changing after each impulsive time. To the best of our knowledge, this is the first model of neural networks of this type studied in the literature. Additionally, we study the general case of variables in time strengths of the $j$-th unit on the $i$-th unit and nonlinear impulsive functions. Both the cases of state coupling controllers and output coupling controllers are considered. Our sufficient conditions naturally depend significantly on the fractional order of the model (compare with sufficient conditions in [11,12]).

## 2. Impulses in Fractional Delay Differential Equations

Let a sequence $\{t_k\}_{k=1}^{\infty}\ :\ 0 \leq t_{k-1} < t_k \leq t_{k+1},\ \lim_{k\to\infty} t_k = \infty$ be given. Let $t_0 \neq t_k$, $k = 1, 2, \ldots$ be the given initial time and $r > 0$. Without loss of generality we can assume $t_0 \in [0, t_1)$. Let $E = C([-r, 0], \mathbb{R}^n)$ with $||\phi||_0 = \max_{s\in[-r,0]} ||\phi(s)||$ for $\phi \in E$, and $||.||$ is a norm in $\mathbb{R}^n$.

In many applications in science and engineering, the fractional order $q$ is often less than 1, so we restrict $q \in (0, 1)$ everywhere in the paper.

**1:** *The Riemann–Liouville (RL) fractional derivative* of order $q \in (0, 1)$ of $m(t)$ is given by (see, for example, [13–15])

$$ {}_{t_0}^{RL}D_t^q m(t) = \frac{1}{\Gamma(1-q)} \frac{d}{dt} \int_{t_0}^{t} (t-s)^{-q} m(s) ds, \quad t \geq t_0, $$

where $\Gamma(.)$ denotes the Gamma function.

**2:** *The Caputo fractional derivative* of order $q \in (0, 1)$ is defined by (see, for example, [13–15])

$$ {}_{t_0}^{C}D_t^q m(t) = \frac{1}{\Gamma(1-q)} \int_{t_0}^{t} (t-s)^{-q} m'(s) ds, \quad t \geq t_0. $$

**3:** *The Grünwald–Letnikov fractional derivative* is given by (see, for example, [13–15]))

$$ {}_{t_0}^{GL}D_t^q m(t) = \lim_{h\to 0+} \frac{1}{h^q} \sum_{r=0}^{[\frac{t-t_0}{h}]} (-1)^r (qCr)\, m(t - rh), \quad t \geq t_0. $$

The Mittag–Leffler function with one parameter is defined as

$$E_\alpha(z) = \sum_{i=0}^{\infty} \frac{z^i}{\Gamma(1+\alpha\, i)}, \quad \alpha > 0, \ z \in C.$$

**Definition 1.** ([16]) The function $m(t) \in C^q([t_0, T], \mathbb{R}^n)$ if $m(t)$ is differentiable on $[t_0, T]$ (i.e., $m'(t)$ exists) and the Caputo derivative ${}_{t_0}^{C}D^q m(t)$ exists for $t \in [t_0, T]$.

Consider the initial value problem (IVP) for the nonlinear *impulsive Caputo fractional delay differential equation* (IFrDDE)

$$\begin{aligned} &{}_{t_0}^{C}D^q x(t) = F(t, x_t) \text{ for } t \geq t_0, \ t \neq t_k, k = 1, \dots, \\ &\Delta x(t)|_{t=t_k} = I_k(x(t_k - 0)) \quad \text{for } k = 1, 2, \dots, \\ &x(t_0 + s) = \phi(s), \ \ s \in [-r, 0], \end{aligned} \tag{1}$$

where $0 < q < 1$, $\Delta x(t)|_{t=t_k} = x(t_k+0) - x(t_k)$, $x_t = x(t+s), s \in [-r, 0]$, $F : [0, \infty) \times \mathbb{R}^n \to \mathbb{R}^n$, $I_k : \mathbb{R}^n \to \mathbb{R}^n$, $(k = 1, 2, 3, \dots)$, $\phi \in E$, where $x(t_k+0) = \lim_{t \to t_k,\, t > t_k} x(t) < \infty$, $x(t_k - 0) = \lim_{t \to t_k,\, t < t_k} x(t) = x(t_k)$.

Denote $\Phi_k(x) = x + I_k(x), k = 1, 2, \dots, \ x \in \mathbb{R}^n$.

We will denote the solution of the IVP for IFrDDE (1) by $x(t; t_0, \phi)$ for $t \geq t_0$. The solution of IFrDDE (1) is a piecewise continuous function. In connection with this, we introduce the following sets of functions:

$$\begin{aligned} PC(a,b) = \Big\{ &x : [a,b] \to \mathbb{R}^n \text{ such that } x(t) \in C([a,b]/\{t_k\}), \\ &x(t_k) = x(t_k - 0) = \lim_{t \to t_k - 0} x(t), \ x(t_k+0) = \lim_{t \to t_k+0} x(t) < \infty \Big\}, \\ PC^1(a,b) = \Big\{ &x \in PC(a,b) \text{ such that } x(t) \in C^1([a,b]/\{t_k\}), \\ &x'(t_k) = x'(t_k - 0) = \lim_{t \to t_k - 0} x'(t), \ x'(t_k+0) = \lim_{t \to t_k+0} x'(t) < \infty \Big\}. \end{aligned}$$

The fractional derivatives depend significantly on their lower limit and it allows different interpretations of piecewise continuous solutions of impulsive differential equations. This phenomena is not characteristic for ordinary derivatives. In the literature, there are two main approaches to interpret the solutions of impulsive fractional delay differential equations:

**First approach to the solutions of (1) (A1 for IFrDDE).**

The solution of the IVP for IFrDDE (1) satisfies the equalities (integral)

$$x(t) = x(t; t_0, \phi) = \begin{cases} \phi(t - t_0), & t \in [t_0 - r, t_0], \\ \phi(0) + \frac{1}{\Gamma(q)} \int_{t_0}^{t} (t-s)^{q-1} F(s, x_s) ds & \\ \quad + \sum_{i=1}^{k} I_i(x(t_i - 0)), & t \in (t_k, t_{k+1}], \ k = 0, 1, 2, \dots. \end{cases} \tag{2}$$

Formula (2) is given and used in [17]. It is a generalization to the formula proved in [18] for the solution of impulsive fractional differential equations without delays.

**Second approach to the solutions of (1) (A2 for IFrDDE).**

The idea of this approach is based on the dependence of the Caputo fractional derivative on the initial time point of the interval of differential equation, i.e., the lower limit of the Caputo fractional derivative is changing at each moment of impulse of the differential equation. Sometimes, Equation (1) in this case is written by

$$
\begin{aligned}
&{}_{t_k}^{C}D^q x(t) = F(t, x_t) \text{ for } t \in (t_k, t_{k+1}],\ k = 0, 1, \ldots, \\
&\Delta x(t)|_{t=t_k} = I_k(x(t_k - 0)) \quad \text{for } k = 1, 2, \ldots, \\
&x(t_0 + s) = \phi(s),\ \ s \in [-r, 0].
\end{aligned} \tag{3}
$$

Then, the solution of the IVP for IFrDDE (1), respectively (3), is given by

$$
x(t) = \begin{cases} \phi(t - t_0), & t \in [t_0 - r, t_0], \\ \phi(0) + \frac{1}{\Gamma(q)} \int_{t_0}^{t} (t-s)^{q-1} F(s, x_s) ds \text{ for } t \in (t_0, t_1], \\ \Phi_k(x(t_k - 0)) + \frac{1}{\Gamma(q)} \int_{t_k}^{t} (t-s)^{q-1} F(s, x_s) ds \\ \qquad \text{for } t \in (t_k, t_{k+1}],\ k = 1, 2, \ldots. \end{cases} \tag{4}
$$

**Remark 1.** *Both Formulas (2) and (4) differ for fractional differential equations and they are generalizations to impulsive ordinary differential equations. Both formulas coincide in the case of the ordinary derivative ($q = 1$) because in this case we have*

$$
\begin{aligned}
&\Phi_k(x(t_k - 0)) + \int_{t_k}^{t} F(s, x_s) ds = x(t_k - 0) + I_k(x(t_k - 0)) + \int_{t_k}^{t} f(s, x_s) ds \\
&= \phi(0) + \int_{t_0}^{t_k} F(s, x_s) ds + \sum_{i=1}^{k-1} I_i(x(t_i - 0)) + I_k(x(t_k - 0)) + \int_{t_k}^{t} f(s, x_s) ds \\
&= \phi(0) + \int_{t_0}^{t} F(s, x_s) ds + \sum_{i=1}^{k} I_i(x(t_i - 0)),
\end{aligned} \tag{5}
$$

*but*

$$
\begin{aligned}
&\Phi_k(x(t_k - 0)) + \frac{1}{\Gamma(q)} \int_{t_k}^{t} (t-s)^{q-1} F(s, x_s) ds \\
&= x(t_k - 0) + I_k(x(t_k - 0)) + \frac{1}{\Gamma(q)} \int_{t_k}^{t} F(s, x_s) ds \\
&= \Big(\phi(0) + \frac{1}{\Gamma(q)} \int_{t_0}^{t_k} (t_k - s)^{q-1} F(s, x_s) ds + \sum_{i=1}^{k-1} I_i(x(t_i - 0))\Big) \\
&\quad + I_k(x(t_k - 0)) + \frac{1}{\Gamma(q)} \int_{t_k}^{t} (t-s)^{q-1} F(s, x_s) ds \\
&\neq \phi(0) + \frac{1}{\Gamma(q)} \int_{t_0}^{t} (t-s)^{q-1} F(s, x_s) ds + \sum_{i=1}^{k} I_i(x(t_i - 0)).
\end{aligned} \tag{6}
$$

**Remark 2.** *In the case $q = 1$, the solution of the impulsive ordinary differential equation on each interval of continuity could be considered as a solution of the same differential equation with a new initial condition, defined by the impulsive function. It allows the application of induction w.r.t. the interval of continuity.*

*This is different for fractional differential equation.*

*If A1 for IFrDDE is applied, then ${}_{t_0}^{C}D^q x(t) \neq {}_{t_k}^{C}D^q x(t)$ for $t \in (t_k, t_{k+1})$ and induction w.r.t. the interval of continuity is not useful.*

*If A2 for IFrDDE is applied, then induction w.r.t. the interval of continuity could be used.*

A detailed explanation of advantages/disadvantages of both the above approaches for equations without delays is given in [19,20]. The definition of the solution $x(t; t_0, \phi)$ of the IVP for IFrDDE (1) depends on your point of view.

In this paper, we will use approach A2 for IFrDDE.

## 3. Lyapunov Functions and Their Fractional Derivatives

In this paper, we will use piecewise continuous Lyapunov functions [21]):

**Definition 1.** *Let $\alpha < \beta \leq \infty$ be given numbers and $\mathcal{D} \subset \mathbb{R}^n$, $0 \in \mathcal{D}$ be a given set. We will say that the function $V(t,x) : [\alpha - r, \beta) \times \mathcal{D} \to \mathbb{R}_+$ belongs to the class $\Lambda([\alpha - r, \beta), \mathcal{D})$ if*

1. *The function $V(t,x)$ is continuous on $[\alpha, \beta)/\{t_k\} \times \mathcal{D}$ and it is locally Lipschitz with respect to its second argument;*
2. *For each $t_k \in (\alpha, \beta)$ and $x \in \mathcal{D}$, there exist finite limits*

$$V(t_k, x) = V(t_k - 0, x) = \lim_{t \uparrow t_k} V(t,x), \quad \text{and } V(t_k + 0, x) = \lim_{t \downarrow t_k} V(t,x).$$

In connection with the Caputo fractional derivative, it is necessary to define in an appropriate way the derivative of the Lyapunov functions among the studied equation. The choice $\frac{dV(t,x)}{dt}$ is adapted from the case of ordinary differential equations, but it is not applicable since it does not depend on the initial time point (such as the Caputo fractional derivative).

We will give a brief overview of the three main types derivatives of Lyapunov functions $V(t,x) \in \Lambda([t_0 - r, T), \mathcal{D})$ among solutions of fractional differential equations in the literature:

- first type– Let $x(t) \in \mathcal{D}$, $t \in [t_0 - r, T)$, be a solution of the IVP for the IFrDDE (3) (according to A2 for IFrDDE). Then, we can consider **the Caputo fractional derivative** of the function $V(t,x) \in \Lambda([t_0 - r, T), \mathcal{D})$ defined by

$$
\begin{gathered}
{}_{t_k}^{c}D^q V(t, x(t)) = \frac{1}{\Gamma(1-q)} \int_{t_k}^{t} (t-s)^{-q} \frac{d}{ds}\Big(V(s, x(s))\Big) ds, \\
t \in J_k = (t_k, t_{k+1}), \quad k = 1, 2, \cdots : \ t_k \in (t_0, T).
\end{gathered} \tag{7}
$$

  This type of derivative is applicable for continuously differentiable Lyapunov functions.

- second type– Let $\psi \in C([-\tau, 0], \mathcal{D})$. Then, **the Dini fractional derivative** of the Lyapunov function $V(t,x) \in \Lambda([t_0 - r, T), \mathcal{D})$ is defined by

$$
\begin{aligned}
& {}_{t_k}D^q_{(3)} V(t, \psi(0), \psi) \\
& = \limsup_{h \to 0} \frac{1}{h^q} \Big[ V(t, \psi(0)) - \sum_{r=1}^{[\frac{t-t_k}{h}]} (-1)^{r+1} {}_qC_r V(t - rh, \psi(0) - h^q F(t, \psi_0)) \Big] \\
& \quad t \in J_k, \quad k = 1, 2, \cdots : \ t_k \in (t_0, T),
\end{aligned} \tag{8}
$$

  where $\psi_0 = \psi(s)$, $s \in [-r, 0]$.

  The Dini fractional derivative (8) keeps the concept of fractional derivatives because it has a memory.

  Note that Dini fractional derivative, defined by (8), is based on the notation

$$D^+ V(t, \psi(0), \psi) = \limsup_{h \to 0} \frac{1}{h^q} \Big[ V(t, \psi(0)) - V(t - h, \psi(0) - h^q F(t, \psi_0)) \Big]. \tag{9}$$

  In [17], the notation (9) is used directly. However, the notation (9) does not depend on the order $q$ of the fractional derivative nor on the initial time $t_0$, which is typical for the Caputo fractional derivative. The operator defined by (9) has no memory. In addition, if $x(t)$ is a solution of (3), then $D^+ V(t, x(t), x) \neq {}_{t_k}^{c}D^q V(t, x(t))$.

For fractional differential equations without any impulses, notation similar to (9) is defined and $V(t-h, x-h^q F(t,x)) = \sum_{r=1}^{[\frac{t-t_0}{h}]} (-1)^{r+1} {}_qC_r V(t-rh, x-h^q F(t,x))$ is applied [16].

- third type—let the initial data $(t_0, \phi_0) \in \mathbb{R}_+ \times C([-\tau, 0], \mathcal{D}))$, $\mathcal{D}) \subset \mathbb{R}^n$, of IVP for IFrDDE (3) and $\psi \in C([-\tau, 0], \mathcal{D})$ be given. Then, **the Caputo fractional Dini derivative** of the Lyapunov function $V(t,x) \in \Lambda([t_0 - r, T), \mathcal{D})$ is defined by:

$$
\begin{aligned}
&{}_{t_k}^{c}D^q_{(3)} V(t, \psi; t_0, \phi_0(0)) \\
&= \limsup_{h \to 0^+} \frac{1}{h^q} \Big\{ V(t, \psi(0)) - V(t_0, \phi_0(0)) \\
&- \sum_{r=1}^{[\frac{t-t_k}{h}]} (-1)^{r+1} {}_qC_r \Big( V(t-rh, \psi(0) - h^q F(t, \psi_0)) - V(t_0, \phi_0(0)) \Big) \Big\}, \\
&\text{for } t \in J_k,
\end{aligned}
\tag{10}
$$

where $\psi_0 = \psi(s)$, $s \in [-r, 0]$,

or its equivalent

$$
\begin{aligned}
&{}_{t_k}^{c}D^q_{(3)} V(t, \psi; t_k, \phi_0(0)) \\
&= \limsup_{h \to 0^+} \frac{1}{h^q} \Big\{ V(t, \psi(0)) + \sum_{r=1}^{[\frac{t-t_k}{h}]} (-1)^r {}_qC_r V(t-rh, \psi(0) - h^q f(t, \psi_0)) \Big\} \\
&- \frac{V(t_0, \phi_0(0))}{(t-t_k)^q \Gamma(1-q)}, \qquad \text{for } t \in J_k.
\end{aligned}
\tag{11}
$$

The derivative ${}_{t_k}^{c}D^q_{(3)} V(t, \psi; t_0, \phi_0(0))$ given by (11) depends significantly on both the fractional order $q$ and the initial data $(t_0, \phi_0)$ of IVP for IFrDDE (3) and it makes this type of derivative close to the idea of the Caputo fractional derivative of a function.

**Remark 3.** *For any initial data $(t_0, \phi_0) \in \mathbb{R}_+ \times C([-\tau, 0], \mathcal{D})$ of the IVP for IFrDDE (3), the relation between the Dini fractional derivative defined by (8) and for any $t \in J_k$, $\psi \in C([-\tau, 0], \mathcal{D})$ and the Caputo fractional Dini derivative defined by (11) is given by*

$$
{}_{t_0}^{c}D^q_{(3)} V(t, \psi; t_0, \phi_0(0)) = {}_{t_k}D^q_{(3)} V(t, \psi(0), \psi) - \frac{V(t_0, \phi_0(0))}{(t-t_k)^q \Gamma(1-q)}
$$

*or*

$$
{}_{t_k}^{c}D^q_{(3)} V(t, \psi; t_0, \phi_0(0)) = {}_{t_k}D^q_{(3)} V(t, \psi(0), \psi) - {}_{t_0}^{RL}D^q \Big( V(t_0, \phi_0(0)) \Big).
$$

In the next example, to simplify the calculations and to emphasize the derivatives and their properties, we will consider the scalar case, i.e., $n = 1$.

**Example 1.** *(Quadratic Lyapunov function). Let $V(t,x) = x^2$, with $x \in \mathbb{R}$.*

*Case 1. Caputo fractional derivative: Let $x(t) = x(t; t_0, \phi_0)$, $t \in [t_0 - \tau, T)$ be a solution of the IVP for IFrDDE (3) with $n = 1$ and $\phi_0 \in C([-\tau, 0], \mathbb{R})$ and we get*

$$
{}_{t_k}^{c}D^q V(t, x(t)) = \frac{2}{\Gamma(1-q)} \int_{t_k}^{t} (t-s)^{-q} x(s) \frac{d}{ds}\Big(x(s)\Big) ds \le 2x(t) F(t, x_t), \quad t \in J_k. \tag{12}
$$

*Case 2. Dini fractional derivative. Consider IVP for IFrDDE (3) with given initial data $(t_0, \phi_0) \in \mathbb{R}_+ \times C([-\tau, 0], \mathbb{R})$. Let $\psi \in C([-\tau, 0], \mathbb{R})$ be any function. Apply (8) and we obtain*

$$ {}_{t_k}D^q_{(3)}V(t,\psi(0),\psi) = 2\psi(0)F(t,\psi_0) + \frac{(\psi(0))^2}{(t-t_k)^q\Gamma(1-q)}, \quad t \in J_k, \tag{13} $$

*where* $\psi_0 = \psi(s), s \in [-r, 0]$.

*Note, if we apply (9) directly, we obtain* $D^+V(t,\psi(0),\psi) = 2\psi(0)F(t,\psi_0)$.

*Case 3. Caputo fractional Dini derivative. Use (11), Case 2 and Remark 3 and we obtain*

$$ \begin{aligned} {}^{c}_{t_k}D^q_{(3)}V(t,\psi;t_0,\phi_0) &= 2\psi(0)F(t,\psi_0) + ((\psi(0))^2 - (\phi_0(0))^2)\frac{(t-t_k)^{-q}}{\Gamma(1-q)} \\ &= {}_{t_k}D^q_{(3)}V(t,\psi(0),\psi) - {}^{RL}_{t_k}D^q\Big(V(t_0,\phi_0(0))\Big), \quad \text{for } t \in J_k,\ \psi \in C([-\tau,0],\mathbb{R}). \end{aligned} \tag{14} $$

**Example 2.** *(Lyapunov function depending directly on the time variable). Let* $V(t,x) = m(t)\, x^2$ *where* $m \in C^1(\mathbb{R}_+, \mathbb{R}_+)$.

*Case 1. Caputo fractional derivative. Let* $x(t) = x(t;t_0,\phi_0)$ *be a solution of the IVP for IFrDDE (3). The fractional derivative*

$$ {}^{c}_{t_k}D^qV(t,x(t)) = {}^{c}_{t_k}D^q\Big(m(t)\,x^2(t)\Big) = \frac{1}{\Gamma(1-q)}\int_{t_k}^{t}\frac{m'(s)x^2(s)+2m(s)x(s)x'(s)}{(t-s)^q}ds $$

*is difficult to obtain in the general case for any solution of (3). In addition, its upper bound is difficult to find.*

*Case 2. Dini fractional derivative. Let* $\psi \in C([-\tau,0],\mathcal{D})$ *be given. Applying (8), we obtain*

$$ {}_{t_k}D^q_{(3)}V(t,\psi(0),\psi) = \psi(0)\, m(t)F(t,\psi_0) + (\psi(0))^2\, {}^{RL}_{t_k}D^q\Big(m(t)\Big). \tag{15} $$

*Note that if we use (9) directly, then* $D^+V(t,\psi(0),\psi) = 2\psi(0)\, m(t)F(t,\psi_0)$*, which unusually is missing the derivative of the function* $m(t)$ *(compare with the case of ordinary derivatives).*

*Case 3. Caputo fractional Dini derivative. Use (11) and we obtain*

$$ \begin{aligned} &{}^{c}_{t_k}D^q_{(3)}V(t,\psi;t_0,\phi_0(0)) \\ &= 2\psi(0)m(t)F(t,\psi_0) + (\psi(0))^2\ {}^{RL}_{t_k}D^q\Big(m(t)\Big) - (\phi_0(0))^2m(t_0)\frac{(t-t_k)^{-q}}{\Gamma(1-q)} \\ &= {}_{t_k}D^q_{(3)}V(t,\psi(0),\psi) - V(t_0,\phi_0(0))\frac{(t-t_k)^{-q}}{\Gamma(1-q)} \\ &= {}_{t_k}D^q_{(3)}V(t,\psi(0),\psi) - {}^{RL}_{t_k}D^q\Big(V(t_0,\phi_0(0))\Big). \end{aligned} \tag{16} $$

## 4. Some Comparison Results for Lyapunov Functions

### *4.1. Comparison Results for Delay Fractional Differential Equations*

First, we will prove several comparison results for fractional delay differential equation without any impulses. We will use Lyapunov function with the Razumikhin condition $V(t+\Theta,\psi(\Theta)) \le p(V(t,\psi(0)))$, $\Theta \in [-r,0]$ for $\psi \in C([-\tau,0],\mathbb{R}^n)$, where $p \in C([0,\infty),\mathbb{R}_+)$, $p(s) > s$ for $s > 0$.

**Remark 4.** *A comparison result is given in Theorem 4.5 [22] by applying definition (9) for the derivative of V and incorrectly replacing it with the Caputo derivative in the proof. Some comparison results applying A1 for IFrDDE are obtained in [17], but induction w.r.t. the interval of continuity is incorrectly used (see Remark 2).*

Consider the IVP for the following delay fractional differential equation (FrDDE)

$$ \begin{aligned} &{}^{C}_{\tau_0}D^qx(t) = F(t,x_t) \text{ for } t \in (\tau_0,\Theta], \\ &x(\tau_0+s) = \phi(s), \quad s \in [-r,0], \end{aligned} \tag{17} $$

where $x \in R^n$, $\phi \in C([-r,0],\mathcal{D})$, $\mathcal{D} \subset \mathbb{R}^n$.

**Lemma 1.** *(Comparison result with the Caputo fractional derivative) Assume:*

1. *The function $x(t) = x(t;\tau_0,\phi) \in C^q([\tau_0,\Theta],\mathcal{D})$ is a solution of the IVP for FrDDE (17).*
2. *The function $V \in \Lambda([\tau_0 - r,\Theta],\mathcal{D})$, $\mathcal{D} \subset \mathbb{R}^n$ is such that there exist positive numbers $p,\alpha : p > \frac{1}{E_q(-\alpha r^q)}$ such that, for any point $t \in [\tau_0,\Theta] : V(t+s,x(t+s)) < pV(t,x(t))$, $s \in [-r,0)$, the fractional derivative ${}^{C}_{\tau_0}D^q V(t,x(t))$ exists and the inequality*

$$ {}^{C}_{\tau_0}D^q V(t,x(t)) \leq -\alpha V(t,x(t)) \tag{18} $$

*holds.*

*Then, $V(t,x(t;\tau_0,\phi)) \leq \max_{s\in[-r,0]} V(t_0+s,\phi(s))E_q(-\alpha(t-t_0)^q)$ for $t \in [\tau_0,\Theta]$.*

**Proof.** Denote $B = \max_{s\in[-r,0]} V(\tau_0+s,\phi_0(s))$. Let $\varepsilon > 0$ be an arbitrary number. Define the functions $m(t) = V(t,x(t))$ for $t \in [\tau_0 - r,\Theta]$. We will prove that

$$ m(t) < BE_q(-\alpha(t-\tau_0)^q) + \varepsilon, \ \ t \geq \tau_0. \tag{19} $$

For $t = \tau_0$, the inequality $m(\tau_0) \leq B < B+\varepsilon$ holds. Assume (19) is not true and, therefore, there exists a point $t^* \in (\tau_0,\Theta)$ such that

$$ m(t) < BE_q(-\alpha(t-\tau_0)^q) + \varepsilon, \ \ t \in [\tau_0,t^*) \ \text{ and } m(t^*) = BE_q(-\alpha(t^*-\tau_0)^q) + \varepsilon. \tag{20} $$

Let $s \in [-r,0]$.
*Case 1.* Let $t^*+s \in [\tau_0,t^*]$.
Then, $t^*+s \geq \tau_0 - r$, $t^* - \tau_0 > 0$ and $0 \leq t^* - \tau_0 - r \leq t^* - \tau_0 + s$. Using the inequality $E_q(-\alpha a^q)E_q(-\alpha(t-a)^q) \leq E_q(-\alpha t^q)$ for $t \geq a$ with $\alpha > 0, a \geq 0$ and the choice of $p$, we obtain

$$ \begin{aligned} p(m(t^*)) &= pBE_q(-\alpha(t^*-\tau_0)^q) + p\varepsilon > B\frac{E_q(-\alpha(t^*-\tau_0)^q)}{E_q(-\alpha r^q)} + \varepsilon \\ &\geq BE_q(-\alpha(t^*-\tau_0-r)^q) + \varepsilon \geq BE_q(-\alpha(t^*+s-\tau_0)^q) + \varepsilon > m(t^*+s). \end{aligned} \tag{21} $$

*Case 2.* Let $t^* - r \leq t^*+s < \tau_0$.
Then, $t^* - \tau_0 < -s \leq r$ and we get

$$ p(m(t^*)) > B\frac{E_q(-\alpha(t^*-\tau_0)^q)}{E_q(-\alpha r^q)} + \varepsilon \geq B+\varepsilon > B \geq m(t^*+s). \tag{22} $$

From (21), (22) and condition 2 of Lemma 1, it follows that the fractional derivative ${}^{C}_{\tau_0}D^q m(t^*)$ exists.
From (20), it follows that

$$ {}^{C}_{\tau_0}D^q \Big(m(t^*) - BE_q(-\alpha(t^*-\tau_0)^q) - \varepsilon\Big) > 0. \tag{23} $$

Then, using ${}^{C}_{\tau_0}D^q E_q(-\alpha(t-\tau_0)^q) = -aE_q(-\alpha(t-\tau_0)^q)$, we get ${}^{C}_{\tau_0}D^q m(t^*) > -B\alpha E_q(-\alpha(t^*-\tau_0)^q)$.
From inequality (18), we get ${}^{C}_{\tau_0}D^q V(t^*,x(t^*)) \leq -\alpha V(t^*,x(t^*))$. Therefore, the inequality $-B\alpha E_q(-\alpha(t^*-\tau_0)^q) < {}^{C}_{\tau_0}D^q m(t^*) \leq -\alpha m(t^*)$ holds or $BE_q(-\alpha(t^*-\tau_0)^q) > m(t^*)$, which contradicts (20). Therefore, inequality (19) holds for an arbitrary $\varepsilon > 0$. Thus, the claim in our Lemma is true. □

**Remark 5.** *If $p = 1$ in condition 2 of Lemma 1 then since $E_q(-\alpha a^q)E_q(-\alpha(t-a)^q) \neq E_q(-\alpha t^q)$ for $t \geq a$ (see [23]), the inequality (21) is not true and the claim of Lemma 1 is not true.*

Note the comparison result for Lyapunov functions is true if the Caputo fractional derivative in Lemma 1 is replaced by any of the other two derivatives. In the proof, we will use the following result.

**Lemma 2.** *[22] Let $m \in C([\tau_0,\Theta],\mathbb{R})$ and there exists $\xi \in (\tau_0,\Theta]$ such that $m(\xi) = 0$ and $m(t) < 0$ for $t \in [\tau_0,\xi)$. Then, ${}_{\tau_0}^{GL}D_+^q m(\xi) > 0$.*

**Lemma 3.** *(Comparison result with the Dini fractional derivative) Assume the function $V \in \Lambda([\tau_0 - r,\Theta],\mathcal{D})$, $\mathcal{D} \subset \mathbb{R}^n$, is such that there exist positive numbers $p,\alpha$ : $p > \frac{1}{E_q(-\alpha r^q)}$ such that for any point $t \in [\tau_0,\Theta]$ and any function $\psi \in C([-r,0],\mathbb{R}^n)$ : $V(t+s,\psi(s)) < pV(t,\psi(0))$, $s \in [-r,0)$ the inequality*

$$ {}_{\tau_0}D^q_{(17)}V(t,\psi(0),\psi) \le -\alpha V(t,\psi(0)) \tag{24}$$

*holds.*

*Then, $V(t,x(t;\tau_0,\phi)) \le \max_{s\in[-r,0]} V(t_0+s,\phi(s))E_q(-\alpha(t-t_0)^q)$ for $t \in [\tau_0,\Theta]$ where $x(t;\tau_0,\phi) \in C^q([\tau_0,\Theta],\mathcal{D}$ is a solution of the IVP for FrDDE (17).*

**Proof.** The proof is similar to that in Lemma 1. The main difference is in connection with inequality (23). Follow the proof in Lemma 1 and in this case we use Lemma 2 and obtain

$$ {}_{\tau_0}^{GL}D_+^q\Big(m(t^*) - BE_q(-\alpha(t^*-\tau_0)^q) - \varepsilon\Big) > 0. \tag{25}$$

Now, using ${}_{\tau_0}^{C}D^q u(t) = {}_{\tau_0}^{GL}D_+^q(u(t)-u(\tau_0)) = {}_{\tau_0}^{GL}D_+^q u(t) - \frac{u(\tau_0)}{(t-\tau_0)^q\Gamma(1-q)}$, we get

$$\begin{aligned} {}_{\tau_0}^{GL}D_+^q m(t^*) &> {}_{\tau_0}^{GL}D_+^q\Big(BE_q(-\alpha(t^*-\tau_0)^q) - B\Big) + \frac{B+\varepsilon}{(t-\tau_0)^q\Gamma(1-q)} \\ &> {}_{\tau_0}^{C}D^q\Big(BE_q(-\alpha(t^*-\tau_0)^q) = -B\alpha E_q(-\alpha(t^*-\tau_0)^q). \end{aligned} \tag{26}$$

It remains to show that we have a contradiction. To see this for any $t \in [\tau_0,t^*]$ and $h > 0$, we let

$$S(x(t),h) = \sum_{r=1}^{[\frac{t-\tau_0}{h}]} (-1)^{r+1}\,{}_qC_r\Big[x(t-rh) - \phi(0)\Big].$$

Now, ${}_{\tau_0}^{C}D_+^q x(t) ={}_{\tau_0}^{GL} D_+^q(x(t)-x(\tau_0)) = \limsup_{h\to 0+}\frac{1}{h^q}\Big[x(t) - x(\tau_0) - S(x(t),h)\Big] = F(t,x_t)$. Therefore, $S\left(x(t),h\right) = x(t) - \phi(0) - h^q F(t,x_t) - \Lambda(h^q)$ or

$$x(t) - h^q F(t,x_t) = S\left(x(t),h\right) + \phi(0) + \Lambda(h^q) \tag{27}$$

with $\frac{||\Lambda(h^q)||}{h^q} \to 0$ as $h \to 0$. Then, for any $t \in [t_0,t^*]$, we obtain

$$\begin{aligned} {}_{\tau_0}^{GL}D_+^q m(t) &= m(t) - \sum_{r=1}^{[\frac{t-\tau_0}{h}]}(-1)^{r+1}\,{}_qC_r m(t-rh) \\ &= \Big\{V(t,x(t)) - \sum_{r=1}^{[\frac{t-\tau_0}{h}]}(-1)^{r+1}\,{}_qC_r V(t-rh,x(t)-h^qF(t,x_t)\Big\} \\ &\quad + \sum_{r=1}^{[\frac{t-\tau_0}{h}]}(-1)^{r+1}\,{}_qC_r\Big\{V(t-rh,S\left(x(t),h\right)+\phi(0)+\Lambda(h^q)) \\ &\quad - V(t-rh,x(t-rh))\Big\}. \end{aligned} \tag{28}$$

Since $V$ is locally Lipschitzian in its second argument with a Lipschitz constant $L > 0$, we obtain

$$\begin{aligned}
&\sum_{r=1}^{[\frac{t-\tau_0}{h}]} (-1)^{r+1}\,{}_qC_r\Big\{V(t-rh, S(x(t),h)+\phi(0)+\Lambda(h^q)) - V(t-rh, x(t-rh))\Big\}\\
&\leq L\Big|\Big|\sum_{r=1}^{[\frac{t-\tau_0}{h}]} (-1)^{r+1}\,{}_qC_r \sum_{j=1}^{[\frac{t-\tau_0}{h}]} (-1)^{j+1}\,{}_qC_j\Big(x(t-jh)-\phi(0)\Big)\\
&\quad - \sum_{r=1}^{[\frac{t-\tau_0}{h}]} (-1)^{r+1}\,{}_qC_r\Big((x(t-rh)-\phi(0))\Big)\Big|\Big| + L||\Lambda(h^q)||\,\Big|\sum_{r=1}^{[\frac{t-\tau_0}{h}]} (-1)^{r+1}\,{}_qC_r\Big| \qquad (29)\\
&= L\Big|\Big|\Big(\sum_{r=0}^{[\frac{t-\tau_0}{h}]} (-1)^{r+1}\,{}_qC_r\Big)\Big(\sum_{j=1}^{[\frac{t-\tau_0}{h}]} (-1)^{j+1}\,{}_qC_j\Big(x(t-jh)-\phi(0)\Big)\Big)\Big|\Big|\\
&\quad + L||\Lambda(h^q)||\,\Big|\sum_{r=1}^{[\frac{t-\tau_0}{h}]} (-1)^{r+1}\,{}_qC_r\Big|.
\end{aligned}$$

Substitute (29) in (28), divide both sides by $h^q$, take the limit as $h \to 0^+$, use $\sum_{r=0}^{\infty} {}_qC_r z^r = (1+z)^q$ if $|z| \leq 1$, and we obtain for any $t \in [\tau_0, t^*]$ the inequality

$$\begin{aligned}
&{}_{\tau_0}^{GL}D_+^q m(t) \leq \lim_{h\to 0+} \frac{1}{h^q}\Big\{V(t,x(t))\\
&\quad - \sum_{r=1}^{[\frac{t-\tau_0}{h}]} (-1)^{r+1}\,{}_qC_r V(t-rh, x(t)-h^q f(t,x^*(t))\Big\}\\
&\quad + L\lim_{h\to 0+}\frac{||\Lambda(h^q)||}{h^q}\lim_{h\to 0+}\Big|\sum_{r=1}^{[\frac{t-\tau_0}{h}]} (-1)^{r+1}\,{}_qC_r\Big| \qquad (30)\\
&\quad + L\limsup_{h\to 0^+}\Big|\sum_{r=1}^{[\frac{t-\tau_0}{h}]} (-1)^{r+1}\,{}_qC_r\Big|\,\Big|\Big|\frac{1}{h^q}\sum_{j=1}^{[\frac{t-\tau_0}{h}]} (-1)^{j+1}\,{}_qC_j\Big(x(t-jh)-\phi(0)\Big)\Big|\Big|\\
&\leq \lim_{h\to 0+}\frac{1}{h^q}\Big\{V(t,x(t)) - \sum_{r=1}^{[\frac{t-\tau_0}{h}]} (-1)^{r+1}\,{}_qC_r V(t-rh, x(t)-h^q f(t,x^*(t))\Big\}.
\end{aligned}$$

Let $t = t^*$. Define the function $\psi(\Theta) = x(t^* + \Theta)$, $\theta \in [-\tau, 0]$. Applying condition 2 to (30) for $t = t^*$, we get

$$\begin{aligned}
&{}_{t_0}^{GL}D_+^q m(t^*) \leq \lim_{h\to 0+}\frac{1}{h^q}\Big\{V(t^*,\psi(0)) - \sum_{r=1}^{[\frac{t^*-\tau_0}{h}]} (-1)^{r+1}\,{}_qC_r V(t^*-rh, \psi(0)-h^q F(t^*,\psi)\Big\} \qquad (31)\\
&= {}_{\tau_0}D^q_{(17)}V(t^*,\psi(0),\psi_0) \leq -\alpha V(t^*,u(0)) = -\alpha m(t^*) = -\alpha B E_q(-\alpha(t^*-\tau_0)).
\end{aligned}$$

Inequality (31) contradicts (26). □

**Remark 6.** *(Comparison result with the Caputo fractional Dini derivative) Lemma 3 remains true if the Dini fractional derivative in inequality (24) is replaced by the Caputo fractional Dini derivative* ${}_{\tau_0}^{c}D^q_{(17)}V(t,\psi;\tau_0,\phi_0(0))$.

### *4.2. Comparison Results for Impulsive Delay Fractional Differential Equations*

Now, we will prove some comparison result for IFrDDE (3) using approach A2 for IFrDDE.

**Lemma 4.** *(Comparison result with the Caputo fractional derivative) Assume:*

1. *The function $x(t) = x(t; t_0, \phi_0) \in PC^q([t_0, \Theta], \mathcal{D})$ is a solution of the IVP for FrDDE (3) with $\phi_0 \in C([-r, 0], \mathcal{D})$.*
2. *The function $V \in \Lambda([t_0 - r, \Theta], \mathcal{D})$, $\mathcal{D} \subset \mathbb{R}^n$ is such that there exist positive numbers $p, \alpha : p > \frac{1}{E_q(-\alpha r^q)}$ :*

   (i) *for any $k = 0, 1, \dots$ and any point $t \in [t_k, t_{k+1}] \bigcap (t_0, \Theta]$ such that $V(t+s, x(t+s)) < pV(t, x(t))$, $s \in [-r, 0)$ the fractional derivative ${}_{t_k}^{c}D^q V(t, x(t))$ exists and the inequality*

$$ {}_{t_k}^{c}D^q V(t, x(t)) \le -\alpha V(t, x(t)) \tag{32} $$

   *holds;*

   (ii) *for all $k = 1, 2, \dots : t_k \in (t_0, \Theta)$ and $x \in \mathcal{D}$, the inequality $V(t_k, x + I_k(x)) \le V(t_k, x)$.*

*Then,*

$$ V(t, x(t; t_0, \phi_0)) \le \max_{s \in [-r,0]} V(t_0 + s, \phi_0(s)) \Big( \prod_{i=1}^{k} E_q(-\alpha(t_i - t_{i-1})^q) \Big) E_q(-\alpha(t - t_k)^q), $$
$$ t \in (t_k, t_{k+1}] \bigcap [t_0, \Theta]. $$

**Proof.** Let $\varepsilon > 0$ be an arbitrary number. Define the functions $m(t) = V(t, x(t))$ for $t \in [\tau_0 - r, \Theta]$.

We use Lemma 1 and induction w.r.t. to the interval (see Remark 2) to prove the claim. For $t \in [t_0, t_1]$, the claim follows directly from Lemma 1, i.e.,

$$ m(t) = V(t, x(t)) \le \max_{s \in [-r,0]} V(t_0 + s, \phi_0(s)) E_q(-\alpha(t - t_0)^q), \quad t \in [t_0, t_1]. \tag{33} $$

Let $t \in (t_1, t_2]$. Denote $B_1 = m(t_1)$. Let $\varepsilon > 0$ be an arbitrary number. According to (33), the inequality $B_1 \le \max_{s \in [-r,0]} V(t_0 + s, \phi_0(s)) E_q(-\alpha(t_1 - t_0)^q)$ holds.

We will prove

$$ m(t) < B_1 E_q(-\alpha(t - t_1)^q) + \varepsilon, \quad t \in (t_1 + 0, t_2]. \tag{34} $$

From condition 2(ii), we get $m(t_1 + 0) \le m(t_1) < m(t_1) + \varepsilon$. Assume (34) is not true on $(t_1, t_2]$, i.e., there exists a point $t^* \in (t_1, t_2)$ such that

$$ m(t) < B_1 E_q(-\alpha(t - t_1)^q) + \varepsilon, \quad t \in [t_1, t^*) \quad \text{and } m(t^*) = B_1 E_q(-\alpha(t^* - t_1)^q) + \varepsilon. \tag{35} $$

Let $s \in [-r, 0]$.

*Case 1.* Let $t^* + s \in [t_1, t^*]$.

Then, $t^* + s \ge t_1 - r$. Then, $t^* - t_1 > 0$ and $0 \le t^* - t_1 - r \le t^* - t_1 + s$. Using the inequality $E_q(-\alpha a^q) E_q(-\alpha(t - a)^q) \le E_q(-\alpha t^q)$ for $t \ge a$ with $\alpha > 0, a \ge 0$ and (35), we get

$$ \begin{aligned} p\, m(t^*) &= p B_1 E_q(-\alpha(t^* - t_1)^q) + p\varepsilon > B_1 \frac{E_q(-\alpha(t^* - t_1)^q)}{E_q(-\alpha r^q)} + \varepsilon \\ &\ge B_1 E_q(-\alpha(t^* - t_1 - r)^q) + \varepsilon \ge B_1 E_q(-\alpha(t^* + s - t_1)^q) + \varepsilon > m(t^* + s). \end{aligned} \tag{36} $$

*Case 2.* Let $t^* - r \le t^* + s < t_1$.

Then, $t^* - t_1 < -s \le r$ and we get

$$ p\, m(t^*) > B_1 \frac{E_q(-\alpha(t^* - t_1)^q)}{E_q(-\alpha r^q)} + \varepsilon \ge B_1 + \varepsilon > B_1 \ge m(t^* + s), \; s \in [-r, 0]. \tag{37} $$

From (36), (22) and condition 2 (i) of Lemma 4, it follows that the fractional derivative ${}_{t_1}^{c}D^q m(t^*)$ exists.

Similar to the proof in Lemma 1 with $\tau_0 = t_1$ and $B = B_1$, we obtain a contradiction.

Therefore, inequality (34) holds for an arbitrary $\varepsilon > 0$. Thus, for $t \in (t_1, t_2]$,

$$m(t) \leq B_1 E_q(-\alpha(t-t_1)^q) \leq \max_{s\in[-r,0]} V(t_0+s, \phi_0(s)) E_q(-\alpha(t_1-t_0)^q) E_q(-\alpha(t-t_1)^q).$$

Using induction, we prove the claim. □

In the case when the Dini fractional derivative, defined by (8), is applied instead of the Caputo fractional derivative by using Lemma 3, we obtain the following result:

**Lemma 5.** *(Comparison result with the Dini fractional derivative) Assume the function* $V \in \Lambda([\tau_0 - r, \Theta], \mathcal{D})$, $\mathcal{D} \subset \mathbb{R}^n$ *is such that there exist positive numbers* $p, \alpha :\ p > \frac{1}{E_q(-\alpha r^q)}$:

*(i)* *for any* $k = 0, 1, \dots$ *and any point* $t \in [t_k, t_{k+1}] \bigcap (t_0, \Theta]$ *and any function* $\psi \in C([-r,0], \mathbb{R}^n)$ : $V(t+s, \psi(s)) < pV(t, \psi(0))$, $s \in [-r, 0)$ *the inequality*

$$ {}_{t_k}D^q_{(3)} V(t, \psi(0), \psi) \leq -\alpha V(t, \psi(0)) \tag{38}$$

*holds;*

*(ii)* *for all* $k = 1, 2, \dots : t_k \in (t_0, \Theta)$ *and* $x \in \mathcal{D}$ *the inequality* $V(t_k, x + I_k(x)) \leq V(t_k, x)$.

*Then,*

$$\begin{aligned} V(t, x(t; t_0, \phi_0)) &\leq \max_{s\in[-r,0]} V(t_0+s, \phi_0(s)) \Big( \prod_{i=1}^{k} E_q(-\alpha(t_i - t_{i-1})^q) \Big) E_q(-\alpha(t-t_k)^q), \\ & t \in (t_k, t_{k+1}] \bigcap [t_0, \Theta], \end{aligned} \tag{39}$$

*where* $x(t; \tau_0, \phi_0) \in PC^q([\tau_0, \Theta], \mathcal{D})$ *is a solution of the IVP for FrDDE (3) with* $\phi_0 \in C([-r,0], \mathcal{D})$.

**Remark 7.** *(Comparison result with the Caputo fractional Dini derivative) Lemma 5 remains true if the Dini fractional derivative in inequality (38) is replaced by Caputo fractional Dini derivative* ${}_{t_k}^{c}D^q_{(3)} V(t, \psi; \tau_0, \phi_0(0))$.

## 5. Application to Neural Networks

### 5.1. Problem Formulation

We will study neural networks modeled by impulsive Caputo fractional differential equations with bounded time dependent delays and distributed delays. We will consider the case when the lower limit of the fractional derivative is changed after each impulse, i.e., we will use approach A2 for IFrDDE. Following the notations in (3), we consider the general model of Hopfield's graded response neural networks with impulses and bounded delays and distributed delays (INND)

$$\begin{aligned} {}_{t_k}^{C}D_t^q x_i(t) &= -c_i(t)x_i(t) + \sum_{j=1}^{n} a_{ij}(t) f_j(x_j(t)) + \sum_{j=1}^{n} b_{ij}(t) g_j(x_j(t-\tau_j(t))) \\ &+ \sum_{j=1}^{n} d_{ij}(t) \int_{-r}^{0} K_{ij}(s) h_j(x_j(t+s)) ds + I_i(t) \\ & \text{for } t \in (t_k, t_{k+1}], k = 0, 1, 2, \dots, \\ \Delta x_i(t)|_{t=t_k} &= I_{k,i}(x_i(t_k - 0)), \quad k = 1, 2, \dots, \\ x_i(t) &= \phi_i^0(t - t_0), \ \ t \in [t_0 - r, t_0], \ \ i = 1, 2, \dots n, \end{aligned} \tag{40}$$

where $n$ represents the number of neurons in the network, $x_i(t)$ is the pseudostate variable denoting the average membrane potential of the $i$-th neuron at time $t$, $x(t) = (x_1(t), x_2(t), \dots, x_n(t)) \in \mathbb{R}^n$, $q \in (0,1)$, $c_i(t) > 0, i = 1,2,\dots,n$, is the self-regulating parameter of the $i$-th unit, they correspond to the rate with which the $i$-th neuron rests its potential in the resting state in isolation, $a_{ij}(t), b_{ij}(t)$, $i,j = 1,2,\dots,n$, correspond to the synaptic connection strength of the $i$-th neuron to the $j$-th neuron at time $t$ and $t-\tau_j(t)$, respectively, $f_j(x), g_j(x), h_j(x)$ are nonlinear activation functions such that $f(x) = (f_1(x_1), f_2(x_2), \dots, f_n(x_n))$, $h(x) = (h_1(x_1), h_2(x_2), \dots, h_n(x_n))$, $g(x) = (g_1(x_1), g_2(x_2), \dots, g_n(x_n))$; $I = (I_1, I_2, \dots, I_n)$ is an external bias vector, $\tau_j(t)$ represents the transmission delay along the axis of the $j$-th unit and satisfies $0 \leq \tau_j(t) \leq r$, the $t_k, k = 1,2,\dots$, are points of acting the state displacements, the functions $\Phi_{k,i}(t,u,v)$, $k = 1,2,\dots$ are the impulsive functions giving the impulsive perturbation of the $i$-th neuron on the point $t_k$, the numbers $x_i(t_k - 0) = x_i(t_k)$ and $x_i(t_k + 0)$ are the state of the $i$-th neuron before and after impulsive perturbation at time $t_k$; $K_{ij}(.)$ is the delay kernel with $\int_{-r}^{0} |K_{ij}(s)|ds = 1$, $\phi_i^0 \in C([-r,0],\mathbb{R})$, $i = 1,2,\dots,n$ are the initial functions.

The slave system is given by

$$\begin{aligned}
{}_{t_k}^{C}D_t^q y_i(t) &= -c_i(t)y_i(t) + \sum_{j=1}^{n} a_{ij}(t) f_j(y_j(t)) + \sum_{j=1}^{n} b_{ij}(t) g_j(y_j(t-\tau_j(t))) \\
&\quad + \sum_{j=1}^{n} d_{ij}(t) \int_{-r}^{0} K_{ij}(s) h_j(y_j(t+s))ds - u_i(t) + I_i(t) \\
&\qquad \text{for } t \in (t_k, t_{k+1}], k = 0,1,2,\dots, \\
\Delta y_i(t)|_{t=t_k} &= I_{k,i}(y_i(t_k - 0)), \quad k = 1,2,\dots, \\
y_i(t) &= \varphi_i^0(t - t_0), \;\; t \in [t_0 - r, t_0], \;\; i = 1,2,\dots n,
\end{aligned} \tag{41}$$

where $u_i(t)$, $i = 1,2,\dots,n$ are the suitable controllers at time $t$, $\varphi_i^0 \in C([-r,0],\mathbb{R})$, $i = 1,2,\dots,n$.

### 5.2. Mittag–Leffler Synchronization

**Definition 2.** *The master impulsive Caputo fractional system (40) and the slave impulsive Caputo fractional system (41) are globally Mittag–Leffler synchronized if for any initial functions $\phi_i^0, \varphi_i^0 \in C([-r,0],\mathbb{R})$ there exist constants $C, K > 0$ such that*

$$\begin{aligned}
||x(t;t_0,\phi^0) - y(t;t_0,\varphi^0)|| &\leq \{m(\phi^0 - \varphi^0)E_q(-C(t-t_k)^q) \prod_{j=0}^{k-1} E_q(-C(t_{j+1} - t_j)^q)\}^K, \\
t \in J_k = (t_k, t_{k=1}], \; k = 0,1,2,\dots,
\end{aligned}$$

*where $m \in C(\mathbb{R}_+^n, \mathbb{R}_+)$ (with $m(0) = 0$) is Lipschitz.*

**Remark 8.** *The synchronization of the problem (40) is studied in [11] and the authors consider the case of constant strengths between the neurons and linear impulsive functions. They used approach A1 for IFrDDE. The main result is based on incorrectly citing and using the Lemma from [17] where they use the derivative (9), which is different than the Caputo fractional derivative (see Remarks 2 and 4).*

The main goal of the paper is to implement appropriate controllers $u_i(t)$, $i = 1,2,\dots,n$ for the response system, such that the controlled response system (41) could be synchronized with the drive system (40).

5.2.1. Output Coupling Controller

Inspired by the ideas in [24], the control inputs in the response system are taken as output coupling $u_j(t) = \sum_{j=1}^{n} m_{ij}(f_j(y_j(t)) - f_j(x_j(t))),\ i = 1,2,\dots,n$. The synchronization via output coupling is important because, in many real systems, only output signals can be measured.

Define the synchronization error $e_i(t) = y_i(t) - x_i(t)$. Therefore, the error dynamics between (40) and (41) can be expressed by

$$\begin{aligned} {}_{t_k}^{C}D_t^q e_i(t) &= -c_i(t)e_i(t) + \sum_{j=1}^{n}(a_{ij}(t) - m_{ij})F_j(e_j(t)) + \sum_{j=1}^{n} b_{ij}(t)G_j(e_j(t-\tau_j(t))) \\ &\quad + \sum_{j=1}^{n} d_{ij}(t)\int_{-r}^{0} K_{ij}(s)H_j(e_j(t+s))ds \text{ for } t \in (t_k, t_{k+1}], k = 0,1,2,\dots, \\ \Delta e_i(t)|_{t=t_k} &= L_{k,i}(e_i(t_k - 0)), \qquad k = 1,2,\dots, \\ e_i(t) &= \Phi_i^0(t - t_0),\ t \in [t_0 - r, t_0],\ i = 1,2,\dots n, \end{aligned} \tag{42}$$

where $F_j(e_j(t)) = f_j(y_j(t)) - f_j(x_j(t))$, $G_j(e_j(t)) = g_j(y_j(t)) - g_j(x_j(t))$, $H_j(e_j(t)) = h_j(y_j(t)) - h_j(x_j(t))$ for $t \in (t_k, t_{k+1}], k = 0,1,2,\dots$, $L_{k,i}(e_j(t_k - 0)) = L_{k,i}(y_j(t_k - 0)) - L_{k,i}(x_j(t_k - 0))$, $i = 1,2,\dots,n, k = 1,2,\dots$, and $\Phi_i^0(s) = \varphi_i^0(s) - \phi_i^0(s)$, $s \in [-r, 0]$.

We assume the following:

**Assumption A1**. The neuron activation functions are Lipschitz, i.e., there exist positive numbers $\lambda_i, \mu_i, \nu_i\ i = 1,2,\dots,n$ such that $|f_i(u) - f_i(v)| \le \lambda_i|u - v|$, $|g_i(u) - g_i(v)| \le \mu_i|u - v|$ and $|h_i(u) - h_i(v)| \le \nu_i|u - v|, i = 1,2,\dots,n$ for $u, v \in \mathbb{R}$.

**Assumption A2**. There exist positive numbers $M_{i,j},\ C_{i,j}, D_{ij}\ i,j = 1,2,\dots,n$ such that $|a_{i,j}(t)| \le M_{i,j}$, $|b_{i,j}(t)| \le C_{i,j}$, $|d_{i,j}(t)| \le D_{i,j}$ for $t > 0$.

**Assumption A3**. There exists a constant $\eta > 0$ such that $c_i(t) \ge \eta,\ i = 1,2,\dots,n,\ t \ge 0$.

**Assumption A4**. The impulsive functions $\Phi_{k,i}(u) = u + I_{k,i}(u)$ are Lipschitz with constants $A_{k,i} \in (0,1]$, i.e., $|\Phi_{k,i}(u) - \Phi_{k,i}(v)| \le A_{k,i}|u - v|, i = 1,2,\dots,n, k = 1,2,\dots$ for $u, v \in \mathbb{R}$.

**Assumption A5**. The inequality

$$\begin{aligned} 2\eta > &\sum_{i=1}^{n}\Big(\max_{j=1,2,\dots,n}(M_{ij} + |m_{ij}|)\lambda_j + \sum_{j=1}^{n}(D_{ij}\nu_j + C_{ij}\mu_j)\Big) \\ &+ \max_{i=1,2,\dots,n}\sum_{j=1}^{n}\Big[(M_{ij} + |m_{ij}|)\lambda_j + C_{ij}\mu_j + D_{ij}\nu_j\Big] \end{aligned} \tag{43}$$

holds.

**Remark 9.** *If assumption A1 is satisfied, then the functions $F, G, H$ in (42) satisfy $|F_j(u)| \le \lambda_j|u|$, $|G_j(u)| \le \mu_j|u|$, $|H_j(u)| \le \nu_j|u|$, $j = 1,2,\dots,n$ for any $u \in \mathbb{R}$. If assumption A4 is satisfied, then the impulsive functions $L_{k,i}$ in (42) satisfy $|u + L_{k,i}| \le A_{k,i}|u|$ $i = 1,2,\dots,n,\ k = 1,2,\dots$ for any $u \in \mathbb{R}$.*

The case of multiple time constant delays (no distributed delays) and the constant synaptic connection strength between neurons is studied in [22] by using quadratic Lyapunov function. We will study the case of variable bounded synaptic connection strength and nonlinear impulsive functions.

**Theorem 1.** *Let assumptions A1–A5 be satisfied.*

*Then, the master impulsive Caputo fractional system (40) and the slave impulsive Caputo fractional system (41) are globally Mittag–Leffler synchronized.*

**Proof.** According to condition A5, there exists a positive constant $\alpha$ such that $\alpha \leq \frac{2\eta - B_1 - B_2}{B_3}$, where

$$B_1 = \sum_{i=1}^{n}\Big(\max_{j=1,2,\dots,n}(M_{ij}+|m_{ij}|)\lambda_j + \sum_{j=1}^{n}(D_{ij}\nu_j + C_{ij}\mu_j)\Big),$$

$$B_2 = \max_{i=1,2,\dots,n}\sum_{j=1}^{n}\Big[C_{ij}\mu_j + D_{ij}\nu_j + (M_{ij}+|m_{ij}|)\lambda_j\Big],$$

$$B_3 = 1 + \frac{r^q}{\Gamma(1+q)}\sum_{i=1}^{n}\sum_{j=1}^{n}(D_{ij}\nu_j + C_{ij}\mu_j).$$

Choose the positive constants $p = 1 + \frac{\alpha r^q}{\Gamma(1+q)}$. Then, $\frac{1}{E_q(-\alpha r^q)} \leq 1 + \frac{\alpha r^q}{\Gamma(1+q)} = p$ ( see (3.8) and (3.11) [23]).

Consider the quadratic functions $V(t,x) = x^T x$. Let the point $t > 0 : t \in (t_m, t_{m+1}]$, $m \geq 0$ being an integer, be such that $\sup_{s\in[-r,0]} V(t+s, e(t+s)) = \sup_{s\in[-r,0]} \sum_{j=1}^{n} e_j^2(t+s) = p\sum_{j=1}^{n} e_j^2(t) = pV(t, e(t))$ where $e(t)$ is a solution of (42). Then, since $\tau_j(t) \in [0, r]$, $t \geq 0$, we have $pV(t,e(t)) = p\sum_{j=1}^{n}(e_j(t))^2 \geq \sum_{j=1}^{n}(e_j(t-\tau_j(t)))^2 \geq (e_i(t-\tau_j(t)))^2, i = 1,2,\dots,n$. In addition, $(e_j(t+s))^2 \leq \sup_{s\in[-r,0]} V(t+s, e(t+s)) = pV(t,e(t))$, $s \in [-r, 0]$.

From conditions A1,A2, A3, A5, the choice of the constants $\alpha, p$, and Remark 9, we get for the chosen above point $t$:

$$\begin{aligned}
&{}^{C}_{t_m}D^q_t V(t,e(t)) = {}^{C}_{t_m}D^q_t e^T(t)e(t) \leq 2\sum_{i=1}^{n} e_i(t)\, {}^{C}_{t_m}D^q_t e_i(t)\\
&\leq -2\sum_{i=1}^{n} c_i e_i^2(t) + 2\sum_{i=1}^{n}\sum_{j=1}^{n}(|a_{ij}(t)| + |m_{ij}|)|F_j(e_j(t))||e_i(t)|\\
&\quad + 2\sum_{i=1}^{n}\sum_{j=1}^{n}|b_{ij}(t)||G_j(e_j(t-\tau(t)))|e_i(t)|\\
&\quad + 2\sum_{i=1}^{n}\sum_{j=1}^{n}|d_{ij}(t)|\int_{-r}^{0}|K_{ij}(s)||H_j(e_j(t+s))|ds|e_i(t)|\\
&\leq -2\sum_{i=1}^{n} c_i e_i^2(t) + \sum_{i=1}^{n}\sum_{j=1}^{n}(M_{ij}+|m_{ij}|)\lambda_j((e_j(t))^2 + (e_i(t))^2)\\
&\quad + \sum_{i=1}^{n}\sum_{j=1}^{n} C_{ij}\mu_j((e_j(t-\tau(t)))^2 + (e_i(t))^2)\\
&\quad + \sum_{i=1}^{n}\sum_{j=1}^{n}|d_{ij}(t)|\int_{-r}^{0}|K_{ij}(s)|\nu_j((e_j(t+s)))^2 + (e_i(t))^2)ds\\
&\leq \Big\{-2\eta + \sum_{i=1}^{n}\max_{j=1,2,\dots,n}(M_{ij}+|m_{ij}|)\lambda_j\\
&\quad + \max_{i=1,2,\dots,n}\sum_{j=1}^{n}\Big[(M_{ij}+|m_{ij}|)\lambda_j + C_{ij}\mu_j + D_{ij}\nu_j\Big]\\
&\quad + p(\sum_{i=1}^{n}\sum_{j=1}^{n}(D_{ij}\nu_j + C_{ij}\mu_j)\Big\}V(t,e(t)) \leq -\alpha V(t,e(t)).
\end{aligned} \tag{44}$$

Let $t = t_k$, $k$ be a natural number and $x \in \mathbb{R}^n$, $x - (x_1, x_2, \dots, x_n)$. Then, according to condition A4 and Remark 9, the inequalities

$$V(t, x + L_k(x)) = \sum_{i=1}^{n}(x_i + L_{k,i}(x))^2 \leq \sum_{i=1}^{n} A_{k,i}^2 x_i^2 \leq V(t,x) \tag{45}$$

hold.

Inequalities (44), (45) and Lemma 4 prove the claim. □

In the case when the conditions A3 and A5 are not satisfied (i.e., the bounds of the functions $c_i(t)$ are not small enough), we introduce:

**Assumption A6.** There exists a continuous positive function $m(t) \in C([0,\infty),(0,\infty))$ such that $0 < \beta \le m(t) \le \gamma$, $\beta, \gamma$ are constants, the fractional derivative ${}_{t_k}^{RL}D_t^q m(t)$ exists for $t \in (t_k, t_{k+1})$, $k = 0,1,2,\ldots$,

$$2\min_{i=1,2,\ldots,n} c_i(t) - \frac{{}_{t_k}^{RL}D^q\Big(m(t)\Big)}{m(t)} \ge \xi > 0 \ \text{ for } t \in (t_k, t_{k+1}),\ k = 0,1,2,\ldots,$$

and

$$\xi > \sum_{i=1}^{n}\Big(\max_{j=1,2,\ldots,n}(M_{ij} + |m_{ij}|)\lambda_j + \frac{1}{\beta}\sum_{j=1}^{n}(D_{ij}\nu_j + C_{ij}\mu_j)\Big)$$
$$+ \max_{i=1,2,\ldots,n}\sum_{j=1}^{n}\Big[(M_{ij} + |m_{ij}|)\lambda_j + C_{ij}\mu_j + D_{ij}\nu_j\Big].$$

**Theorem 2.** *Let ssumptions A1, A2, A4 and A6 be satisfied.*

*Then, the master impulsive Caputo fractional system (40) and the slave impulsive Caputo fractional system (41) are globally Mittag–Leffler synchronized.*

**Proof.** According to condition A6, there exists a positive constant $\alpha$ such that $\alpha \le \frac{\xi - B_4 - B_2}{B_5}$ where $B_2$ is defined in Theorem 1 and

$$B_4 = \sum_{i=1}^{n}\Big(\max_{j=1,2,\ldots,n}(M_{ij} + |m_{ij}|)\lambda_j + \frac{1}{\beta}\sum_{j=1}^{n}(D_{ij}\nu_j + C_{ij}\mu_j)\Big),$$

$$B_5 = 1 + \frac{r^q}{\beta\Gamma(1+q)}\sum_{i=1}^{n}\sum_{j=1}^{n}(D_{ij}\nu_j + C_{ij}\mu_j).$$

Choose the positive constants $p = 1 + \frac{\alpha r^q}{\Gamma(1+q)}$ and consider the Lyapunov function $V(t,x) = m(t)\sum_{i=1}^{n} x_i^2$, $x = (x_1, x_2, \ldots, x_n)$ where the function $m(t)$ is defined in Assumption A6.

Let $k$ be any nonegative integer, the point $t \in (t_k, t_{k+1})$ and the function $\psi \in C([-r,0],\mathbb{R}^n)$ be such that $V(t+s,\psi(s)) = m(t+s)\sum_{i=1}^{n}(\psi_i(s))^2 < pm(t)\sum_{i=1}^{n}(\psi_i(0))^2 = pV(t,\psi(0))$, $s \in [-r,0)$. Then, we have $(\psi_i(s))^2 \le \frac{1}{\beta}\beta(\psi_i(s))^2 \le \frac{1}{\beta}m(t+s)(\psi_i(s))^2 \le \frac{1}{\beta}m(t+s)\sum_{i=1}^{n}(\psi_i(s))^2 = \frac{1}{\beta}V(t+s,\psi(s)) < p\frac{1}{\beta}V(t,\psi(0)) = p\frac{1}{\beta}m(t)\sum_{i=1}^{n}(\psi_i(0))^2$, $s \in [-r,0]$.

From conditions A1, A2, A6, the choice of the constants $\alpha, p$, Example 2, Case 2 and Equation (15), we get for the chosen above point $t$ and function $\psi$:

$$
\begin{aligned}
&{}_{t_k}D^q_{(3)}V(t,\psi(0),\psi)\\
&\leq m(t)\sum_{i=1}^{n}\Big\{-2\sum_{i=1}^{n}c_i(t)|\psi_i(0)|^2+2\sum_{i=1}^{n}\sum_{j=1}^{n}|a_{ij}(t)+m_{ij}||F_j(\psi_j(0))|\psi_i(0)|\\
&\quad+2\sum_{i=1}^{n}\sum_{j=1}^{n}|b_{ij}(t)||G_j(\psi_j(\tau(0)))|\psi_i(0)|+{}^{RL}_{t_k}D^q\Big(m(t)\Big)\sum_{i=1}^{n}(\psi_(0))^2\\
&\quad+2\sum_{i=1}^{n}\sum_{j=1}^{n}|d_{ij}(t)|\int_{-r}^{0}|K_{ij}(s)||H_j(\psi_j(s))|ds|\psi_i(0)|\Big\}\\
&\leq\Big\{-2\min_{i=1,2,\dots,n}c_i(t)+\sum_{i=1}^{n}\max_{j=1,2,\dots,n}(M_{ij}+|m_{ij}|)\lambda_j+\frac{{}^{RL}_{t_k}D^q\Big(m(t)\Big)}{m(t)}\\
&\quad+\max_{i=1,2,\dots,n}\sum_{j=1}^{n}\Big[(M_{ij}+|m_{ij}|)\lambda_j+C_{ij}\mu_j+D_{ij}\nu_j\Big]\\
&\quad+p\frac{1}{\beta}(\sum_{i=1}^{n}\sum_{j=1}^{n}(D_{ij}\nu_j+C_{ij}\mu_j)\Big\}V(t,\psi(0))<-\alpha V(t,\psi(0)).
\end{aligned}
\tag{46}
$$

For any natural number $k$ and $x\in\mathbb{R}^n, x=(x_1,x_2,\dots,x_n)$ according to condition A4 and Remark 9, we have $V(t_k,x+L_k(x))\leq\sum_{i=1}^{n}A_{k,i}^2x_i^2\leq V(t,x)$.

According to Lemma 5, the claim of Theorem 2 follows. □

**Example 3.** *Consider the master impulsive Caputo fractional system (40) with $n=3$, $c_i(t)\equiv c_i$, with the activation functions $f_j(s)=g_j(s)=h_j(s)=0.5\tanh(s)$, the delays $\tau_i(t)=|sin(t)|\leq 1$, i.e., $r=1$ and $|a_{ij}(t)|\leq M_{ij}$, $|b_{ij}(t)|\leq C_{ij}$, $|d_{ij}(t)|\leq D_{ij}$ $i,j=1,2,3$, $t\geq 0$, where $M=\{M_{ij}\}, C=\{C_{ij}\}$ are given by*

$$
M=\begin{pmatrix}0.1&0.5&0.3\\0.2&0.3&0.2\\0.4&0.2&0.1\end{pmatrix},\quad C=\begin{pmatrix}0.1&0.1&0.2\\0.3&0.2&0.1\\0.2&0.5&0.3\end{pmatrix},\qquad D=\begin{pmatrix}0.2&0.1&0.1\\0.1&0.2&0.1\\0.2&0.3&0.1\end{pmatrix}.
$$

*Let the output coupling controller be $u_j(t)=(tanh(y_j(t))-tanh(x_j(t))\sum_{j=1}^{3}m_{ij}$, $i=1,2,3$ with*

$$
m=\begin{pmatrix}0.1&0.2&0.1\\0.2&0.1&0.1\\0.1&0.2&0.1\end{pmatrix}.
$$

*Then, $\lambda_i=\mu_i=\nu_i=0.5$ and $\sum_{i=1}^{3}\Big(\max_{j=1,2,3}(M_{ij}+m_{ij})\lambda_j+\sum_{j=1}^{3}(D_{ij}\nu_j+C_{ij}\mu_j)\Big)=2.5$ and $\max_{i=1,2,3}\sum_{j=1}^{3}\Big[(M_{ij}+m_{ij})\lambda_j+C_{ij}\mu_j+D_{ij}\nu_j\Big]=0.5\max_{i=1,2,3}\sum_{j=1}^{3}\Big[(M_{ij}+m_{ij})\lambda_j+C_{ij}\mu_j+D_{ij}\nu_j\Big]=1.35$.*

*Therefore, if $c_i>\frac{3.85}{2}=1.925$, $i=1,2,3$ then, according to Theorem 1, the master impulsive Caputo fractional system (40) and the slave impulsive Caputo fractional system (41) are globally Mittag–Leffler synchronized.*

**Example 4.** *Consider the master impulsive Caputo fractional system (40) with $n=2$, $q=0.3$, $t_k=k, k=0,1,2,\dots$, $c_i(t)=\frac{0.55}{(t-k)^{0.3}\Gamma(0.7)}$ for $t\in(k,k+1]$, with the activation functions $f_j(s)=g_j(s)=h_j(s)=\frac{1}{1+e^{-s}}$, the delays $\tau_i(t)=|sin(t)|\leq 1$, i.e., $r=1$ and $|a_{ij}(t)|\leq M_{ij}$, $|b_{ij}(t)|\leq C_{ij}$, $|d_{ij}(t)|\leq D_{ij}$ $i,j=1,2$, $t\geq 0$ where $M=\{M_{ij}\}, C=\{C_{ij}\}$ are given by*

$$
M=\begin{pmatrix}0.1&0.03\\0.02&0.3\end{pmatrix},\quad C=\begin{pmatrix}0.001&0.002\\0.003&0.001\end{pmatrix},\qquad D=\begin{pmatrix}0.002&0.001\\0.001&0.002\end{pmatrix}.
$$

*Let the output coupling controller be* $u_j(t) = (\frac{1}{1+e^{-y_j(t)}} - \frac{1}{1+e^{-x_j(t)}})\sum_{j=1}^{3} m_{ij},\ i = 1,2$ *with*

$$m = \begin{pmatrix} 0.1 & 0.02 \\ 0.02 & 0.1 \end{pmatrix} \quad M + m = \begin{pmatrix} 0.2 & 0.05 \\ 0.04 & 0.4 \end{pmatrix}.$$

*Let* $m(t) = E_{0.3}(-(t-k)^{0.3}) + 0.1$ *with* $\beta = 0.1$. *Then,* ${}_{t_k}^{RL}D^q\Big(m(t)\Big) = -E_{0.3}(-(t-k)^{0.3}) + \frac{1.1}{(t-k)^{0.3}\Gamma(0.7)}$ *and* $2\min_{i=1,2,\ldots,n} c_i(t) - \frac{{}_{t_k}^{RL}D^q\big(m(t)\big)}{m(t)} = E_{0.3}(-(t-k)^{0.3}) > E_{0.3}(-1) = \xi = 0.456594 > 0$.

*Then,* $\lambda_i = \mu_i = \nu_i = 0.25$ *and* $0.25\sum_{i=1}^{2}\Big(\max_{j=1,2}(M_{ij} + |m_{ij}|) + 10\sum_{j=1}^{2}(D_{ij} + C_{ij})\Big) = 0.1775$, $0.25\max_{i=1,2}\sum_{j=1}^{2}\Big[(M_{ij} + |m_{ij}|) + C_{ij} + D_{ij}\Big] = 0.114$. *Then,* $E_{0.3}(-1) > 0.1775 + 0.114$ *and according to Theorem 2 the master impulsive Caputo fractional system (40) and the slave impulsive Caputo fractional system (41) are globally Mittag–Leffler synchronized.*

#### 5.2.2. State Coupling Controllers

Note that in [25] the state coupling was used to achieve the exponential lag synchronization of chaotic neural networks with impulsive effects. Now, we will consider the case when the control inputs are $u_j(t) = N_j(y_j(t)) - x_j(t)),\ j = 1,2,\ldots,n$. Then, the synchronization error $e_i(t) = y_i(t) - x_i(t)$ will satisfy

$$\begin{aligned}
{}_{t_k}^{C}D_t^q e_i(t) &= -c_i(t)e_i(t) + \sum_{j=1}^{n} a_{ij}(t)F_j(e_j(t)) + \sum_{j=1}^{n} b_{ij}(t)G_j(e_j(t-\tau_j(t))) \\
&\quad + \sum_{j=1}^{n} d_{ij}(t)\int_{-r}^{0} K_{ij}(s)H_j(e_j(t+s))ds - N_i e_i(t) \text{ for } t \in (t_k, t_{k+1}], k = 0,1,2,\ldots, \\
\Delta e_i(t)|_{t=t_k} &= L_{k,i}(e_i(t_k - 0)), \qquad k = 1,2,\ldots \\
e_i(t) &= \Phi_i^0(t - t_0),\ \ t \in [t_0 - r, t_0],\ \ i = 1,2,\ldots n.
\end{aligned} \tag{47}$$

In this case, we can derive the following result (its proof is similar to the one in Theorem 1 and we omit it). We assume the following:

**Assumption A7.** The inequality

$$\begin{aligned}
2(\eta + \min_{i=1,2,\ldots,n} N_i) &> \sum_{i=1}^{n}\Big(\max_{j=1,2,\ldots,n} M_{ij}\lambda_j + \sum_{j=1}^{n}(D_{ij}\nu_j + C_{ij}\mu_j)\Big) \\
&\quad + \max_{i=1,2,\ldots,n}\sum_{j=1}^{n}\Big[M_{ij}\lambda_j + C_{ij}\mu_j + D_{ij}\nu_j\Big]
\end{aligned} \tag{48}$$

holds.

**Theorem 3.** *Let assumptions A1–A4 and A7 be satisfied.*

*Then, the master impulsive Caputo fractional system (40) and the slave impulsive Caputo fractional system (41) are globally Mittag–Leffler synchronized.*

In the case when assumptions A3 and A7 are not satisfied, we introduce the following:

**Assumption A8.** There exists a continuous positive function $m(t) \in C([0,\infty),(0,\infty))$ such that $0 < \beta \le m(t) \le \gamma$, $\beta, \gamma$ are constants, the fractional derivative ${}_{t_k}^{RL}D_t^q m(t)$ exists for $t \in (t_k, t_{k+1})$, $k = 0,1,2,\ldots,$

$$2 \min_{i=1,2,\ldots,n} (c_i(t) + N_i) - \frac{{}^{RL}_{t_k}D^q\Big(m(t)\Big)}{m(t)} \geq \xi > 0 \text{ for } t \in (t_k, t_{k+1}),\ k = 0,1,2,\ldots,$$

and

$$\xi > \sum_{i=1}^{n} \Big( \max_{j=1,2,\ldots,n} M_{ij}\lambda_j + \frac{1}{\beta}\sum_{j=1}^{n}(D_{ij}\nu_j + C_{ij}\mu_j)\Big) + \max_{i=1,2,\ldots,n} \sum_{j=1}^{n} \Big[M_{ij}\lambda_j + C_{ij}\mu_j + D_{ij}\nu_j\Big].$$

**Theorem 4.** *Let assumptions A1, A2, A4 and A8 be satisfied.*

*Then, the master impulsive Caputo fractional system (40) and the slave impulsive Caputo fractional system (41) are globally Mittag–Leffler synchronized.*

## 6. Conclusions

The paper presents sufficient conditions for the global Mittag–Leffler synchronization of a fractional-order neural network with time-varying and distributed delay and with impulsive effects. We consider the case of two types of controllers, output coupling controller and state coupling controller. The study is based on the application of the fractional generalization of the Lyapunov–Razumikhin technique. We study the case of the time varying rate with which the $i$-th neuron resets its potential to the resting state in isolation and time varying synaptic connection strength of the $i$-th neuron to the $j$-th neuron. The case when the lower bound of the Caputo fractional derivative is changeable at each point of impulse is investigated. Consequently, our results are significant for various applications in engineering and technology.

It would be interesting to extend our results to the case of non-Lipschitz discontinuous activation functions applying both approaches for the interpretation of solutions of fractional equations with impulses. This would lead to wider possibilities for appropriate modeling of the connections between neurons in the networks. This topic goes beyond the scope of this paper and will be a challenging issue for future research.

**Author Contributions:** Conceptualization, R.A., S.H. and D.O.; Methodology, R.A., S.H. and D.O.; Formal Analysis, R.A., S.H. and D.O.; Investigation, R.A., S.H. and D.O.;Writing—Original Draft Preparation, R.A., S.H. and D.O.; Writing—Review and Editing, R.A., S.H. and D.O.; Supervision, R.A., S.H. and D.O.; Funding Acquisition, S.H.

**Funding:** This research was partially supported by Fund of Plovdiv University FP17FMI008.

**Conflicts of Interest:** The authors declare no conflict of interest.

## References

1. Nouh, M.I.; Abed El Salam, E.A.B. Analytical solution to the fractional polytropic gas spheres. *Eur. Phys. J. Plus* **2018**, *149*, 133–149. [CrossRef]
2. Srivastava, T.; Singh, A.P.; Agarwal, H. Modeling the under-Aactuated mechanical system with fractional order derivative. *Progr. Fract. Differ. Appl.* **2015**, *1*, 57–64.
3. Longhi, S. Fractional Schrödinger equation in optics. *Opt. Lett.* **2015**, *40*, 1117–1120. [CrossRef] [PubMed]
4. Sadati, S.J.; Ghaderi, R.; Ranjbar, A. Some fractional comparison results and stability theorem for fractional time delay systems. *Romanian Rep. Phys.* **2013**, *65*, 94–102.
5. Wang, J.R.; Zhang, Y. Ulam-Hyers-Mittag–Leffler stability of fractional-order delay differential equations. *Optim. J. Math. Program. Oper. Res.* **2014**, *63*, 1181–1190. [CrossRef]
6. Hristova, S.; Stamova, I. On the Mittag–Leffler stability of impulsive fractional neural networks with finite delays. *Int. J. Pure Appl. Math.* **2016**, *109*, 105–117. [CrossRef]
7. Li, Y.; Chen, Y.; Podlubny, I. Mittag—Leffler stability of fractional order nonlinear dynamic systems. *Automatica* **2009**, *45*, 1965–1969. [CrossRef]

8. Sadati, S.J.; Baleanu, D.; Ranjbar, A.; Ghaderi, R.; Abdeljawad (Maraaba), T. Mittag–Leffler stability theorem for fractional nonlinear systems with delay. *Abstr. Appl. Anal.* **2010**, *2010*, 108651. [CrossRef]
9. Yang, X.; Li, C.; Huang, T.; Song, Q. Mittag–Leffler stability analysis of nonlinear fractional-order systems with impulses. *Appl. Math. Comput.* **2017**, *293*, 416–422. [CrossRef]
10. Peng, X.; Wu, H.; Song, K.; Shi, J. Global synchronization in finite time for fractional-order neural networks with discontinuous activations and time delays. *Neural Net.* **2017**, *94*, 46–54. [CrossRef] [PubMed]
11. Rifhat, R.; Muhammadhaji, A.; Teng, Z. Global Mittag–Leffler synchronization for impulsive fractional-order neural networks with delays. *Int. J. Nonlinear Sci. Numer. Simul.* **2018**, *19*, 205–213. [CrossRef]
12. Zhang, W.; Wu, R.; Cao, J.; Alsaedid, A.; Hayat, T. Synchronization of a class of fractional-order neural networks with multiple time delays by comparison principles. *Nonliinear Anal. Model. Control* **2017**, *22*, 636–645. [CrossRef]
13. Das, S. *Functional Fractional Calculus*; Springer: Berlin/Heidelberg, Germany, 2011.
14. Diethelm, K. *The Analysis of Fractional Differential Equations*; Springer: Berlin/Heidelberg, Germany, 2010.
15. Podlubny, I. *Fractional Differential Equations*; Academic Press: San Diego, CA, USA, 1999.
16. Devi, J.V.; Rae, F.A.M.; Drici, Z. Variational Lyapunov method for fractional differential equations. *Comput. Math. Appl.* **2012**, *64*, 2982–2989. [CrossRef]
17. Stamova, I.; Stamov, G. Stability analysis of impulsive functional systems of fractional order. *Commun. Nonlinear Sci. Numer. Simul.* **2014**, *19*, 702–709. [CrossRef]
18. Feckan, M.; Zhou, Y.; Wang, J.R. On the concept and existence of solution for impulsive fractional differential equations. *Commun. Nonlinear Sci. Numer. Simul.* **2012**, *17*, 3050–3060. [CrossRef]
19. Agarwal, R.; Hristova, S.; O'Regan, D. A survey of Lyapunov functions, stability and impulsive Caputo fractional differential equations. *Fract. Calc. Appl. Anal.* **2016**, *19*, 290–318. [CrossRef]
20. Wang, J.R.; Feckan, M.; Zhou, Y. A survey on impulsive fractional differential equations. *Fract. Calc. Appl. Anal.* **2016**, **19**, 806–831. [CrossRef]
21. Lakshmikantham, V.; Bainov, D.D.; Simeonov, P.S. *Theory of Impulsive Differential Equations*; World Scientiffic: Singapore, 1989.
22. Laskin, N. Principles of fractional quantum mechanics. *Fract. Dyn.* **2011**, 393–427. [CrossRef]
23. Mainardi, F. On some properties of the Mittag–Leffler function $E_\alpha(-t^\alpha)$, completely monotone for $t > 0$ with $0 < \alpha < 1$. *Discrt. Contin. Dyn. Syst. Ser. B* **2014**, *19*, 2267–2278.
24. Kaslik, E.; Sivasundaram, S. Nonlinear dynamics and chaos in fractional-order neural networks. *Neural Netw.* **2012**, *32*, 245–256. [CrossRef] [PubMed]
25. Yang, Y.; Cao, J. Exponential lag synchronization of a class of chaotic delayed neural networks with impulsive effects. *Phys. A* **2007**, *386*, 492–502. [CrossRef]

*symmetry*

MDPI

*Article*

# On Conformable Double Laplace Transform and One Dimensional Fractional Coupled Burgers' Equation

**Hassan Eltayeb [1], Imed Bachar [1] and Adem Kılıçman [2,3,*]**

1 Mathematics Department, College of Science, King Saud University, P.O. Box 2455, Riyadh 11451, Saudi Arabia; hgadain@ksu.edu.sa (H.E.); abachar@ksu.edu.sa (I.B.)
2 Department of Mathematics and Institute for Mathematical Research, Universiti Putra Malaysia, Serdang 43400 UPM, Selangor, Malaysia
3 Department of Electrical and Electronic Engineering, Istanbul Gelisim University, Avcilar, Istanbul 34310, Turkey
* Correspondence: akilic@upm.edu.my; Tel.: +60-3-8946-6813

Received: 24 February 2019; Accepted: 18 March 2019; Published: 21 March 2019

**Abstract:** In the present work we introduced a new method and name it the conformable double Laplace decomposition method to solve one dimensional regular and singular conformable functional Burger's equation. We studied the existence condition for the conformable double Laplace transform. In order to obtain the exact solution for nonlinear fractional problems, then we modified the double Laplace transform and combined it with the Adomian decomposition method. Later, we applied the new method to solve regular and singular conformable fractional coupled Burgers' equations. Further, in order to illustrate the effectiveness of present method, we provide some examples.

**Keywords:** conformable fractional derivative; conformable partial fractional derivative; conformable double Laplace decomposition method; conformable Laplace transform; singular one dimensional coupled Burgers' equation

## 1. Introduction

The fractional partial differential equations play a crucial role in mathematical and physical sciences. In [1], the authors studied the solution of some time-fractional partial differential equations by using a method known as simplest equation method. In this work, we deal with Burgers' equation, these type of equations have appeared in the area of applied sciences such as fluid mechanics and mathematical modeling. In fact, Burgers' equation was first proposed in [2], where the steady state solutions were discussed. Later it was modified by Burger, in order to solve the descriptive certain viscosity of flows. Today in the literature it is widely known as Burgers' equation, see [3]. Several researchers focused and concentrated to study the exact as well as the numerical solutions of this type of equation. In the present work, we considered and modified the conformable double Laplace transform method which was introduced in [4] in order to solve the fractional partial differential equations. The authors in [5] applied the first integral method to establish the exact solutions for time-fractional Burgers' equation. In [6], the researchers applied the generalized two-dimensional differential transform method (DTM) and obtained the solution for the coupled Burgers' equation with space- and time-fractional derivatives. Recently in [7], the conformable fractional Laplace transform method was applied to solve the coupled system of conformable fractional differential equations. Thus the aim of this study is to propose an analytic solution for the one dimensional regular and singular conformable fractional coupled Burgers' equation by using conformable double Laplace decomposition method (CDLDM). In [8], the following space-time fractional order coupled Burgers' equation, were considered

$$\begin{aligned} \frac{\partial^\beta u}{\partial t^\beta} - \frac{\partial^{2\alpha} u}{\partial x^{2\alpha}} + \eta u \frac{\partial^\alpha}{\partial x^\alpha} u + \zeta \frac{\partial^\alpha}{\partial x^\alpha}(uv) &= f\left(\frac{x^\alpha}{\alpha}, \frac{t^\beta}{\beta}\right) \\ \frac{\partial^\beta v}{\partial t^\beta} - \frac{\partial^{2\alpha} v}{\partial x^{2\alpha}} + \eta v \frac{\partial^\alpha}{\partial x^\alpha} v + \mu \frac{\partial^\alpha}{\partial x^\alpha}(uv) &= g\left(\frac{x^\alpha}{\alpha}, \frac{t^\beta}{\beta}\right). \end{aligned} \tag{1}$$

Conformable fractional derivatives were studied in [9] and extended in [10]. Next, we recall the definition of conformable fractional derivatives, which are used in this study.

**Definition 1.** *Let $f : (0, \infty) \to R$ then the conformable fractional derivative of $f$ order $\beta$ is defined by*

$$\frac{d^\beta}{dt^\beta} f\left(\frac{t^\beta}{\beta}\right) = \lim_{\epsilon \to 0} \frac{f\left(\frac{t^\beta}{\beta} + \epsilon t^{1-\beta}\right) - f\left(\frac{t^\beta}{\beta}\right)}{\epsilon}, \quad \frac{t^\beta}{\beta} > 0,\ 0 < \beta \leq 1,$$

*see [9,11,12].*

**Conformable Partial Derivatives:**

**Definition 2.** *([13]): Given a function $f\left(\frac{x^\alpha}{\alpha}, \frac{t^\beta}{\beta}\right) : R \times (0, \infty) \to R$. Then, the conformable space fractional partial derivative of order $\alpha$ a function $f\left(\frac{x^\alpha}{\alpha}, \frac{t^\beta}{\beta}\right)$ is defined as:*

$$\frac{\partial^\alpha}{\partial x^\alpha} f\left(\frac{x^\alpha}{\alpha}, \frac{t^\beta}{\beta}\right) = \lim_{\epsilon \to 0} \frac{f\left(\frac{x^\alpha}{\alpha} + \epsilon x^{1-\alpha}, t\right) - f\left(\frac{x^\alpha}{\alpha}, \frac{t^\beta}{\beta}\right)}{\epsilon}, \quad \frac{x^\alpha}{\alpha}, \frac{t^\beta}{\beta} > 0, 0 < \alpha, \beta \leq 1.$$

**Definition 3.** *([13]): Given a function $f\left(\frac{x^\alpha}{\alpha}, \frac{t^\beta}{\beta}\right) : R \times (0, \infty) \to R$. Then, the conformable time fractional partial derivative of order $\beta$ a function $f\left(\frac{x^\alpha}{\alpha}, \frac{t^\beta}{\beta}\right)$ is defined as:*

$$\frac{\partial^\beta}{\partial t^\beta} f\left(\frac{x^\alpha}{\alpha}, \frac{t^\beta}{\beta}\right) = \lim_{\sigma \to 0} \frac{f\left(\frac{x^\alpha}{\alpha}, \frac{t^\beta}{\beta} + \sigma t^{1-\beta}\right) - f\left(\frac{x^\alpha}{\alpha}, \frac{t^\beta}{\beta}\right)}{\sigma}, \quad \frac{x^\alpha}{\alpha}, \frac{t^\beta}{\beta} > 0,\ 0 < \alpha, \beta \leq 1.$$

**Conformable fractional derivatives of certain functions:**

**Example 1.** *We have the following*

$$\begin{aligned} \frac{\partial^\alpha}{\partial x^\alpha}\left(\frac{x^\alpha}{\alpha}\right)\left(\frac{t^\beta}{\beta}\right) &= \left(\frac{t^\beta}{\beta}\right), \quad \frac{\partial^\alpha}{\partial x^\alpha}\left(\frac{x^\alpha}{\alpha}\right)^n\left(\frac{t^\beta}{\beta}\right) = n\left(\frac{x^\alpha}{\alpha}\right)^{n-1}\left(\frac{t^\beta}{\beta}\right) \\ \frac{\partial^\beta}{\partial t^\beta}\left(\frac{x^\alpha}{\alpha}\right)\left(\frac{t^\beta}{\beta}\right) &= \left(\frac{x^\alpha}{\alpha}\right), \quad \frac{\partial^\beta}{\partial t^\beta}\left(\frac{x^\alpha}{\alpha}\right)^n\left(\frac{t^\beta}{\beta}\right)^m = m\left(\frac{x^\alpha}{\alpha}\right)^n\left(\frac{t^\beta}{\beta}\right)^{m-1} \\ \frac{\partial^\beta}{\partial t^\beta}\left(\sin\left(\frac{x^\alpha}{\alpha}\right)\sin\left(\frac{t^\beta}{\beta}\right)\right) &= \sin\left(\frac{x^\alpha}{\alpha}\right)\cos\left(\frac{t^\beta}{\beta}\right), \\ \frac{\partial^\alpha}{\partial x^\alpha}\left(\sin a\left(\frac{x^\alpha}{\alpha}\right)\sin\left(\frac{t^\beta}{\beta}\right)\right) &= a\cos\left(\frac{x^\alpha}{\alpha}\right)\sin\left(\frac{t^\beta}{\beta}\right) \\ \frac{\partial^\alpha}{\partial x^\alpha}\left(e^{\lambda\frac{x^\alpha}{\alpha} + \frac{\tau t^\beta}{\beta}}\right) &= \lambda e^{\lambda\frac{x^\alpha}{\alpha} + \frac{\tau t^\beta}{\beta}}, \quad \frac{\partial^\beta}{\partial t^\beta}\left(e^{\lambda\frac{x^\alpha}{\alpha} + \frac{\tau t^\beta}{\beta}}\right) = \tau e^{\lambda\frac{x^\alpha}{\alpha} + \frac{\tau t^\beta}{\beta}}. \end{aligned}$$

**Conformable Laplace transform:**

**Definition 4.** *([14]): Let $f : [0,\infty) \to \mathbb{R}$ be a real valued function. The conformable Laplace transform of $f$ is defined by*

$$L_t^{\beta}\left(f\left(\frac{t^{\beta}}{\beta}\right)\right) = \int_0^{\infty} e^{-s\frac{t^{\beta}}{\beta}} f\left(\frac{t^{\beta}}{\beta}\right) t^{\beta-1}dt$$

*for all values of s, provided the integral exists.*

**Definition 5.** *([4]): Let $u\left(\frac{x^{\alpha}}{\alpha},\frac{t^{\beta}}{\beta}\right)$ be a piecewise continuous function on the interval $[0,\infty)\times[0,\infty)$ having exponential order. Consider for some $a,b \in \mathbb{R}$ $\sup \frac{x^{\alpha}}{\alpha},\frac{t^{\beta}}{\beta} > 0$, $\frac{\left|u\left(\frac{x^{\alpha}}{\alpha},\frac{t^{\beta}}{\beta}\right)\right|}{e^{\frac{ax^{\alpha}}{\alpha}+\frac{bt^{\beta}}{\beta}}}$. Under these conditions the conformable double Laplace transform is given by*

$$L_x^{\alpha}L_t^{\beta}\left(u\left(\frac{x^{\alpha}}{\alpha},\frac{t^{\beta}}{\beta}\right)\right) = U(p,s) = \int_0^{\infty}\int_0^{\infty} e^{-p\frac{x^{\alpha}}{\alpha}-s\frac{t^{\beta}}{\beta}} u\left(\frac{x^{\alpha}}{\alpha},\frac{t^{\beta}}{\beta}\right) t^{\beta-1}x^{\alpha-1}dtdx$$

*where $p,s \in \mathbb{C}$, $0<\alpha,\beta\leq 1$ and the integrals are by means of conformable fractional with respect to $\frac{x^{\alpha}}{\alpha}$ and $\frac{t^{\beta}}{\beta}$ respectively.*

**Example 2.** *The double fractional Laplace transform for certain functions given by*

1. $L_x^{\alpha}L_t^{\beta}\left[\left(\frac{x^{\alpha}}{\alpha}\right)^n\left(\frac{t^{\beta}}{\beta}\right)^m\right] = L_xL_t\left[(x)^n(t)^m\right] = \frac{n!m!}{p^{n+1}s^{m+1}}.$
2. $L_x^{\alpha}L_t^{\beta}\left[e^{\lambda\frac{x^{\alpha}}{\alpha}+\frac{\tau t^{\beta}}{\beta}}\right] = L_xL_t\left[e^{\lambda x+\tau t}\right] = \frac{1}{(p-\lambda)(s-\tau)}.$
3. $L_x^{\alpha}L_t^{\beta}\left[\left(\sin(\lambda\frac{x^{\alpha}}{\alpha}\right)\sin\left(\tau\frac{t^{\beta}}{\beta}\right)\right] = L_xL_t\left[\sin(x)\sin(t)\right] = \frac{1}{p^2+\lambda^2}\frac{1}{s^2+\tau^2}.$
4. *If $a(>-1)$ and $b(>-1)$ are real numbers, then double fractional Laplace transform of the function $f\left(\frac{x^{\alpha}}{\alpha},\frac{t^{\beta}}{\beta}\right) = \left(\frac{x^{\alpha}}{\alpha}\right)^a\left(\frac{t^{\beta}}{\beta}\right)^b$ is given by*

$$L_x^{\alpha}L_t^{\beta}\left[(\frac{x^{\alpha}}{\alpha})^a(\frac{t^{\beta}}{\beta})^b\right] = \frac{\Gamma(a+1)\Gamma(b+1)}{p^{a+1}s^{b+1}}.$$

**Theorem 1.** *Let $0<\alpha,\beta\leq 1$ and $m,n\in\mathbb{N}$ such that $u\left(\frac{x^{\alpha}}{\alpha},\frac{t^{\beta}}{\beta}\right)\in C^l(\mathbb{R}^+\times\mathbb{R}^+)$, $l=\max(m,n)$. Further let the conformable Laplace transforms of the functions given as $u\left(\frac{x^{\alpha}}{\alpha},\frac{t^{\beta}}{\beta}\right)$, $\frac{\partial^{m\alpha}u}{\partial x^{m\alpha}}$ and $\frac{\partial^{n\beta}u}{\partial t^{n\beta}}$. Then*

$$L_x^{\alpha}L_t^{\beta}\left(\frac{\partial^{m\alpha}u}{\partial x^{m\alpha}}\right) = p^mU(p,s) - p^{m-1}U(0,s) - \sum_{i=1}^{m-1} p^{m-1-i}L_t^{\beta}\left(\frac{\partial^{i\alpha}}{\partial x^{i\alpha}}u\left(0,\frac{t^{\beta}}{\beta}\right)\right)$$

$$L_x^{\alpha}L_t^{\beta}\left(\frac{\partial^{n\beta}}{\partial t^{n\beta}}u\left(\frac{x^{\alpha}}{\alpha},\frac{t^{\beta}}{\beta}\right)\right) = s^nU(p,s) - s^{n-1}U(p,0) - \sum_{j=1}^{n-1} s^{n-1-j}L_x^{\alpha}\left(\frac{\partial^{j\beta}}{\partial t^{j\beta}}u\left(\frac{x^{\alpha}}{\alpha},0\right)\right)$$

*where $\frac{\partial^{m\alpha}u}{\partial x^{m\alpha}}$ and $\frac{\partial^{n\beta}v}{\partial t^{n\beta}}$ denotes $m,n$ times conformable fractional derivatives of function $u\left(\frac{x^{\alpha}}{\alpha},\frac{t^{\beta}}{\beta}\right)$, for more details see [4].*

In the following theorem, we study double Laplace transform of the function $\left(\frac{x^{\alpha}}{\alpha}\right)^n\frac{\partial^{\beta}}{\partial t^{\beta}}f\left(\frac{x^{\alpha}}{\alpha},\frac{t^{\beta}}{\beta}\right)$ as follows:

**Theorem 2.** *If conformable double Laplace transform of the partial derivatives* $\frac{\partial^\beta}{\partial t^\beta} f\left(\frac{x^\alpha}{\alpha}, \frac{t^\beta}{\beta}\right)$ *is given by Equation (27), then double Laplace transform of* $\left(\frac{x^\alpha}{\alpha}\right)^n \frac{\partial^\beta}{\partial t^\beta} f\left(\frac{x^\alpha}{\alpha}, \frac{t^\beta}{\beta}\right)$ *and* $\left(\frac{x^\alpha}{\alpha}\right)^n g\left(\frac{x^\alpha}{\alpha}, \frac{t^\beta}{\beta}\right)$ *are given by*

$$(-1)^n \frac{d^n}{dp^n}\left(L_x^\alpha L_t^\beta\left[\frac{\partial^\beta}{\partial t^\beta} f\left(\frac{x^\alpha}{\alpha}, \frac{t^\beta}{\beta}\right)\right]\right) = L_x^\alpha L_t^\beta\left[\left(\frac{x^\alpha}{\alpha}\right)^n \frac{\partial^\beta}{\partial t^\beta} f\left(\frac{x^\alpha}{\alpha}, \frac{t^\beta}{\beta}\right)\right] \tag{2}$$

*and*

$$(-1)^n \frac{d^n}{dp^n}\left(L_x^\alpha L_t^\beta\left[g\left(\frac{x^\alpha}{\alpha}, \frac{t^\beta}{\beta}\right)\right]\right) = L_x^\alpha L_t^\beta\left[\left(\frac{x^\alpha}{\alpha}\right)^n g\left(\frac{x^\alpha}{\alpha}, \frac{t^\beta}{\beta}\right)\right], \tag{3}$$

*where* $n = 1, 2, 3, \ldots$.

**Proof.** Using the definition of double Laplace transform of the fractional partial derivatives one gets

$$L_x^\alpha L_t^\beta\left[\frac{\partial^\beta}{\partial t^\beta} f(\frac{x^\alpha}{\alpha}, \frac{t^\beta}{\beta})\right] = \int_0^\infty \int_0^\infty e^{-p\frac{x^\alpha}{\alpha} - s\frac{t^\beta}{\beta}} \left(\frac{\partial^\beta}{\partial t^\beta} f\left(\frac{x^\alpha}{\alpha}, \frac{t^\beta}{\beta}\right)\right) t^{\beta-1} x^{\alpha-1} dt\, dx, \tag{4}$$

by taking the ${}^n$th derivative with respect to $p$ for both sides of Equation (4), we have

$$\begin{aligned}
\frac{d^n}{dp^n}\left(L_x^\alpha L_t^\beta\left[\frac{\partial^\beta}{\partial t^\beta} f\left(\frac{x^\alpha}{\alpha}, \frac{t^\beta}{\beta}\right)\right]\right) &= \int_0^\infty \int_0^\infty \frac{d^n}{dp^n}\left(e^{-p\frac{x^\alpha}{\alpha} - s\frac{t^\beta}{\beta}} \frac{\partial^\beta}{\partial t^\beta} f\left(\frac{x^\alpha}{\alpha}, \frac{t^\beta}{\beta}\right)\right) t^{\beta-1} x^{\alpha-1} dt dx \\
&= (-1)^n \int_0^\infty \int_0^\infty \left(\frac{x^\alpha}{\alpha}\right)^n e^{-p\frac{x^\alpha}{\alpha} - s\frac{t^\beta}{\beta}} t^{\beta-1} x^{\alpha-1} \frac{\partial^\beta}{\partial t^\beta} f\left(\frac{x^\alpha}{\alpha}, \frac{t^\beta}{\beta}\right) dt\, dx \\
&= (-1)^n L_x^\alpha L_t^\beta\left[\left(\frac{x^\alpha}{\alpha}\right)^n \frac{\partial^\beta}{\partial t^\beta} f(\frac{x^\alpha}{\alpha}, \frac{t^\beta}{\beta})\right],
\end{aligned}$$

thus we obtain

$$(-1)^n \frac{d^n}{dp^n}\left(L_x^\alpha L_t^\beta\left[\frac{\partial^\beta}{\partial t^\beta} f\left(\frac{x^\alpha}{\alpha}, \frac{t^\beta}{\beta}\right)\right]\right) = L_x^\alpha L_t^\beta\left[\left(\frac{x^\alpha}{\alpha}\right)^n \frac{\partial^\beta}{\partial t^\beta} f\left(\frac{x^\alpha}{\alpha}, \frac{t^\beta}{\beta}\right)\right].$$

Similarly, we can prove Equation (3). □

**Existence Condition for the conformable double Laplace transform:**

If $f\left(\frac{x^\alpha}{\alpha}, \frac{t^\beta}{\beta}\right)$ is an exponential order $a$ and $b$ as $\frac{x^\alpha}{\alpha} \to \infty$, $\frac{t^\beta}{\beta} \to \infty$, if there exists a positive constant $K$ such that for all $x > X$ and $t > T$

$$\left|f\left(\frac{x^\alpha}{\alpha}, \frac{t^\beta}{\beta}\right)\right| \le K e^{a\frac{x^\alpha}{\alpha} + b\frac{t^\beta}{\beta}}, \tag{5}$$

it is easy to get,

$$f\left(\frac{x^\alpha}{\alpha}, \frac{t^\beta}{\beta}\right) = O\left(e^{a\frac{x^\alpha}{\alpha} + b\frac{t^\beta}{\beta}}\right) \text{ as } \frac{x^\alpha}{\alpha} \to \infty, \frac{t^\beta}{\beta} \to \infty.$$

Or, equivalently,

$$\lim_{\substack{\frac{x^\alpha}{\alpha} \to \infty \\ \frac{t^\beta}{\beta} \to \infty}} e^{-\mu\frac{x^\alpha}{\alpha} - \eta\frac{t^\beta}{\beta}} \left|f\left(\frac{x^\alpha}{\alpha}, \frac{t^\beta}{\beta}\right)\right| = K \lim_{\substack{\frac{x^\alpha}{\alpha} \to \infty \\ \frac{t^\beta}{\beta} \to \infty}} e^{-(\mu - a)\frac{x^\alpha}{\alpha} - (\eta - b)\frac{t^\beta}{\beta}} = 0,$$

where $\mu > a$ and $\eta > b$. The function $f(\frac{x^\alpha}{\alpha}, \frac{t^\beta}{\beta})$ is called an exponential order as $\frac{x^\alpha}{\alpha} \to \infty$, $\frac{t^\beta}{\beta} \to \infty$, and clearly, it does not grow faster than $Ke^{a\frac{x^\alpha}{\alpha}+b\frac{t^\beta}{\beta}}$ as $\frac{x^\alpha}{\alpha} \to \infty$, $\frac{t^\beta}{\beta} \to \infty$.

**Theorem 3.** *If a function $f\left(\frac{x^\alpha}{\alpha}, \frac{t^\beta}{\beta}\right)$ is a continuous function in every finite intervals $(0, X)$ and $(0, T)$ and of exponential order $e^{a\frac{x^\alpha}{\alpha}+b\frac{t^\beta}{\beta}}$, then the conformable double Laplace transform of $f(\frac{x^\alpha}{\alpha}, \frac{t^\beta}{\beta})$ exists for all $Re(p) > \mu, Re(s) > \eta$.*

**Proof.** From the definition of the conformable double Laplace transform of $f(\frac{x^\alpha}{\alpha}, \frac{t^\beta}{\beta})$, we have

$$\begin{aligned} |U(p,s)| &= \left| \int_0^\infty \int_0^\infty e^{-p\frac{x^\alpha}{\alpha}-s\frac{t^\beta}{\beta}} f(\tfrac{x^\alpha}{\alpha}, \tfrac{t^\beta}{\beta}) t^{\beta-1} x^{\alpha-1} dt\, dx \right| \\ &\leq K \left| \int_0^\infty \int_0^\infty e^{-(p-a)\frac{x^\alpha}{\alpha}-(s-b)\frac{t^\beta}{\beta}} t^{\beta-1} x^{\alpha-1} dt\, dx \right| \\ &= \frac{K}{(p-a)(s-b)}. \end{aligned} \tag{6}$$

For $Re(p) > \mu, Re(s) > \eta$, from Equation (6), we have

$$\lim_{\substack{p\to\infty \\ s\to\infty}} |U(p,s)| = 0 \text{ or } \lim_{\substack{p\to\infty \\ s\to\infty}} U(p,s) = 0.$$

□

## 2. One Dimensional Fractional Coupled Burgers' Equation

In this section, we discuss the solution of regular and singular one dimensional conformable fractional coupled Burgers' equation by using conformable double Laplace decomposition methods (CDLDM). We note that if $\alpha = 1$ and $\beta = 1$ in the following problems, one can obtain the problems which was studied in [15]:

**The first problem:** One dimensional conformable fractional coupled Burgers' equation is given by

$$\begin{aligned} \frac{\partial^\beta u}{\partial t^\beta} - \frac{\partial^{2\alpha} u}{\partial x^{2\alpha}} + \eta u \frac{\partial^\alpha}{\partial x^\alpha} u + \zeta \frac{\partial^\alpha}{\partial x^\alpha}(uv) &= f\left(\frac{x^\alpha}{\alpha}, \frac{t^\beta}{\beta}\right) \\ \frac{\partial^\beta v}{\partial t^\beta} - \frac{\partial^{2\alpha} v}{\partial x^{2\alpha}} + \eta v \frac{\partial^\alpha}{\partial x^\alpha} v + \mu \frac{\partial^\alpha}{\partial x^\alpha}(uv) &= g\left(\frac{x^\alpha}{\alpha}, \frac{t^\beta}{\beta}\right), \end{aligned} \tag{7}$$

subject to

$$u\left(\frac{x^\alpha}{\alpha}, 0\right) = f_1\left(\frac{x^\alpha}{\alpha}\right), \quad v\left(\frac{x^\alpha}{\alpha}, 0\right) = g_1\left(\frac{x^\alpha}{\alpha}\right). \tag{8}$$

for $t > 0$. Here, $f\left(\frac{x^\alpha}{\alpha}, \frac{t^\beta}{\beta}\right)$, $g\left(\frac{x^\alpha}{\alpha}, \frac{t^\beta}{\beta}\right)$, $f_1\left(\frac{x^\alpha}{\alpha}\right)$ and $g_1\left(\frac{x^\alpha}{\alpha}\right)$ are given functions, $\eta$, $\zeta$ and $\mu$ are arbitrary constants depend on the system parameters such as; Peclet number, Stokes velocity of particles due to gravity and Brownian diffusivity, see [16]. By taking conformable double Laplace transform for both sides of Equation (7) and conformable single Laplace transform for Equation (8), we have

$$U(p,s) = \frac{F_1(p)}{s} + \frac{F(p,s)}{s} + \frac{1}{s} L_x^\alpha L_t^\beta \left[ \frac{\partial^{2\alpha} u}{\partial x^{2\alpha}} - \eta u \frac{\partial^\alpha}{\partial x^\alpha} u - \zeta \frac{\partial^\alpha}{\partial x^\alpha}(uv) \right], \tag{9}$$

and

$$V(p,s) = \frac{G_1(p)}{s} + \frac{G(p,s)}{s} + \frac{1}{s} L_x^\alpha L_t^\beta \left[ \frac{\partial^{2\alpha} v}{\partial x^{2\alpha}} - \eta v \frac{\partial^\alpha}{\partial x^\alpha} v - \mu \frac{\partial^\alpha}{\partial x^\alpha}(uv) \right]. \tag{10}$$

The conformable double Laplace decomposition methods (CDLDM) defines the solution of one dimensional conformable fractional coupled Burgers' equation as $u\left(\frac{x^{\alpha}}{\alpha},\frac{t^{\beta}}{\beta}\right)$ and $v\left(\frac{x^{\alpha}}{\alpha},\frac{t^{\beta}}{\beta}\right)$ by the infinite series

$$u\left(\frac{x^{\alpha}}{\alpha},\frac{t^{\beta}}{\beta}\right)=\sum_{n=0}^{\infty}u_n\left(\frac{x^{\alpha}}{\alpha},\frac{t^{\beta}}{\beta}\right),\quad v\left(\frac{x^{\alpha}}{\alpha},\frac{t^{\beta}}{\beta}\right)=\sum_{n=0}^{\infty}v_n\left(\frac{x^{\alpha}}{\alpha},\frac{t^{\beta}}{\beta}\right). \tag{11}$$

We can give Adomian's polynomials $A_n$, $B_n$ and $C_n$ respectively as follows

$$A_n=\sum_{n=0}^{\infty}u_nu_{xn},\ B_n=\sum_{n=0}^{\infty}v_nv_{xn},\ C_n=\sum_{n=0}^{\infty}u_nv_n. \tag{12}$$

In particular, the Adomian polynomials for the nonlinear terms $uu_x$, $vv_x$ and $uv$ can be computed by the following equations

$$\begin{aligned}
A_0 &= u_0u_{0x}\\
A_1 &= u_0u_{1x}+u_1u_{0x}\\
A_2 &= u_0u_{2x}+u_1u_{1x}+u_2u_{0x},\\
A_3 &= u_0u_{3x}+u_1u_{2x}+u_2u_{1x}+u_3u_{0x},\\
A_4 &= u_0u_{4x}+u_1u_{3x}+u_2u_{2x}+u_3u_{1x}+u_4u_{0x},
\end{aligned} \tag{13}$$

$$\begin{aligned}
B_0 &= v_0v_{0x}\\
B_1 &= v_0v_{1x}+v_1v_{0x},\\
B_2 &= v_0v_{2x}+v_1v_{1x}+v_2v_{0x},\\
B_3 &= v_0v_{3x}+v_1v_{2x}+v_2v_{1x}+v_3v_{0x},\\
B_4 &= v_0v_{4x}+v_1v_{3x}+v_2v_{2x}+v_3v_{1x}+v_4v_{0x}.
\end{aligned} \tag{14}$$

and

$$\begin{aligned}
C_0 &= u_0v_0\\
C_1 &= u_0v_1+u_1v_0\\
C_2 &= u_0v_2+u_1v_1+u_2v_0.\\
C_3 &= u_0v_3+u_1v_2+u_2v_1+u_3v_0,\\
C_4 &= u_0v_4+u_1v_3+u_2v_2+u_3v_1+u_4v_0.
\end{aligned} \tag{15}$$

By applying the inverse conformable double Laplace transform on both sides of Equations (9) and (10), making use of Equation (12), we have

$$\begin{aligned}
\sum_{n=0}^{\infty}u_n\left(\tfrac{x^{\alpha}}{\alpha},\tfrac{t^{\beta}}{\beta}\right) &= f_1(x)+L_p^{-1}L_s^{-1}\left[\tfrac{F(p,s)}{s}\right]+L_p^{-1}L_s^{-1}\left[\tfrac{1}{s}L_x^{\alpha}L_t^{\beta}\left[\tfrac{\partial^{2\alpha}u_n}{\partial x^{2\alpha}}\right]\right]\\
&\quad -L_p^{-1}L_s^{-1}\left[\tfrac{1}{s}L_x^{\alpha}L_t^{\beta}\left[\eta A_n\right]\right]-L_p^{-1}L_s^{-1}\left[\tfrac{1}{s}L_x^{\alpha}L_t^{\beta}\left[\zeta\left(C_n\right)\right]\right],
\end{aligned} \tag{16}$$

and

$$\begin{aligned}
\sum_{n=0}^{\infty}v_n\left(\tfrac{x^{\alpha}}{\alpha},\tfrac{t^{\beta}}{\beta}\right) &= g_1(x)+L_p^{-1}L_s^{-1}\left[\tfrac{G(p,s)}{s}\right]+L_p^{-1}L_s^{-1}\left[\tfrac{1}{s}L_x^{\alpha}L_t^{\beta}\left[\tfrac{\partial^{2\alpha}v_n}{\partial x^{2\alpha}}\right]\right]\\
&\quad -L_p^{-1}L_s^{-1}\left[\tfrac{1}{s}L_x^{\alpha}L_t^{\beta}\left[\eta B_n\right]\right]-L_p^{-1}L_s^{-1}\left[\tfrac{1}{s}L_x^{\alpha}L_t^{\beta}\left[\mu\left(C_n\right)\right]\right].
\end{aligned} \tag{17}$$

On comparing both sides of the Equations (16) and (17) we have

$$\begin{aligned} u_0 &= f_1(x) + L_p^{-1}L_s^{-1}\left[\tfrac{F(p,s)}{s}\right], \\ v_0 &= g_1(x) + L_p^{-1}L_s^{-1}\left[\tfrac{G(p,s)}{s}\right]. \end{aligned} \tag{18}$$

In general, the recursive relation is given by the following equations

$$u_{n+1} = L_p^{-1}L_s^{-1}\left[\frac{1}{s}L_x^{\alpha}L_t^{\beta}\left[\frac{\partial^{2\alpha}u_n}{\partial x^{2\alpha}}\right]\right] - L_p^{-1}L_s^{-1}\left[\frac{1}{s}L_x^{\alpha}L_t^{\beta}\left[\eta A_n\right]\right] - L_p^{-1}L_s^{-1}\left[\frac{1}{s}L_x^{\alpha}L_t^{\beta}\left[\zeta\left(C_n\right)\right]\right], \tag{19}$$

and

$$v_{n+1} = L_p^{-1}L_s^{-1}\left[\frac{1}{s}L_x^{\alpha}L_t^{\beta}\left[\frac{\partial^{2\alpha}v_n}{\partial x^{2\alpha}}\right]\right] - L_p^{-1}L_s^{-1}\left[\frac{1}{s}L_x^{\alpha}L_t^{\beta}\left[\eta B_n\right]\right] - L_p^{-1}L_s^{-1}\left[\frac{1}{s}L_x^{\alpha}L_t^{\beta}\left[\mu\left(C_n\right)\right]\right], \tag{20}$$

provided that the double inverse Laplace transform with respect to $p$ and $s$ exist in the above equations. In order to illustrate this method for one dimensional conformable fractional coupled Burgers' equation we provide the following example:

**Example 3.** *Consider the homogeneous one dimensional conformable fractional coupled Burgers' equation*

$$\begin{aligned} \tfrac{\partial^{\beta}u}{\partial t^{\beta}} - \tfrac{\partial^{2\alpha}u}{\partial x^{2\alpha}} - 2u\,\tfrac{\partial^{\alpha}}{\partial x^{\alpha}}u + \tfrac{\partial^{\alpha}}{\partial x^{\alpha}}(uv) &= 0 \\ \tfrac{\partial^{\beta}v}{\partial t^{\beta}} - \tfrac{\partial^{2\alpha}v}{\partial x^{2\alpha}} - 2v\,\tfrac{\partial^{\alpha}}{\partial x^{\alpha}}v + \tfrac{\partial^{\alpha}}{\partial x^{\alpha}}(uv) &= 0, \end{aligned} \tag{21}$$

*with initial condition*

$$u\left(\frac{x^{\alpha}}{\alpha},0\right) = \sin\left(\frac{x^{\alpha}}{\alpha}\right),\quad v\left(\frac{x^{\alpha}}{\alpha},0\right) = \sin\left(\frac{x^{\alpha}}{\alpha}\right). \tag{22}$$

*By using Equations (18)–(20) we have*

$$\begin{aligned} u_0 &= \sin\left(\frac{x^{\alpha}}{\alpha}\right),\ v_0 = \sin\left(\frac{x^{\alpha}}{\alpha}\right) \\ u_1 &= L_p^{-1}L_s^{-1}\left[\frac{1}{s}L_x^{\alpha}L_t^{\beta}\left[\frac{\partial^{2\alpha}u_0}{\partial x^{2\alpha}} + 2u_0\frac{\partial^{\alpha}u_0}{\partial x^{\alpha}} - \frac{\partial^{\alpha}}{\partial x^{\alpha}}(u_0v_0)\right]\right] \\ &= L_p^{-1}L_s^{-1}\left[\frac{1}{s}L_x^{\alpha}L_t^{\beta}\left[-\sin\left(\frac{x^{\alpha}}{\alpha}\right)\right]\right] = L_p^{-1}L_s^{-1}\left[\frac{1}{s^2\left(p^2+1\right)}\right] = -\frac{t^{\beta}}{\beta}\sin\left(\frac{x^{\alpha}}{\alpha}\right), \\ v_1 &= L_p^{-1}L_s^{-1}\left[\frac{1}{s}L_x^{\alpha}L_t^{\beta}\left[\frac{\partial^{2\alpha}v_0}{\partial x^{2\alpha}} + 2v_0\frac{\partial^{\alpha}v_0}{\partial x^{\alpha}} - \frac{\partial^{\alpha}}{\partial x^{\alpha}}(u_0v_0)\right]\right] \\ &= L_p^{-1}L_s^{-1}\left[\frac{1}{s}L_x^{\alpha}L_t^{\beta}\left[-\sin\left(\frac{x^{\alpha}}{\alpha}\right)\right]\right] = L_p^{-1}L_s^{-1}\left[\frac{1}{s^2\left(p^2+1\right)}\right] = -\frac{t^{\beta}}{\beta}\sin\left(\frac{x^{\alpha}}{\alpha}\right) \end{aligned}$$

$$\begin{aligned} u_2 &= L_p^{-1}L_s^{-1}\left[\frac{1}{s}L_x^{\alpha}L_t^{\beta}\left[\frac{\partial^{2\alpha}u_1}{\partial x^{2\alpha}} + 2\left(u_0\frac{\partial^{\alpha}u_1}{\partial x^{\alpha}} + u_1\frac{\partial^{\alpha}u_0}{\partial x^{\alpha}}\right) - \frac{\partial^{\alpha}}{\partial x^{\alpha}}(u_0v_1 + u_1v_0)\right]\right] \\ &= L_p^{-1}L_s^{-1}\left[\frac{1}{s}L_x^{\alpha}L_t^{\beta}\left[\frac{t^{\beta}}{\beta}\sin\left(\frac{x^{\alpha}}{\alpha}\right)\right]\right] = L_p^{-1}L_s^{-1}\left[\frac{1}{s^3\left(p^2+1\right)}\right] = \frac{\left(\frac{t^{\beta}}{\beta}\right)^2}{2}\sin\left(\frac{x^{\alpha}}{\alpha}\right), \end{aligned}$$

$$\begin{aligned} v_2 &= L_p^{-1}L_s^{-1}\left[\frac{1}{s}L_x^{\alpha}L_t^{\beta}\left[\frac{\partial^{2\alpha}v_1}{\partial x^{2\alpha}} + 2\left(v_0\frac{\partial^{\alpha}v_1}{\partial x^{\alpha}} + v_1\frac{\partial^{\alpha}v_0}{\partial x^{\alpha}}\right) - \frac{\partial^{\alpha}}{\partial x^{\alpha}}(u_0v_1 + u_1v_0)\right]\right] \\ &= L_p^{-1}L_s^{-1}\left[\frac{1}{s}L_x^{\alpha}L_t^{\beta}\left[\frac{t^{\beta}}{\beta}\sin\left(\frac{x^{\alpha}}{\alpha}\right)\right]\right] = L_p^{-1}L_s^{-1}\left[\frac{1}{s^3\left(p^2+1\right)}\right] = \frac{\left(\frac{t^{\beta}}{\beta}\right)^2}{2}\sin\left(\frac{x^{\alpha}}{\alpha}\right), \end{aligned}$$

*and*

$$\begin{aligned} u_3 &= L_p^{-1}L_s^{-1}\left[\frac{1}{s}L_x^{\alpha}L_t^{\beta}\left[\frac{\partial^{2\alpha}u_2}{\partial x^{2\alpha}}+2\left(u_0\frac{\partial^{\alpha}u_2}{\partial x^{\alpha}}+u_1\frac{\partial^{\alpha}u_1}{\partial x^{\alpha}}+u_2\frac{\partial^{\alpha}}{\partial x^{\alpha}}u_0\right)\right]\right] \\ &\quad -L_p^{-1}L_s^{-1}\left[\frac{1}{s}L_x^{\alpha}L_t^{\beta}\left[\frac{\partial^{\alpha}}{\partial x^{\alpha}}\left(u_0v_2+u_1v_1+u_2v_0\right)\right]\right] \\ &= -\frac{\left(\frac{t^{\beta}}{\beta}\right)^3}{6}\sin\left(\frac{x^{\alpha}}{\alpha}\right), \\ v_3 &= L_p^{-1}L_s^{-1}\left[\frac{1}{s}L_x^{\alpha}L_t^{\beta}\left[\frac{\partial^{2\alpha}v_2}{\partial x^{2\alpha}}+2\left(v_0\frac{\partial^{\alpha}v_2}{\partial x^{\alpha}}+v_1\frac{\partial^{\alpha}v_1}{\partial x^{\alpha}}+v_2\frac{\partial^{\alpha}}{\partial x^{\alpha}}v_0\right)\right]\right] \\ &\quad -L_p^{-1}L_s^{-1}\left[\frac{1}{s}L_x^{\alpha}L_t^{\beta}\left[\frac{\partial^{\alpha}}{\partial x^{\alpha}}\left(u_0v_2+u_1v_1+u_2v_0\right)\right]\right] \\ &= -\frac{\left(\frac{t^{\beta}}{\beta}\right)^3}{6}\sin\left(\frac{x^{\alpha}}{\alpha}\right), \end{aligned}$$

*and similar to the other components. Therefore, by using Equation (11), the series solutions are given by*

$$\begin{aligned} u\left(\frac{x^{\alpha}}{\alpha},\frac{t^{\beta}}{\beta}\right) &= u_0+u_2+u_3+\ldots=\left(1-\left(\frac{t^{\beta}}{\beta}\right)+\frac{\left(\frac{t^{\beta}}{\beta}\right)^2}{2!}-\frac{\left(\frac{t^{\beta}}{\beta}\right)^3}{3!}+\ldots\right)\sin\left(\frac{x^{\alpha}}{\alpha}\right) \\ v\left(\frac{x^{\alpha}}{\alpha},\frac{t^{\beta}}{\beta}\right) &= v_0+v_2+v_3+\ldots=\left(1-\left(\frac{t^{\beta}}{\beta}\right)+\frac{\left(\frac{t^{\beta}}{\beta}\right)^2}{2!}-\frac{\left(\frac{t^{\beta}}{\beta}\right)^3}{3!}+\ldots\right)\sin\left(\frac{x^{\alpha}}{\alpha}\right) \end{aligned}$$

*and hence the exact solutions become*

$$u\left(\frac{x^{\alpha}}{\alpha},\frac{t^{\beta}}{\beta}\right)=e^{-\frac{t^{\beta}}{\beta}}\sin\left(\frac{x^{\alpha}}{\alpha}\right),\quad v\left(\frac{x^{\alpha}}{\alpha},\frac{t^{\beta}}{\beta}\right)=e^{-\frac{t^{\beta}}{\beta}}\sin\left(\frac{x^{\alpha}}{\alpha}\right).$$

*By taking $\alpha=1$ and $\beta=1$, the fractional solution become*

$$u\left(\frac{x^{\alpha}}{\alpha},\frac{t^{\beta}}{\beta}\right)=e^{-t}\sin x,\quad v\left(\frac{x^{\alpha}}{\alpha},\frac{t^{\beta}}{\beta}\right)=e^{-t}\sin x.$$

**The second problem:** Now consider the singular one dimensional conformable fractional coupled Burgers' equation with Bessel operator

$$\begin{aligned} \frac{\partial^{\beta}u}{\partial t^{\beta}}-\frac{\alpha}{x^{\alpha}}\frac{\partial^{\alpha}}{\partial x^{\alpha}}\left(\frac{x^{\alpha}}{\alpha}\frac{\partial^{\alpha}}{\partial x^{\alpha}}u\right)+\eta u\frac{\partial^{\alpha}}{\partial x^{\alpha}}u+\zeta\frac{\partial^{\alpha}}{\partial x^{\alpha}}(uv) &= f\left(\frac{x^{\alpha}}{\alpha},\frac{t^{\beta}}{\beta}\right) \\ \frac{\partial^{\beta}v}{\partial t^{\beta}}-\frac{\alpha}{x^{\alpha}}\frac{\partial^{\alpha}}{\partial x^{\alpha}}\left(\frac{x^{\alpha}}{\alpha}\frac{\partial^{\alpha}}{\partial x^{\alpha}}v\right)+\eta u\frac{\partial^{\alpha}}{\partial x^{\alpha}}v+\mu\frac{\partial^{\alpha}}{\partial x^{\alpha}}(uv) &= g\left(\frac{x^{\alpha}}{\alpha},\frac{t^{\beta}}{\beta}\right), \end{aligned} \tag{23}$$

and with initial conditions

$$u\left(\frac{x^{\alpha}}{\alpha},0\right)=f_1\left(\frac{x^{\alpha}}{\alpha}\right),\quad v\left(\frac{x^{\alpha}}{\alpha},0\right)=g_1\left(\frac{x^{\alpha}}{\alpha}\right), \tag{24}$$

where the linear terms $\frac{\alpha}{x^{\alpha}}\frac{\partial^{\alpha}}{\partial x^{\alpha}}\left(\frac{x^{\alpha}}{\alpha}\frac{\partial^{\alpha}}{\partial x^{\alpha}}\right)$ is known as conformable Bessel operator where $\zeta$, $\mu$ and $\eta$ are real constants. Now to obtain the solution of Equation (23), First, we multiply both sides of Equation (23) by $\frac{x^{\alpha}}{\alpha}$ and obtain

$$\begin{aligned}\frac{x^\alpha}{\alpha}\frac{\partial^\beta u}{\partial t^\beta}-\frac{\partial^\alpha}{\partial x^\alpha}\left(\frac{x^\alpha}{\alpha}\frac{\partial^\alpha}{\partial x^\alpha}u\right)+\eta\frac{x^\alpha}{\alpha}u\frac{\partial^\alpha}{\partial x^\alpha}u+\zeta\frac{x^\alpha}{\alpha}\frac{\partial^\alpha}{\partial x^\alpha}(uv) &= \frac{x^\alpha}{\alpha}f\left(\frac{x^\alpha}{\alpha},\frac{t^\beta}{\beta}\right)\\ \frac{x^\alpha}{\alpha}\frac{\partial^\beta v}{\partial t^\beta}-\frac{\partial^\alpha}{\partial x^\alpha}\left(\frac{x^\alpha}{\alpha}\frac{\partial^\alpha}{\partial x^\alpha}v\right)+\eta\frac{x^\alpha}{\alpha}v\frac{\partial^\alpha}{\partial x^\alpha}v+\mu\frac{x^\alpha}{\alpha}\frac{\partial^\alpha}{\partial x^\alpha}(uv) &= \frac{x^\alpha}{\alpha}g\left(\frac{x^\alpha}{\alpha},\frac{t^\beta}{\beta}\right).\end{aligned} \tag{25}$$

Second: we apply conformable double Laplace transform on both sides of Equation(25) and single conformable Laplace transform for initial condition, we get

$$\begin{aligned}L_x^\alpha L_t^\beta\left[\frac{x^\alpha}{\alpha}\frac{\partial^\beta u}{\partial t^\beta}\right] &= L_x^\alpha L_t^\beta\left[\frac{\partial^\alpha}{\partial x^\alpha}\left(\frac{x^\alpha}{\alpha}\frac{\partial^\alpha}{\partial x^\alpha}u\right)-\eta\frac{x^\alpha}{\alpha}u\frac{\partial^\alpha}{\partial x^\alpha}u-\zeta\frac{x^\alpha}{\alpha}\frac{\partial^\alpha}{\partial x^\alpha}(uv)+\frac{x^\alpha}{\alpha}f\left(\frac{x^\alpha}{\alpha},\frac{t^\beta}{\beta}\right)\right],\\ L_x^\alpha L_t^\beta\left[\frac{x^\alpha}{\alpha}\frac{\partial^\beta v}{\partial t^\beta}\right] &= L_x^\alpha L_t^\beta\left[\frac{\partial^\alpha}{\partial x^\alpha}\left(\frac{x^\alpha}{\alpha}\frac{\partial^\alpha}{\partial x^\alpha}v\right)-\eta\frac{x^\alpha}{\alpha}v\frac{\partial^\alpha}{\partial x^\alpha}v-\mu\frac{x^\alpha}{\alpha}\frac{\partial^\alpha}{\partial x^\alpha}(uv)+\frac{x^\alpha}{\alpha}g\left(\frac{x^\alpha}{\alpha},\frac{t^\beta}{\beta}\right)\right]\end{aligned} \tag{26}$$

by applying Theorems 1 and 2, we have

$$\begin{aligned}-s\frac{d}{dp}U(p,s)+\frac{d}{dp}L_x^\alpha[f_1(x)] &= L_x^\alpha L_t^\beta\left[\frac{\partial^\alpha}{\partial x^\alpha}\left(\frac{x^\alpha}{\alpha}\frac{\partial^\alpha}{\partial x^\alpha}u\right)-\eta\frac{x^\alpha}{\alpha}u\frac{\partial^\alpha}{\partial x^\alpha}u-\zeta\frac{x^\alpha}{\alpha}\frac{\partial^\alpha}{\partial x^\alpha}(uv)\right]\\ &\quad -\frac{d}{dp}\left(L_x^\alpha L_t^\beta\left[f\left(\frac{x^\alpha}{\alpha},\frac{t^\beta}{\beta}\right)\right]\right),\\ -s\frac{d}{dp}V(p,s)+\frac{d}{dp}L_x^\alpha[g_1(x)] &= L_x^\alpha L_t^\beta\left[\frac{\partial^\alpha}{\partial x^\alpha}\left(\frac{x^\alpha}{\alpha}\frac{\partial^\alpha}{\partial x^\alpha}v\right)-\eta\frac{x^\alpha}{\alpha}v\frac{\partial^\alpha}{\partial x^\alpha}v-\mu\frac{x^\alpha}{\alpha}\frac{\partial^\alpha}{\partial x^\alpha}(uv)\right]\\ &\quad -\frac{d}{dp}\left(L_x^\alpha L_t^\beta\left[g\left(\frac{x^\alpha}{\alpha},\frac{t^\beta}{\beta}\right)\right]\right),\end{aligned} \tag{27}$$

simplifying Equation (27), we obtain

$$\begin{aligned}\frac{d}{dp}U(p,s) &= \frac{1}{s}\frac{d}{dp}L_x^\alpha[f_1(x)]-\frac{1}{s}L_x^\alpha L_t^\beta\left[\frac{\partial^\alpha}{\partial x^\alpha}\left(\frac{x^\alpha}{\alpha}\frac{\partial^\alpha}{\partial x^\alpha}u\right)-\eta\frac{x^\alpha}{\alpha}u\frac{\partial^\alpha}{\partial x^\alpha}u-\zeta\frac{x^\alpha}{\alpha}\frac{\partial^\alpha}{\partial x^\alpha}(uv)\right]\\ &\quad +\frac{1}{s}\frac{d}{dp}\left(L_x^\alpha L_t^\beta\left[f\left(\frac{x^\alpha}{\alpha},\frac{t^\beta}{\beta}\right)\right]\right).\\ \frac{d}{dp}V(p,s) &= \frac{1}{s}\frac{d}{dp}L_x^\alpha[g_1(x)]-\frac{1}{s}L_x^\alpha L_t^\beta\left[\frac{\partial^\alpha}{\partial x^\alpha}\left(\frac{x^\alpha}{\alpha}\frac{\partial^\alpha}{\partial x^\alpha}v\right)-\eta\frac{x^\alpha}{\alpha}v\frac{\partial^\alpha}{\partial x^\alpha}v-\mu\frac{x^\alpha}{\alpha}\frac{\partial^\alpha}{\partial x^\alpha}(uv)\right]\\ &\quad +\frac{1}{s}\frac{d}{dp}\left(L_x^\alpha L_t^\beta\left[g\left(\frac{x^\alpha}{\alpha},\frac{t^\beta}{\beta}\right)\right]\right).\end{aligned} \tag{28}$$

Third: integrating both sides of Equation (28) from 0 to $p$ respect to $p$, we have

$$\begin{aligned}U(p,s) &= \frac{1}{s}\int_0^p\left(\frac{d}{dp}L_x^\alpha[f_1(x)]\right)dp-\frac{1}{s}\int_0^p L_x^\alpha L_t^\beta\left[\frac{\partial^\alpha}{\partial x^\alpha}\left(\frac{x^\alpha}{\alpha}\frac{\partial^\alpha}{\partial x^\alpha}u\right)-\eta\frac{x^\alpha}{\alpha}N_1-\zeta\frac{x^\alpha}{\alpha}N_2\right]dp\\ &\quad +\frac{1}{s}\int_0^p\left(\frac{d}{dp}\left(L_x^\alpha L_t^\beta\left[f\left(\frac{x^\alpha}{\alpha},\frac{t^\beta}{\beta}\right)\right]\right)\right)dp,\\ V(p,s) &= \frac{1}{s}\int_0^p\left(\frac{d}{dp}L_x^\alpha[g_1(x)]\right)dp-\frac{1}{s}\int_0^p L_x^\alpha L_t^\beta\left[\frac{\partial^\alpha}{\partial x^\alpha}\left(\frac{x^\alpha}{\alpha}\frac{\partial^\alpha}{\partial x^\alpha}v\right)-\eta\frac{x^\alpha}{\alpha}N_3-\mu\frac{x^\alpha}{\alpha}N_2\right]dp\\ &\quad +\frac{1}{s}\int_0^p\left(\frac{d}{dp}\left(L_x^\alpha L_t^\beta\left[g\left(\frac{x^\alpha}{\alpha},\frac{t^\beta}{\beta}\right)\right]\right)\right)dp.\end{aligned} \tag{29}$$

Using conformable double Laplace decomposition method to define a solution of the system as $u\left(\frac{x^\alpha}{\alpha},\frac{t^\beta}{\beta}\right)$ and $v(\frac{x^\alpha}{\alpha},\frac{t^\beta}{\beta})$ by infinite series

$$u\left(\frac{x^\alpha}{\alpha},\frac{t^\beta}{\beta}\right)=\sum_{n=0}^{\infty}u_n\left(\frac{x^\alpha}{\alpha},\frac{t^\beta}{\beta}\right),\quad v\left(\frac{x^\alpha}{\alpha},\frac{t^\beta}{\beta}\right)=\sum_{n=0}^{\infty}v_n\left(\frac{x^\alpha}{\alpha},\frac{t^\beta}{\beta}\right). \tag{30}$$

Here the nonlinear operators can be defined as

$$N_1=\sum_{n=0}^{\infty}A_n,\ \ N_2=\sum_{n=0}^{\infty}C_n\,N_3=\sum_{n=0}^{\infty}B_n \tag{31}$$

$$\begin{aligned}\sum_{n=0}^{\infty} u_n\left(\frac{x^\alpha}{\alpha},\frac{t^\beta}{\beta}\right) &= f_1(x)+L_p^{-1}L_s^{-1}\left[\frac{1}{s}\int_0^p dF(p,s)\right]\\ &-L_p^{-1}L_s^{-1}\left[\frac{1}{s}\int_0^p\left(L_x^\alpha L_t^\beta\left[\frac{\partial^\alpha}{\partial x^\alpha}\left(\frac{x^\alpha}{\alpha}\frac{\partial^\alpha}{\partial x^\alpha}\left(\sum_{n=0}^{\infty}u_n\right)\right)\right]\right)dp\right]\\ &+L_p^{-1}L_s^{-1}\left[\frac{1}{s}\int_0^p\left(L_x^\alpha L_t^\beta\left[\eta\frac{x^\alpha}{\alpha}\sum_{n=0}^{\infty}A_n\right]\right)dp\right]\\ &+L_p^{-1}L_s^{-1}\left[\frac{1}{s}\int_0^p\left(L_x^\alpha L_t^\beta\left[\zeta\frac{x^\alpha}{\alpha}\sum_{n=0}^{\infty}C_n\right]\right)dp\right],\end{aligned}\tag{32}$$

and

$$\begin{aligned}\sum_{n=0}^{\infty} v_n\left(\frac{x^\alpha}{\alpha},\frac{t^\beta}{\beta}\right) &= g_1(x)+L_p^{-1}L_s^{-1}\left[\frac{1}{s}\int_0^p dG(p,s)\right]\\ &-L_p^{-1}L_s^{-1}\left[\frac{1}{s}\int_0^p\left(L_x^\alpha L_t^\beta\left[\frac{\partial^\alpha}{\partial x^\alpha}\left(\frac{x^\alpha}{\alpha}\frac{\partial^\alpha}{\partial x^\alpha}\left(\sum_{n=0}^{\infty}v_n\right)\right)\right]\right)dp\right]\\ &+L_p^{-1}L_s^{-1}\left[\frac{1}{s}\int_0^p\left(L_x^\alpha L_t^\beta\left[\eta\frac{x^\alpha}{\alpha}\sum_{n=0}^{\infty}B_n\right]\right)dp\right]\\ &+L_p^{-1}L_s^{-1}\left[\frac{1}{s}\int_0^p\left(L_x^\alpha L_t^\beta\left[\mu\frac{x^\alpha}{\alpha}\sum_{n=0}^{\infty}C_n\right]\right)dp\right].\end{aligned}\tag{33}$$

The first few components can be written as

$$\begin{aligned}u_0 &= f_1(x)+L_p^{-1}L_s^{-1}\left[\frac{1}{s}\int_0^p dF(p,s)\right],\\ v_0 &= g_1(x)+L_p^{-1}L_s^{-1}\left[\frac{1}{s}\int_0^p dG(p,s)\right],\end{aligned}\tag{34}$$

and

$$\begin{aligned}u_{n+1}\left(\frac{x^\alpha}{\alpha},\frac{t^\beta}{\beta}\right) &= -L_p^{-1}L_s^{-1}\left[\frac{1}{s}\int_0^p\left(L_x^\alpha L_t^\beta\left[\frac{\partial^\alpha}{\partial x^\alpha}\left(\frac{x^\alpha}{\alpha}\frac{\partial^\alpha}{\partial x^\alpha}\left(\sum_{n=0}^{\infty}u_n\right)\right)\right]\right)dp\right]\\ &+L_p^{-1}L_s^{-1}\left[\frac{1}{s}\int_0^p\left(L_x^\alpha L_t^\beta\left[\eta\frac{x^\alpha}{\alpha}\sum_{n=0}^{\infty}A_n\right]\right)dp\right]\\ &+L_p^{-1}L_s^{-1}\left[\frac{1}{s}\int_0^p\left(L_x^\alpha L_t^\beta\left[\zeta\frac{x^\alpha}{\alpha}\sum_{n=0}^{\infty}C_n\right]\right)dp\right],\end{aligned}\tag{35}$$

and

$$\begin{aligned}v_{n+1}\left(\frac{x^\alpha}{\alpha},\frac{t^\beta}{\beta}\right) &= -L_p^{-1}L_s^{-1}\left[\frac{1}{s}\int_0^p\left(L_x^\alpha L_t^\beta\left[\frac{\partial^\alpha}{\partial x^\alpha}\left(\frac{x^\alpha}{\alpha}\frac{\partial^\alpha}{\partial x^\alpha}\left(\sum_{n=0}^{\infty}v_n\right)\right)\right]\right)dp\right]\\ &+L_p^{-1}L_s^{-1}\left[\frac{1}{s}\int_0^p\left(L_x^\alpha L_t^\beta\left[\eta\frac{x^\alpha}{\alpha}\sum_{n=0}^{\infty}B_n\right]\right)dp\right]\\ &+L_p^{-1}L_s^{-1}\left[\frac{1}{s}\int_0^p\left(L_x^\alpha L_t^\beta\left[\zeta\frac{x^\alpha}{\alpha}\sum_{n=0}^{\infty}C_n\right]\right)dp\right].\end{aligned}\tag{36}$$

Provided the double inverse Laplace transform with respect to $p$ and $s$ exist for Equations (34)–(36).

**Example 4.** *Singular one dimensional conformable fractional coupled Burgers' equation*

$$\begin{aligned}\frac{\partial^\beta u}{\partial t^\beta}-\frac{\alpha}{x^\alpha}\frac{\partial^\alpha}{\partial x^\alpha}\left(\frac{x^\alpha}{\alpha}\frac{\partial^\alpha}{\partial x^\alpha}u\right)-2u\frac{\partial^\alpha}{\partial x^\alpha}u+\frac{\partial^\alpha}{\partial x^\alpha}(uv) &= \left(\frac{x^\alpha}{\alpha}\right)^2 e^{\frac{t^\beta}{\beta}}-4e^{\frac{t^\beta}{\beta}}\\ \frac{\partial^\beta v}{\partial t^\beta}-\frac{\alpha}{x^\alpha}\frac{\partial^\alpha}{\partial x^\alpha}\left(\frac{x^\alpha}{\alpha}\frac{\partial^\alpha}{\partial x^\alpha}v\right)-2v\frac{\partial^\alpha}{\partial x^\alpha}v+\frac{\partial^\alpha}{\partial x^\alpha}(uv) &= \left(\frac{x^\alpha}{\alpha}\right)^2 e^{\frac{t^\beta}{\beta}}-4e^{\frac{t^\beta}{\beta}},\end{aligned}\tag{37}$$

*subject to*

$$u(x,0)=\left(\frac{x^\alpha}{\alpha}\right)^2,\quad v(x,0)=\left(\frac{x^\alpha}{\alpha}\right)^2.\tag{38}$$

*By following similar steps, we obtain*

$$\sum_{n=0}^{\infty} u_n\left(\frac{x^\alpha}{\alpha},\frac{t^\beta}{\beta}\right) = \left(\frac{x^\alpha}{\alpha}\right)^2 e^{\frac{t^\beta}{\beta}} - 4e^{\frac{t^\beta}{\beta}} + 4 -L_p^{-1}L_s^{-1}\left[\frac{1}{s}\int_0^p\left(L_x^\alpha L_t^\beta\left[\frac{\partial^\alpha}{\partial x^\alpha}\left(\frac{x^\alpha}{\alpha}\frac{\partial^\alpha}{\partial x^\alpha}\left(\sum_{n=0}^{\infty} v_n\right)\right)\right]\right)dp\right] -L_p^{-1}L_s^{-1}\left[\frac{1}{s}\int_0^p\left(L_x^\alpha L_t^\beta\left[2\frac{x^\alpha}{\alpha}\sum_{n=0}^{\infty} A_n\right]\right)dp\right] +L_p^{-1}L_s^{-1}\left[\frac{1}{s}\int_0^p\left(L_x^\alpha L_t^\beta\left[\frac{x^\alpha}{\alpha}\frac{\partial^\alpha}{\partial x^\alpha}\left(\sum_{n=0}^{\infty} C_n\right)\right]\right)dp\right], \quad (39)$$

*and*

$$\sum_{n=0}^{\infty} v_n\left(\frac{x^\alpha}{\alpha},\frac{t^\beta}{\beta}\right) = \left(\frac{x^\alpha}{\alpha}\right)^2 e^{\frac{t^\beta}{\beta}} - 4e^{\frac{t^\beta}{\beta}} + 4 -L_p^{-1}L_s^{-1}\left[\frac{1}{s}\int_0^p\left(L_x^\alpha L_t^\beta\left[\frac{\partial^\alpha}{\partial x^\alpha}\left(\frac{x^\alpha}{\alpha}\frac{\partial^\alpha}{\partial x^\alpha}\left(\sum_{n=0}^{\infty} v_n\right)\right)\right]\right)dp\right] -L_p^{-1}L_s^{-1}\left[\frac{1}{s}\int_0^p\left(L_x^\alpha L_t^\beta\left[2\frac{x^\alpha}{\alpha}\sum_{n=0}^{\infty} B_n\right]\right)dp\right] +L_p^{-1}L_s^{-1}\left[\frac{1}{s}\int_0^p\left(L_x^\alpha L_t^\beta\left[\frac{x^\alpha}{\alpha}\sum_{n=0}^{\infty} C_n\right]\right)dp\right] \quad (40)$$

*where $A_n, B_n$ and $C_n$ are defined in Equations (14)–(16) respectively. On using Equations (34)–(36) the components are given by*

$$u_0 = \left(\frac{x^\alpha}{\alpha}\right)^2 e^{\frac{t^\beta}{\beta}} - 4e^{\frac{t^\beta}{\beta}} + 4,\ v_0 = \left(\frac{x^\alpha}{\alpha}\right)^2 e^{\frac{t^\beta}{\beta}} - 4e^{\frac{t^\beta}{\beta}} + 4,$$

$$u_1 = -L_p^{-1}L_s^{-1}\left[\frac{1}{s}\int_0^p L_x^\alpha L_t^\beta\left[\frac{\partial^\alpha}{\partial x^\alpha}\left(\frac{x^\alpha}{\alpha}\frac{\partial^\alpha u_0}{\partial x^\alpha}\right) + 2\frac{x^\alpha}{\alpha}u_0\frac{\partial^\alpha u_0}{\partial x^\alpha} - \frac{x^\alpha}{\alpha}\frac{\partial^\alpha}{\partial x^\alpha}(u_0v_0)\right]dp\right]$$

$$u_1 = -L_p^{-1}L_s^{-1}\left[\frac{1}{s}\int_0^p L_x^\alpha L_t^\beta\left[\left(4\frac{x^\alpha}{\alpha}e^{\frac{t^\beta}{\beta}}\right)\right]dp\right] = 4e^{\frac{t^\beta}{\beta}} - 4,$$

$$v_1 = -L_p^{-1}L_s^{-1}\left[\frac{1}{s}\int_0^p L_x^\alpha L_t^\beta\left[\frac{\partial^\alpha}{\partial x^\alpha}\left(\frac{x^\alpha}{\alpha}\frac{\partial^\alpha v_0}{\partial x^\alpha}\right) + 2\frac{x^\alpha}{\alpha}v_0\frac{\partial^\alpha v_0}{\partial x^\alpha} - \frac{x^\alpha}{\alpha}\frac{\partial^\alpha}{\partial x^\alpha}(u_0v_0)\right]dp\right]$$

$$v_1 = -L_p^{-1}L_s^{-1}\left[\frac{1}{s}\int_0^p L_x^\alpha L_t^\beta\left[\left(4\frac{x^\alpha}{\alpha}e^{\frac{t^\beta}{\beta}}\right)\right]dp\right] = 4e^{\frac{t^\beta}{\beta}} - 4.$$

*In a similar way, we obtain*

$$u_2 = -L_p^{-1}L_s^{-1}\left[\frac{1}{s}\int_0^p L_x^\alpha L_t^\beta\left[\frac{\partial^\alpha}{\partial x^\alpha}\left(\frac{x^\alpha}{\alpha}\frac{\partial^\alpha u_0}{\partial x^\alpha}\right)\right]dp\right] -L_p^{-1}L_s^{-1}\left[\frac{1}{s}\int_0^p L_x^\alpha L_t^\beta\left[2\frac{x^\alpha}{\alpha}\left(u_0\frac{\partial^\alpha u_1}{\partial x^\alpha} + u_1\frac{\partial^\alpha u_0}{\partial x^\alpha}\right)\right]dp\right] +L_p^{-1}L_s^{-1}\left[\frac{1}{s}\int_0^p L_x^\alpha L_t^\beta\left[\frac{x^\alpha}{\alpha}\frac{\partial^\alpha}{\partial x^\alpha}(u_0v_1 + u_1v_0)\right]dp\right]$$

$$u_2 = 0,$$

$$v_2 = 0.$$

*Thus it is obvious that the self-canceling some terms appear among various components and following terms, then we have,*

$$u\left(\frac{x^\alpha}{\alpha},\frac{t^\beta}{\beta}\right) = u_0 + u_1 + u_2 + \ldots,\quad v\left(\frac{x^\alpha}{\alpha},\frac{t^\beta}{\beta}\right) = v_0 + v_1 + v_2 + \ldots$$

*Therefore, the exact solution is given by*

$$u\left(\frac{x^\alpha}{\alpha},\frac{t^\beta}{\beta}\right) = \left(\frac{x^\alpha}{\alpha}\right)^2 e^{\frac{t^\beta}{\beta}} \text{ and } v\left(\frac{x^\alpha}{\alpha},\frac{t^\beta}{\beta}\right) = \left(\frac{x^\alpha}{\alpha}\right)^2 e^{\frac{t^\beta}{\beta}}.$$

*By taking $\alpha = 1$ and $\beta = 1$, the fractional solution becomes*

$$u\left(\frac{x^{\alpha}}{\alpha}, \frac{t^{\beta}}{\beta}\right) = x^2 e^t,$$
$$v\left(\frac{x^{\alpha}}{\alpha}, \frac{t^{\beta}}{\beta}\right) = x^2 e^t.$$

## 3. Conclusions

In this work some properties and conditions for existence of solutions for the conformable double Laplace transform are discussed. We give a solution to the one dimensional regular and singular conformable fractional coupled Burgers' equation by using the conformable double Laplace decomposition method, which is the combination between the conformable double Laplace and Adomian decomposition methods. Further, two examples were given to validate the present method. This method can also be applied to solve some nonlinear time-fractional differential equations having conformable derivatives. The present method can also be used to approximate the solutions of the nonlinear differential equations with the linearization of non-linear terms by using Adomian polynomials.

**Author Contributions:** The authors contributed equally and all authors read the manuscript and approved the final submission.

**Funding:** The authors would like to extend their sincere appreciation to the Deanship of Scientific Research at King Saud University for its funding this Research Group No. (RG-1440-030).

**Acknowledgments:** The authors would like to thanks the referees for the valuable comments that helped us to improve the manuscript.

**Conflicts of Interest:** It is hereby the authors declare that there is no conflict of interest.

## References

1. Chen, C.; Jiang, Y.-L. Simplest equation method for some time-fractional partial differential equations with conformable derivative. *Comput. Math. Appl.* **2018**, *75*, 2978–2988. [CrossRef]
2. Bateman, H. Some Recent Researches on the Motion of Fluids. *Mon. Weather Rev.* **1915**, *43*, 163–170. [CrossRef]
3. Burgers, J.M. A Mathematical Model Illustrating the Theory of Turbulence. *Adv. Appl. Mech.* **1948**, *1*, 171–199.
4. Özkan, O.; Kurt, A. On conformable double Laplace transform. *Opt. Quant. Electron.* **2018**, *50*, 103. [CrossRef]
5. Çenesiz, Y.; Baleanu, D.; Kurt, A.; Tasbozan, O. New exact solutions of Burgers' type equations with conformable derivative. *Wave Random Complex Media* **2017**, *27*, 103–116. [CrossRef]
6. Liu, J.; Hou, G. Numerical solutions of the space-and time-fractional coupled Burgers equations by generalized differential transform method. *Appl. Math. Comput.* **2011**, *217*, 7001–7008. [CrossRef]
7. Hashemi, M.S. Invariant subspaces admitted by fractional differential equations with conformable derivatives. *Chaos Solitons Fractals* **2018**, *107*, 161–169. [CrossRef]
8. Younis, M.; Zafar, A.; Haq, K.U.; Rahman, M. Travelling wave solutions of fractional order coupled Burger's equations by $(G'/G)$-expansion method. *Am. J. Comput. Appl. Math.* **2013**, *3*, 81.
9. Khalil, R.; Al-Horani, M.; Yousef, A.; Sababheh, M. A new definition of fractional derivative. *J. Comput. Appl. Math.* **2014**, *264*, 65–70. [CrossRef]
10. Abdeljawad, T. On conformable fractional calculus. *J. Comput. Appl. Math.* **2015**, *279*, 57–66. [CrossRef]
11. Eslami, M. Exact traveling wave solutions to the fractional coupled nonlinear Schrödinger equations. *Appl. Math. Comput.* **2016**, *285*, 141–148. [CrossRef]
12. Abdeljawad, T.; Al-Horani, M.; Khalil, R. Conformable fractional semigroups of operators. *J. Semigroup Theory Appl.* **2015**, *2015*, 7.
13. Thabet, H.; Kendre, S. Analytical solutions for conformable space-time fractional partial differential equations via fractional differential transform. *Chaos Solitons Fractals* **2018**, *109*, 238–245. [CrossRef]

14. Iskender Eroglu, B.B.; Avcı D.; Özdemir, N. Optimal Control Problem for a Conformable Fractional Heat Conduction Equation. *Acta Phys. Polonica A* **2017**, *132*, 658–662. [CrossRef]
15. Eltayeb, H.; Mesloub, S.; Kılıçman, A. A note on a singular coupled Burgers equation and double Laplace transform method. *J. Nonlinear Sci. Appl.* **2018**, *11*, 635–643. [CrossRef]
16. Nee, J.; Duan, J. Limit set of trajectories of the coupled viscous Burgers' equations. *Appl. Math. Lett.* **1998**, *11*, 57–61. [CrossRef]

symmetry

MDPI

Article

# Positive Solutions of a Fractional Thermostat Model with a Parameter

**Xinan Hao * and Luyao Zhang**

School of Mathematical Sciences, Qufu Normal University, Qufu 273165, China; shuxuequfu@126.com
* Correspondence: haoxinan2004@163.com

Received: 19 December 2018; Accepted: 17 January 2019; Published: 21 January 2019

**Abstract:** We study the existence, multiplicity, and uniqueness results of positive solutions for a fractional thermostat model. Our approach depends on the fixed point index theory, iterative method, and nonsymmetry property of the Green function. The properties of positive solutions depending on a parameter are also discussed.

**Keywords:** positive solution; fractional thermostat model; fixed point index; dependence on a parameter

## 1. Introduction

In this paper, we investigate a fractional nonlocal boundary value problem (BVP)

$$\begin{cases} {}^cD^{\alpha}_{0+}x(t) + \lambda g(t)f(x(t)) = 0, \quad t \in (0,1), \\ x'(0) = 0,\ \beta\, {}^cD^{\alpha-1}_{0+}x(1) + x(\eta) = 0, \end{cases} \tag{1}$$

where $1 < \alpha \leq 2$, $\beta > 0$, $0 \leq \eta \leq 1$, $\beta\Gamma(\alpha) > (1-\eta)^{\alpha-1}$, ${}^cD^{\alpha}_{0+}$ is the Gerasimov–Caputo fractional derivative of order $\alpha$, $\lambda > 0$ is a parameter, $f \in C([0,+\infty),[0,+\infty))$, $g \in C((0,1),[0,+\infty))$, and $0 < \int_0^1 g(t)dt < +\infty$.

One motivation is that the thermostat model

$$\begin{cases} x''(t) + g(t)f(t,x(t)) = 0, \quad t \in (0,1), \\ x'(0) = 0,\ \beta x'(1) + x(\eta) = 0, \end{cases} \tag{2}$$

which is a special case with $\alpha = 2$ and $\lambda = 1$, has been discussed by Infante and Webb [1,2]. They established multiplicity results of BVP (2). These types of problems have been investigated by various scholars, see References [3–17].

Recently, the thermostat model was extended to the fractional case

$$\begin{cases} {}^cD^{\alpha}_{0+}x(t) + f(t,x(t)) = 0, \quad t \in (0,1),\ \alpha \in (1,2], \\ x'(0) = 0,\ \beta\, {}^cD^{\alpha-1}_{0+}x(1) + x(\eta) = 0, \end{cases} \tag{3}$$

where $\beta > 0$, $0 \leq \eta \leq 1$, $f \in C([0,1] \times [0,+\infty),[0,+\infty))$. Nieto and Pimentel [18] proved the existence of positive solutions based on the Krasnosel'skii fixed point theorem. Cabada and Infante [19] discussed the multiplicity results of positive solutions for BVP (3).

In Reference [20], Shen, Zhou, and Yang studied a fractional thermostat model

$$\begin{cases} {}^cD^{\alpha}_{0+}x(t) + \lambda f(t,x(t)) = 0, \quad t \in (0,1),\ 1 < \alpha \leq 2, \\ x'(0) = 0,\ \beta\, {}^cD^{\alpha-1}_{0+}x(1) + x(\eta) = 0, \end{cases}$$

where $\beta > 0$, $0 \leq \eta \leq 1$, $\beta\Gamma(\alpha) > (1-\eta)^{\alpha-1}$, $\lambda > 0$, $f : [0,1] \times [0,+\infty) \to [0,+\infty)$ is continuous. The authors obtained intervals of parameter $\lambda$ that correspond to at least one and no positive solutions. Similar fractional thermostat problems have been studied in References [21–24].

In this paper, we deal with positive solutions for the fractional thermostat model (1). The existence, multiplicity, and uniqueness results are established by the fixed point index theory and iterative method. The properties of positive solutions depending on a parameter are also discussed. Some of the ideas in this paper are from References [25,26]. Let us remark that the definition of the Gerasimov–Caputo derivative was first introduced and applied by Gerasimov in 1947 and then by Caputo in 1967, see for example, the overview by Novozhenova in Reference [27]. For details on the theory and applications of the fractional derivatives and integrals and fractional differential equations, see References [28–31].

## 2. Preliminaries

**Lemma 1** ([20]). *Given $u(t) \in C(0,1) \cap L^1(0,1)$, the solution of the problem*

$$\begin{cases} {}^cD_{0+}^{\alpha}x(t) + u(t) = 0, \quad t \in (0,1), \\ x'(0) = 0, \ \beta\, {}^cD_{0+}^{\alpha-1}x(1) + x(\eta) = 0 \end{cases}$$

*is*

$$x(t) = \int_0^1 G(t,s)u(s)ds, \quad t \in [0,1],$$

*where*

$$G(t,s) = \begin{cases} \beta - \dfrac{(t-s)^{\alpha-1}}{\Gamma(\alpha)} + \dfrac{(\eta-s)^{\alpha-1}}{\Gamma(\alpha)}, & 0 \leq s \leq \eta,\ s \leq t, \\ \beta + \dfrac{(\eta-s)^{\alpha-1}}{\Gamma(\alpha)}, & 0 \leq s \leq \eta,\ s \geq t, \\ \beta - \dfrac{(t-s)^{\alpha-1}}{\Gamma(\alpha)}, & \eta \leq s \leq 1,\ s \leq t, \\ \beta, & \eta \leq s \leq 1,\ s \geq t, \end{cases}$$

*and $G(t,s)$ satisfies:*

*(i)* $G(t,s) : [0,1] \times [0,1] \to (0,+\infty)$ *is continuous;*
*(ii)* $\frac{\partial}{\partial t}G(t,s) \leq 0,\ t,s \in [0,1]$;
*(iii)* $\gamma\overline{G} = \underline{G} \leq G(1,s) \leq G(t,s) \leq G(0,s) \leq \overline{G}, \quad t,s \in [0,1],$

*where*

$$\gamma = \frac{\beta\Gamma(\alpha) - (1-\eta)^{\alpha-1}}{\beta\Gamma(\alpha) + \eta^{\alpha-1}}, \quad \underline{G} = \frac{\beta\Gamma(\alpha) - (1-\eta)^{\alpha-1}}{\Gamma(\alpha)}, \quad \overline{G} = \frac{\beta\Gamma(\alpha) + \eta^{\alpha-1}}{\Gamma(\alpha)}.$$

Denote $E = C[0,1]$ and $\|x\| = \sup_{t\in[0,1]} |x(t)|$. We define the cone

$$P = \{x \in E : x(t) \geq 0, \ \inf_{t\in[0,1]} x(t) \geq \gamma\|x\|\}.$$

For any $0 < r < +\infty$, let $P_r = \{x \in P : \|x\| < r\}$. We define $T : (0,+\infty) \times E \to E$ as

$$T(\lambda, x)(t) = \lambda \int_0^1 G(t,s)g(s)f(x(s))ds, \quad t \in [0,1].$$

It is obvious from Lemma 1 that if $x \in P$ is a fixed point of operator $T$, then $x$ is a positive solution of Problem (1). By regularity arguments, we can show that $T$ is completely continuous and $T(P) \subset P$.

Define the linear operator $L : E \to E$ by

$$Lx(t) = \int_0^1 G(t,s)g(s)x(s)ds, \quad t \in [0,1].$$

By the Krein–Rutman theorem, we see that the spectral radius $r(L)$ of the operator $L$ is positive, and $L$ has positive eigenfunction $\varphi_1$ corresponding to its first eigenvalue $\mu_1 = (r(L))^{-1}$.

**Lemma 2** ([32]). *Let $P$ be a cone in Banach space $E$. Suppose that $T : P \to P$ is a completely continuous operator. (i) If $Tu \neq \mu u$ for any $u \in \partial P_r$ and $\mu \geq 1$, then $i(T, P_r, P) = 1$. (ii) If $Tu \neq u$ and $\|Tu\| \geq \|u\|$ for any $u \in \partial P_r$, then $i(T, P_r, P) = 0$.*

*Denote*

$$f_0 = \lim_{s\to 0} \frac{f(s)}{s}, \quad f_\infty = \lim_{s\to\infty} \frac{f(s)}{s}, \quad A = \int_0^1 G(0,s)g(s)ds, \quad l = \min_{s\in(0,\infty)} \frac{f(s)}{s}.$$

*We assume that:*

$(H_1)$ *$f$ is nondecreasing on $[0,+\infty)$;*
$(H_2)$ *there exists a function $\phi : (0,1] \to [0,1]$ continuous nondecreasing, such that $f(\kappa x) \geq \phi(\kappa)f(x)$ for $0 < \kappa < 1$, $x > 0$, and $F(\kappa) := \frac{\kappa}{\phi(\kappa)}$ is strictly increasing on $(0,1]$ and $F(1) = 1$.*

**Lemma 3.** *Suppose that $(H_1)$ holds, $f_0 = \infty$ and $l > 0$. If $0 < \lambda_1 < \lambda_2 < \frac{1}{lA}$, then there exist $x_1, x_2 \in P \setminus \{\theta\}$, $x_1 \leq x_2$, such that $T(\lambda_1, x_1)(t) = x_1(t)$ and $T(\lambda_2, x_2)(t) = x_2(t)$.*

**Proof.** Assume $s_0 \in (0,\infty)$ such that $f(s_0) = ls_0$. Since $0 < \lambda_1 < \lambda_2 < \frac{1}{lA}$, we have $l < \frac{1}{\lambda_2 A} < \frac{1}{\lambda_1 A}$. We define

$$x_0(t) = \frac{s_0}{A}\int_0^1 G(t,s)g(s)ds, \quad t \in [0,1],$$

then

$$\|x_0\| = x_0(0) = s_0, \quad x_0(t) \geq \frac{s_0}{A}\int_0^1 \gamma G(0,s)g(s)ds = \gamma\|x_0\|, \quad t \in [0,1].$$

Therefore, $x_0 \in P$ and $\|x_0\| = s_0$. Direct computations yield

$$\begin{aligned} T(\lambda_1, x_0)(t) =&\lambda_1 \int_0^1 G(t,s)g(s)f(x_0(s))ds \leq \lambda_1 \int_0^1 G(t,s)g(s)f(\|x_0\|)ds \\ =&\lambda_1 l s_0 \int_0^1 G(t,s)g(s)ds < \frac{s_0}{A}\int_0^1 G(t,s)g(s)ds = x_0(t), \quad t \in [0,1]. \end{aligned}$$

Define

$$x_1^1(t) = T(\lambda_1, x_0)(t),\ x_1^j(t) = T(\lambda_1, x_1^{j-1})(t) = T^j(\lambda_1, x_0)(t), \quad j = 2,3,\cdots,\ t \in [0,1].$$

Direct calculations show that $x_0 > x_1^1 > x_1^2 > \cdots > x_1^j > x_1^{j+1} > \cdots \geq \theta$. Hence, sequence $\{x_1^j\}_{j=1}^\infty$ is decreasing and bounded from below, $\lim_{j\to\infty} x_1^j(t)$ exists and convergence is uniform for $t \in [0,1]$. Assume that $\lim_{j\to\infty} x_1^j = x_1$, we claim that $x_1(t) > 0$. Otherwise, since $x_1 \in P$, $x_1(t) = 0$, i.e., $\lim_{j\to\infty} x_1^j(t) = 0$, $t \in [0,1]$, and hence from $x_1^j \in P$, we deduce $\|x_1^j\| \to 0$. Since $f_0 = \infty$, for any $H > \frac{1}{\lambda_1\gamma A}$, there is integral $Z > 0$ such that for $j > Z$, we have $f(x_1^j(t)) > Hx_1^j(t)$, and hence

$$\begin{aligned} x_1^{j+1}(0) =& \lambda_1 \int_0^1 G(0,s)g(s)f(x_1^j(s))ds \\ >& \lambda_1 H\gamma \int_0^1 G(0,s)g(s)\|x_1^j\|ds \\ \geq& x_1^j(0)\lambda_1 H\gamma A > x_1^j(0). \end{aligned}$$

The contradiction shows that $x_1 \in P \setminus \{\theta\}$ and $x_1 = T(\lambda_1, x_1)$.

Similarly, from $x_2^1(t) = T(\lambda_2, x_0)(t)$ and $x_2^j(t) = T(\lambda_2, x_2^{j-1})(t)$, $j = 2, 3, \cdots$, we deduce

$$x_0 > x_2^1 > x_2^2 > \cdots > x_2^j > x_2^{j+1} > \cdots \geq \theta,$$

$\lim_{j\to\infty} x_2^j = x_2 \in P \setminus \{\theta\}$, and $x_2 = T(\lambda_2, x_2)$. It follows from $x_1^1 = T(\lambda_1, x_0) < T(\lambda_2, x_0) = x_2^1$ and the monotonicity of $f$ that $x_1^j \leq x_2^j$, $j = 2, 3, \cdots$. Therefore, $x_1 \leq x_2$. □

**Lemma 4.** *If $f_\infty = \infty$, then for any $\mu > 0$, the set $S_\mu = \{x \in P : T(\lambda, x) = x, \lambda \in [\mu, \infty)\}$ is bounded.*

**Proof.** Otherwise, there exists $x_n \in S_\mu$ corresponding to $\lambda_n \in [\mu, \infty)$ such that

$$T(\lambda_n, x_n) = x_n, \quad \lim_{n\to\infty} \|x_n\| = \infty.$$

Because $f_\infty = \infty$, there is $X > 0$ such that $f(s) > Hs$ for $s > X$, where $H > \frac{1}{\mu\gamma A}$. Since $\lim_{n\to\infty} \|x_n\| = \infty$, there exists $N_0 > 0$ such that $\|x_n\| > \frac{X}{\gamma}$ for $n > N_0$, and $x_n(t) \geq \gamma\|x_n\| > X$, $t \in [0,1]$. Then, for any $n > N_0$, we obtain

$$\|x_n\| > \lambda_n \int_0^1 G(0,s)g(s)Hx_n(s)ds > \mu H\gamma\|x_n\|A > \|x_n\|,$$

which is absurd, and hence $S_\mu$ is bounded. □

**Lemma 5.** *Assume that $(H_1)$ holds, and that $f_0 = f_\infty = \infty$. Then, $T$ admits a fixed point for $\lambda = \frac{1}{IA}$.*

**Proof.** Choosing a sequence $0 < \lambda_1 < \lambda_2 < \cdots < \lambda_n < \lambda_{n+1} < \cdots < \frac{1}{IA}$ such that $\lim_{n\to\infty} \lambda_n = \frac{1}{IA}$. By Lemma 3, there exists a nondecreasing sequence $\{x_n\}_{n=1}^\infty \subset P \setminus \{\theta\}$ such that $x_n = T(\lambda_n, x_n)$. By Lemma 4, we know that $\{x_n\}_{n=1}^\infty$ is uniformly bounded and equicontinuous. By using the Arzela–Ascoli theorem, we can prove that there exists $\{x_{n_k}\}_{k=1}^\infty \subset \{x_n\}_{n=1}^\infty$ such that $x_{n_k} \to \tilde{x} \in E$ uniformly on $[0,1]$. Therefore, $x_{n_k}$ satisfies

$$x_{n_k}(t) = T(\lambda_{n_k}, x_{n_k})(t) = \lambda_{n_k} \int_0^1 G(t,s)g(s)f(x_{n_k}(s))ds, \quad t \in [0,1].$$

Passing to the limit as $k \to \infty$, we obtain

$$\tilde{x}(t) = \frac{1}{IA} \int_0^1 G(t,s)g(s)f(\tilde{x}(s))ds, \quad t \in [0,1].$$

Hence, $\tilde{x} = T\left(\frac{1}{IA}, \tilde{x}\right)$. □

**Lemma 6.** *Assume that $(H_1)$ holds, and that $f(0) > 0$. Then, for any $x \in P$, there exist $U_x \geq V > 0$ such that*

$$VK_\lambda \leq T(\lambda, x)(t) \leq U_x K_\lambda, \quad t \in [0,1],$$

*where*

$$K_\lambda = \lambda \int_0^1 g(t)dt.$$

**Proof.** By $(H_1)$, for any $x \in P$ and $t \in [0,1]$, we have

$$T(\lambda, x)(t) \geq \underline{G}f(0)\lambda \int_0^1 g(t)dt := VK_\lambda,$$

and

$$T(\lambda, x)(t) \leq \overline{G}f(\|x\|)\lambda \int_0^1 g(t)dt := U_x K_\lambda.$$

□

## 3. Main Results

**Theorem 1.** *Assume that $f_\infty = \infty$ and $0 < f_0 < \infty$. Then, for any $0 < \lambda < \frac{\mu_1}{f_0}$, BVP (1) admits a positive solution.*

**Proof.** Since $0 < \lambda < \frac{\mu_1}{f_0}$, there exist $\varepsilon > 0$ small enough and $r > 0$ such that $\lambda(f_0 + \varepsilon) < \mu_1$, and $\frac{f(s)}{s} < f_0 + \varepsilon$ for $s \in (0, r]$. We claim that

$$T(\lambda, x) \neq \mu x, \quad x \in \partial P_r, \ \mu \geq 1.$$

Otherwise, there exist $x_0 \in \partial P_r$ and $\mu_0 \geq 1$ such that $T(\lambda, x_0) = \mu_0 x_0$. Since $0 < \gamma r \leq x_0(t) \leq \|x_0\| = r$, we have

$$\mu_0 x_0(t) \leq \lambda(f_0 + \varepsilon) \int_0^1 G(t,s)g(s)x_0(s)ds = \lambda(f_0 + \varepsilon)Lx_0(t),$$

then $Lx_0(t) \geq \frac{\mu_0}{\lambda(f_0+\varepsilon)} x_0(t)$. Thus, $r(L) \geq \frac{\mu_0}{\lambda(f_0+\varepsilon)} \geq \frac{1}{\lambda(f_0+\varepsilon)}$. It follows that $\mu_1 \leq \lambda(f_0 + \varepsilon)$, which is a contradiction. Then, $i(T, P_r, P) = 1$.

Next, we prove that $i(T, P_R, P) = 0$ for some $R > r$. In fact, $f_\infty = \infty$ implies that $f(s) > Ms$ for some large $R_1 > 0$ and $s \geq R_1$, where $M > (\lambda\gamma A)^{-1}$. Let $R > \max\{r, \frac{R_1}{\gamma}\}$. For $x \in \partial P_R$, we have $x(t) \geq \gamma\|x\| = \gamma R > R_1$, $t \in [0,1]$, then

$$\|T(\lambda, x)\| \geq \lambda M \int_0^1 G(0,s)g(s)x(s)ds \geq \lambda M\gamma\|x\|A > \|x\|.$$

Hence, $i(T, P_R, P) = 0$, and $i(T, P_R \setminus \overline{P}_r, P) = -1$. Therefore, $T$ admits a fixed point $x^* \in P_R \setminus \overline{P}_r$. □

**Theorem 2.** *Assume that $(H_1)$ holds, and that $f_0 = f_\infty = \infty$. Then, BVP (1) has at least one and two positive solutions for $\lambda = \frac{1}{lA}$ and $\lambda \in (0, \frac{1}{lA})$, respectively.*

**Proof.** By Lemma 5, BVP (1) admits a positive solution for $\lambda = \frac{1}{lA}$. For $\lambda \in (0, \frac{1}{lA})$, by Lemmas 3 and 5, there exist $\tilde{x}$, $x_\lambda \in P \setminus \{\theta\}$, $x_\lambda \leq \tilde{x}$ such that

$$T\left(\frac{1}{lA}, \tilde{x}\right)(t) = \tilde{x}(t), \quad T(\lambda, x_\lambda)(t) = x_\lambda(t), \quad t \in [0,1].$$

If $x_\lambda = \tilde{x}$, we have

$$T(\lambda, x_\lambda) = x_\lambda = \tilde{x} = T\left(\frac{1}{lA}, \tilde{x}\right) = T\left(\frac{1}{lA}, x_\lambda\right).$$

This contradiction shows that $x_\lambda < \tilde{x}$.

Define $\Omega_1 = \{x \in E : -r < x(t) < \tilde{x}(t),\ t \in [0,1]\}$, where $r > 0$ is the same as in the first part of Theorem 1. For any $x \in P \cap \partial\Omega_1$, we obtain $\|x\| = \|\tilde{x}\|$, and

$$\|T(\lambda, x)\| < \frac{1}{lA}\int_0^1 G(0,s)g(s)f(\tilde{x}(s))ds = \tilde{x}(0) = \|\tilde{x}\|.$$

Therefore,

$$\|T(\lambda, x)\| < \|x\|, \quad x \in P \cap \partial\Omega_1.$$

As in the proof in Theorem 1, there is $R > 0$ large enough such that

$$\|T(\lambda, x)\| > \|x\|, \quad x \in P \cap \partial\Omega_2,$$

where $\Omega_2 = \{x \in E : \|x\| < R\}$. By compression expansion fixed point theorem, we see that $T$ has a fixed point $\overline{x}_\lambda \in P \cap (\Omega_2 \setminus \overline{\Omega}_1)$. Since $x_\lambda \in \Omega_1$, $x_\lambda \neq \overline{x}_\lambda$, problem (1) has a second positive solution. □

**Theorem 3.** *Assume that* $(H_1)$ *and* $(H_2)$ *hold, and that* $f(0) > 0$. *Then, for any* $\lambda \in (0, \infty)$, *BVP (1) admits a unique positive solution* $\dot{x}_\lambda(t)$, *and* $\dot{x}_\lambda(t)$ *satisfies:*

(i) $\dot{x}_\lambda(t)$ *is nondecreasing with respect to* $\lambda$;
(ii) $\lim_{\lambda\to 0^+} \|\dot{x}_\lambda\| = 0$, $\lim_{\lambda\to\infty} \|\dot{x}_\lambda\| = \infty$;
(iii) $\|\dot{x}_\lambda - \dot{x}_{\lambda_0}\| \to 0$ *as* $\lambda \to \lambda_0$.

**Proof.** Since $T$ is nondecreasing, for $u \in P$, we have

$$T(\lambda, \kappa x)(t) \geq \phi(\kappa)\lambda \int_0^1 G(t,s)g(s)f(x(s))ds = \phi(\kappa)T(\lambda, x)(t), \quad t \in [0,1]. \tag{4}$$

Define $\widehat{x}(t) = K_\lambda$, where $K_\lambda$ is given by Lemma 6, then $\widehat{x} \in P$ and $VK_\lambda \leq T(\lambda, \widehat{x})(t) \leq U_x K_\lambda$. Denote

$$\overline{V} = \sup\{\mu : \mu K_\lambda \leq T(\lambda, \widehat{x})(t)\}, \quad \overline{U} = \inf\{\mu : \mu K_\lambda \geq T(\lambda, \widehat{x})(t)\},$$

then $\overline{V} \geq V$ and $\overline{U} \leq U_x$. Select $\widetilde{V}$ and $\widetilde{U}$ so that

$$0 < \widetilde{V} < \min\{1, F^{-1}(\overline{V})\}, \quad 0 < \frac{1}{\widetilde{U}} < \min\left\{1, F^{-1}\left(\frac{1}{\overline{U}}\right)\right\}.$$

We define

$$x_1(t) = \widetilde{V}K_\lambda,\ x_{k+1}(t) = T(\lambda, x_k)(t), \quad t \in [0,1],\ k = 1, 2, \cdots,$$
$$y_1(t) = \widetilde{U}K_\lambda,\ y_{k+1}(t) = T(\lambda, y_k)(t), \quad t \in [0,1],\ k = 1, 2, \cdots.$$

Combining the properties of $T$ and (4), we get

$$\widetilde{V}K_\lambda = x_1(t) \leq x_2(t) \leq \cdots \leq x_k(t) \leq \cdots \leq y_k(t) \leq \cdots \leq y_2(t) \leq y_1(t) = \widetilde{U}K_\lambda. \tag{5}$$

Let $d = \frac{\widetilde{V}}{\widetilde{U}}$, obviously $0 < d < 1$. We claim that

$$x_k(t) \geq \phi_{k-1}(d)y_k(t), \quad t \in [0,1],\ k = 1, 2, \cdots, \tag{6}$$

where $\phi_0(d) = d$, $\phi_k(d) = \phi(\phi_{k-1}(d))$, $k = 1, 2, \cdots$. In fact, $x_1(t) = dy_1(t) = \phi_0(d)y_1(t)$, $t \in [0,1]$. Suppose $x_n(t) \geq \phi_{n-1}(d)y_n(t)$ for $t \in [0,1]$, then

$$x_{n+1}(t) \geq T(\lambda, \phi_{n-1}(d)y_n)(t) \geq \phi(\phi_{n-1}(d))T(\lambda, y_n)(t) = \phi_n(d)y_{n+1}(t).$$

Hence, it follows by induction that (6) is true. According to (5) and (6), one has

$$0 \leq x_{n+m}(t) - x_n(t) \leq y_n(t) - x_n(t) \leq (1 - \phi_{n-1}(d))y_1(t) = (1 - \phi_{n-1}(d))\widetilde{U}K_\lambda,$$

where $m \geq 0$ is an integer. Thus,

$$\|x_{n+m} - x_n\| \leq \|y_n - x_n\| \leq (1 - \phi_{n-1}(d))\widetilde{U}K_\lambda. \tag{7}$$

We claim that $\lim_{n\to\infty} \phi_n(d) = 1$. From $(H_2)$ and $0 < d < 1$, we see that $\phi(d) \in (d, 1)$ and $d = \phi_0(d) < \phi_1(d) < \cdots < \phi_n(d) < \cdots < 1$. Sequence $\{\phi_n(d)\}_{n=1}^{\infty}$ is increasing and bounded, there is $p \in [d, 1]$ such that $\lim_{n\to\infty} \phi_n(d) = p$. By the continuity of $\phi$ and $\phi_n(d) = \phi(\phi_{n-1}(d))$, we conclude that $p = \phi(p)$, i.e., $F(p) = 1$. It follows that $p = 1$. Inequality (7) implies that there exists $\overline{x} \in P$ such that $\lim_{n\to\infty} x_n(t) = \lim_{n\to\infty} y_n(t) = \overline{x}(t)$ for $t \in [0, 1]$. Clearly, $\overline{x}(t)$ is a positive solution of problem (1).

Suppose that $\bar{x}_1(t)$ and $\bar{x}_2(t)$ are positive solutions of problem (1), then $T(\lambda, \bar{x}_1)(t) = \bar{x}_1(t)$ and $T(\lambda, \bar{x}_2)(t) = \bar{x}_2(t)$, $t \in [0, 1]$. Define $\tilde{\delta} = \sup\{\delta : \bar{x}_1(t) \geq \delta\bar{x}_2(t)\}$, then $\bar{x}_1(t) \geq \tilde{\delta}\bar{x}_2(t)$. We claim that $\tilde{\delta} \geq 1$. Otherwise, $\tilde{\delta} < 1$. Assumption $(H_2)$ implies that $f(\tilde{\delta}\bar{x}_2(t)) > \varphi(\tilde{\delta})f(\bar{x}_2(t))$, $t \in [0, 1]$. Since $f$ is nondecreasing,

$$\bar{x}_1(t) = T(\lambda, \bar{x}_1)(t) \geq T(\lambda, \tilde{\delta}\bar{x}_2)(t) > \phi(\tilde{\delta})T(\lambda, \bar{x}_2)(t) > \tilde{\delta}\bar{x}_2(t), \quad t \in [0, 1],$$

a contradiction. Then, $\bar{x}_1(t) \geq \bar{x}_2(t)$ for $t \in [0, 1]$. Similarly, $\bar{x}_2(t) \geq \bar{x}_1(t)$. Therefore, $\bar{x}_1(t) = \bar{x}_2(t)$, $t \in [0, 1]$. This proves the uniqueness result.

Next, we show that $(i) - (iii)$ hold. Let

$$(Hx)(t) = \int_0^1 G(t, s)g(s)f(x(s))ds, \quad t \in [0, 1],$$

then $T(\lambda, x) = \lambda Hx$. Since $P^o = \{x \in P : x(t) > 0,\ t \in [0, 1]\}$ is nonempty, the operator $H : P^o \to P^o$ is increasing, and $H(\kappa x) \geq \phi(\kappa)Hx$ for $0 < \kappa < 1$. Let $\omega = \frac{1}{\lambda}$. We now write $Hx_\omega = \omega x_\omega$ instead of $\lambda Hx_\lambda = x_\lambda$. Assume $0 < \omega_1 < \omega_2$, then $x_{\omega_1} \geq x_{\omega_2}$. Indeed, denote $\overline{\omega} = \sup\{t > 0 : x_{\omega_1} \geq tx_{\omega_2}\}$, then $\overline{\omega} \geq 1$. Otherwise $0 < \overline{\omega} < 1$. Direct computations yield $\omega_1 x_{\omega_1} = Hx_{\omega_1} \geq H(\overline{\omega}x_{\omega_2}) \geq \phi(\overline{\omega})Hx_{\omega_2} = \phi(\overline{\omega})\omega_2 x_{\omega_2}$, then $x_{\omega_1} \geq \frac{\omega_2}{\omega_1}\phi(\overline{\omega})x_{\omega_2} > \overline{\omega}x_{\omega_2}$. This is a contradiction to the definition of $\overline{\omega}$. Thus, $\overline{\omega} \geq 1$, $x_{\omega_1} \geq x_{\omega_2}$, and further

$$x_{\omega_1} = \frac{1}{\omega_1}Hx_{\omega_1} \geq \frac{1}{\omega_1}Hx_{\omega_2} = \frac{\omega_2}{\omega_1}x_{\omega_2} \gg x_{\omega_2}, \quad 0 < \omega_1 < \omega_2. \tag{8}$$

Then, $x_\omega(t)$ is strong decreasing in $\omega$, that is, $x_\lambda(t)$ is strong increasing in $\lambda$. Let $\omega_2 = \omega$ and fix $\omega_1$ in (8), for $\omega > \omega_1$, we have $x_{\omega_1} \geq \frac{\omega}{\omega_1}x_\omega$, and

$$\|x_\omega\| \leq \frac{N\omega_1}{\omega}\|x_{\omega_1}\|,$$

where $N > 0$ is a normal constant of cone $P$. Because $\omega = \frac{1}{\lambda}$, then $\lim_{\lambda\to 0^+} \|x_\lambda\| = 0$. Let $\omega_1 = \omega$ and fix $\omega_2$ in (8), we obtain $\lim_{\lambda\to+\infty} \|x_\lambda\| = +\infty$.

Finally, for given $\omega_0$, by (8), we have

$$x_\omega \ll x_{\omega_0}, \quad \omega > \omega_0. \tag{9}$$

Let $t_\omega = \sup\{t > 0 : x_\omega \geq tx_{\omega_0},\ \omega > \omega_0\}$, then $0 < t_\omega < 1$ and $x_\omega \geq t_\omega x_{\omega_0}$. Direct computations yield $\omega x_\omega = Hx_\omega \geq H(t_\omega x_{\omega_0}) \geq \phi(t_\omega)Hx_{\omega_0} = \phi(t_\omega)\omega_0 x_{\omega_0}$. By the definition of $t_\omega$, we have $\frac{\omega_0}{\omega}\phi(t_\omega) \leq t_\omega$, and

$$t_\omega \geq F^{-1}\left(\frac{\omega_0}{\omega}\right), \quad \forall \omega > \omega_0. \tag{10}$$

Combining (9) with (10), one has that

$$\|x_{\omega_0} - x_\omega\| \leq N\left[1 - F^{-1}\left(\frac{\omega_0}{\omega}\right)\right] \|x_{\omega_0}\| \to 0, \quad \omega \to \omega_0 + 0.$$

Similarly, $\|x_{\omega_0} - x_\omega\| \to 0$, $\omega \to \omega_0 - 0$. Hence, $\|x_{\omega_0} - x_\omega\| \to 0$ as $\omega \to \omega_0$. □

**Author Contributions:** Both authors have contributed equally to this paper. Writing-original draft, X.H. and L.Z.; Writing-review and editing, X.H. and L.Z.

**Funding:** Supported financially by the National Natural Science Foundation of China (11501318, 11871302), the China Postdoctoral Science Foundation (2017M612230), the Natural Science Foundation of Shandong Province of China (ZR2017MA036) and the International Cooperation Program of Key Professors by Qufu Normal University.

**Conflicts of Interest:** The authors declare no conflict of interest.

## References

1. Infante, G.; Webb, J.R.L. Loss of positivity in a nonlinear scalar heat equation. *NoDEA Nonlinear Differ. Equ. Appl.* **2006**, *13*, 249–261. [CrossRef]
2. Webb J.R.L. Multiple positive solutions of some nonlinear heat flow problems. *Discret. Contin. Dyn. Syst.* **2005**, *2005*, 895–903.
3. Guidotti, P.; Merino, S. Gradual loss of positivity and hidden invariant cones in a scalar heat equation. *Differ. Integral Equ.* **2000**, *13*, 1551–1568.
4. Webb, J.R.L. Existence of positive solutions for a thermostat model. *Nonlinear Anal. Real World Appl.* **2012**, *13*, 923–938. [CrossRef]
5. Webb, J.R.L. Remarks on a non-local boundary value problem. *Nonlinear Anal.* **2010**, *72*, 1075–1077. [CrossRef]
6. Infante, G. Positive solutions of nonlocal boundary value problems with singularities. *Discret. Contin. Dyn. Syst.* **2009**, *2009*, 377–384.
7. Infante, G.; Webb J.R.L. Nonlinear non-local boundary-value problems and perturbed Hammerstein integral equations. *Proc. Edinb. Math. Soc.* **2006**, *49*, 637–656. [CrossRef]
8. Palamides, P.K.; Infante, G.; Pietramala, P. Nontrivial solutions of a nonlinear heat flow problem via Sperner's lemma. *Appl. Math. Lett.* **2009**, *22*, 1444–1450. [CrossRef]
9. Infante, G.; Pietramala, P.; Tenuta, M. Existence and localization of positive solutions for a nonlocal BVP arising in chemical reactor theory. *Commun. Nonlinear Sci. Numer. Simul.* **2014**, *19*, 2245–2251. [CrossRef]
10. Cianciaruso, F.; Infante, G.; Pietramala, P. Solutions of perturbed Hammerstein integral equations with applications. *Nonlinear Anal. Real World Appl.* **2017**, *33*, 317–347. [CrossRef]
11. Calamai, A.; Infante, G. Nontrivial solutions of boundary value problems for second-order functional differential equations. *Ann. Mat. Pura Appl.* **2016**, *195*, 741–756. [CrossRef]
12. Infante, G.; Pietramala, P.; Tojo, F.A.F. Non-trivial solutions of local and non-local Neumann boundary-value problems. *Proc. R. Soc. Edinb. Sect. A* **2016**, *146*, 337–369. [CrossRef]
13. Cabada, A.; Infante, G.; Tojo, F.A.F. Nonzero solutions of perturbed Hammerstein integral equations with deviated arguments and applications. *Topol. Methods Nonlinear Anal.* **2016**, *47*, 265–287. [CrossRef]
14. Li, Z. Existence of positive solutions of superlinear second-order Neumann boundary value problem. *Nonlinear Anal.* **2010**, *72*, 3216–3221. [CrossRef]
15. Ma, R.; An, Y. Global structure of positive solutions for superlinear second order $m$-point boundary value problems. *Topol. Methods Nonlinear Anal.* **2009**, *34*, 279–290. [CrossRef]
16. Cahlon, B.; Schmidt, D.; Shillor, M.; Zou, X. Analysis of thermostat models. *Eur. J. Appl. Math.* **1997**, *8*, 437–455. [CrossRef]
17. Ji, D.; Bai, Z.; Ge, W. The existence of countably many positive solutions for singular multipoint boundary value problems. *Nonlinear Anal.* **2010**, *72*, 955–964. [CrossRef]
18. Nieto, J.J.; Pimentel, J. Positive solutions of a fractional thermostat model. *Bound. Value Probl.* **2013**, *2013*, 5. [CrossRef]

19. Cabada, A.; Infante, G. Positive solutions of a nonlocal Caputo fractional BVP. *Dyn. Syst. Appl.* **2014**, *23*, 715–722.
20. Shen, C.; Zhou, H.; Yang, L. Existence and nonexistence of positive solutions of a fractional thermostat model with a parameter. *Math. Methods Appl. Sci.* **2016**, *39*, 4504–4511. [CrossRef]
21. Senapati, T.; Dey, L.K. Relation-theoretic metrical fixed-point results via $w$-distance with applications. *J. Fixed Point Theory Appl.* **2017**, *19*, 2945–2961. [CrossRef]
22. Cabrera, I.J.; Rocha, J.; Sadarangani, K.B. Lyapunov type inequalities for a fractional thermostat model. *Rev. Real Acad. Cienc. Exactas Fis. Nat. Ser. A Math.* **2018**, *112*, 17–24. [CrossRef]
23. Hao, X.; Sun, H.; Liu L. Existence results for fractional integral boundary value problem involving fractional derivatives on an infinite interval. *Math. Methods Appl. Sci.* **2018**, *41*, 6984–6996. [CrossRef]
24. Hao, X. Positive solution for singular fractional differential equations involving derivatives. *Adv. Differ. Equ.* **2016**, *2016*, 139. [CrossRef]
25. Graef, J.R.; Kong, L.; Wang, H. Existence, multiplicity, and dependence on a parameter for a periodic boundary value problem. *J. Differ. Equ.* **2008**, *245*, 1185–1197. [CrossRef]
26. Hao, X.; Wang, H. Positive solutions of semipositone singular fractional differential systems with a parameter and integral boundary conditions. *Open Math.* **2018**, *16*, 581–596. [CrossRef]
27. Novozhenova, O.G. Life and science of Alexey N. Gerasimov. On the linear operators, elastic viscosity, elevterous and fractional derivatives. *arXiv* **2018**, arXiv:1808.04397.
28. Samko, S.G.; Kilbas, A.A.; Marichev, O.I. *Fractional Integrals and Derivatives: Theory and Applications*; Gordon and Breach: New York, NY, USA, 1993.
29. Kilbas, A.A.; Srivastava, H.M.; Trujillo, J.J. *Theory and Applications of Fractional Differential Equations*; Elsevier: Amsterdam, The Netherlands, 2006.
30. Kiryakova, V. *Generalized Fractional Calculus and Applications*; Longman: Harlow, UK, 1994.
31. Podlubny, I. *Fractional Differential Equations*; Academic Press: New York, NY, USA, 1999.
32. Guo, D.; Lakshmikantham, V. *Nonlinear Problems in Abstract Cones*; Academic Press: Boston, MA, USA, 1988.

symmetry

MDPI

Article

# Oscillatory Behavior of Three Dimensional $\alpha$-Fractional Delay Differential Systems

**Adem Kilicman [1,2,*], Vadivel Sadhasivam [3] and Muthusamy Deepa [3] and Nagamanickam Nagajothi [3]**

1 Department of Mathematics and Institute for Mathematical Research, Universiti Putra Malaysia, Serdang 43400 UPM, Selangor, Malaysia
2 Department of Electrical and Electronic Engineering, Istanbul Gelisim University, Avcilar, Istanbul 34310, Turkey
3 Post Graduate and Research Department of Mathematics, Thiruvalluvar Government Arts College (Periyar University), Rasipuram 637 401, Namakkal Dt., India; ovsadha@gmail.com (V.S.); mdeepa.maths@gmail.com (M.D.); nagajothi006@gmail.com (N.N.)
* Correspondence: akilic@upm.edu.my; Tel.: +603-89466813

Received: 18 October 2018; Accepted: 11 December 2018; Published: 18 December 2018

**Abstract:** In the present work we study the oscillatory behavior of three dimensional $\alpha$-fractional nonlinear delay differential system. We establish some sufficient conditions that will ensure all solutions are either oscillatory or converges to zero, by using the inequality technique and generalized Riccati transformation. The newly derived criterion are also used to establish a new class of systems with delay which are not covered in the existing study of literature. Further, we constructed some suitable illustrations.

**Keywords:** oscillation; nonlinear differential system; delay differential system; $\alpha$-fractional derivative

## 1. Introduction

In the literature there are many advanced strategies in the expansion of ordinary and partial differential equations of fractional order and they have been used as excellent sources and tools in order to model many phenomena in the different fields of engineering, science and technology. Further, these tools are also used in fields such as chemical processes, polymer rheology, mathematical biology, industrial robotics, viscoelasticity, and many more, see the monographs [1–7].

At the end of the nineteenth century, Henry Poincare initiated the method and used the qualitative analysis of nonlinear systems of integer order differential equations. Since then, there has been significant development in the theory of oscillation of integer order differential systems [8–18].

In a study [19], Vreeke et al. applied the differential systems in the application of physics in order to solve the problem of a nuclear reactor which involved two temperature feedback. In the current literature there are many established results in the oscillation theory of classical differential systems (see [20–24]). However, in the nonlinear fractional differential system development is relatively slow due to the occurrence of nonlocal behavior of fractional derivatives that possess weakly singular kernels.

In 2014, Khalil et al. introduced the idea of conformable fractional derivative as a kind of local derivative with no memory (see [25–27]). By following the idea of Khalil, an interesting application of the conformable fractional derivative in physics was discussed and the action principle for particles under the frictional forces were formulated, see [28].

The idea of conformable fractional derivatives was generalized by Katugampola, and today it is known as the Katugampola fractional derivative. Nowadays, many researchers have interest in this type of derivative for their useful properties (see [29–31]). In this respect, we list the contributions of

Spanikova [32], Sadhasivam [33] and Chatzarakis [34] where the oscillation of $\alpha$-fractional nonlinear three dimensional delay differential systems were also studied.

Now we study oscillatory behavior of the following system having the form

$$\begin{aligned} D^{\alpha}(u(t)) &= p(t)g(v(\sigma(t))), \\ D^{\alpha}(v(t)) &= -q(t)h(w(t))), \\ D^{\alpha}(w(t)) &= r(t)f(u(\delta(t))),\ t \geq t_0, \end{aligned} \tag{1}$$

where $0 < \alpha \leq 1$, $D^{\alpha}$ denotes the $\alpha$-fractional derivative respect to $t$.

Based on the following assumptions:

$(A_1)$ $p(t) \in C^{2\alpha}([t_0,\infty),\mathbb{R}^+)$, $q(t) \in C^{\alpha}([t_0,\infty),\mathbb{R}^+)$, $r(t) \in C([t_0,\infty),\mathbb{R}^+)$, $p(t)$, $q(t)$ and $r(t)$ are not identically zero on any interval of $[T_0,\infty)$, $T_0 \geq t_0$, $r(t)$ and $q(t)$ are decreasing and positive;
$(A_2)$ $g \in C^{\alpha}(\mathbb{R},\mathbb{R}), vg(v) > 0, D^{\alpha}g(v) \geq l' > 0$, $h \in C^{\alpha}(\mathbb{R},\mathbb{R}), wh(w) > 0, D^{\alpha}h(w) \geq m' > 0$, $f \in C(\mathbb{R},\mathbb{R}), yf(y) > 0$ and $\frac{f(y)}{y} \geq k > 0$ for $y \neq 0$;
$(A_3)$ $\sigma(t) \leq t$ with $D^{\alpha}\sigma(t) \geq l > 0$, $\delta(t) \leq t$ and satisfies $\lim_{t\to\infty}\sigma(t) = \infty$, $\lim_{t\to\infty}\delta(t) = \infty$;
$(A_4)$ The case will be considered as

$$\int_{t_0}^{\infty} s^{\alpha-1}\frac{1}{b(s)}ds = \infty, \int_{t_0}^{\infty} s^{\alpha-1}\frac{1}{a(s)}ds = \infty,$$

where $b(t) = \frac{1}{q(t)}$, $a(t) = \frac{1}{p(t)}$ and $c(t) = l^2 l' m' r(t)$, $a(t), b(t)$ and $c(t)$ are positive real-valued continuous functions with $b(t)t^{1-\alpha} < 1$.

The solution implies that, it is a vector-valued function such that $U(t) = (u(t), v(t), w(t))$ with $T_1 = \min\{\delta(t_1), \sigma(t_1)\}$ for some $t_1 \geq t_0$ which has the property such that $b(t)D^{\alpha}(a(t)D^{\alpha}u(t)) \in C^{\alpha}([T_1,\infty),\mathbb{R})$ and satisfies the system (1) on $[T_1,\infty)$. Denote by $\mathbb{P}$, the set of all solutions $U(t)$ of (1) which exist on some half line $[T_1,\infty)$, $T_1 > t_0$. The researchers only focus to the nontrivial solutions of system (1) and satisfy $\sup\{|u(\xi)| + |v(\xi)| + |w(\xi)|,\ t \leq \xi < \infty\} > 0$ for any $t \geq T_1$. We make a standing hypothesis that (1) has such a solution.

A proper solution $U(t) \in \mathbb{P}$ for the system (1) is called oscillatory if all the components are oscillatory, otherwise it is nonoscillatory. Further, the system (1) is said to be oscillatory if all proper solutions oscillate.

The main goal of this paper is to establish some new oscillation criteria for the system (1) by making use of the generalized Riccati transformation and inequality technique. The study is structured as follows. In Section 2, we recall some preliminary concepts relative to the $\alpha$- fractional derivative. In Section 3, some new conditions for the oscillatory behavior of the solutions of system (1) were presented. Illustrative examples are included in the final part of the paper in order to demonstrate the efficiency of new theorems.

## 2. Preliminaries

We begin this section with the following definition of the operator $D^{\alpha}$.

**Definition 1.** *[30] Let $y : [0,\infty) \to \mathbb{R}$, then $\alpha$-fractional derivative of $y$ is defined by*

$$D^{\alpha}(y)(t) := \lim_{\epsilon\to 0}\frac{y(te^{\epsilon t^{-\alpha}}) - y(t)}{\epsilon} \quad for \quad t > 0 \text{ and } \alpha \in (0,1]. \tag{2}$$

*If $y$ is differentiable $\alpha$-times in some $(0,a)$ with $a > 0$, $\lim_{t\to 0^+} D^{\alpha}(y)(t)$ exists, then we have*

$$D^{\alpha}(y)(0) := \lim_{t\to 0^+} D^{\alpha}(y)(t).$$

**$\alpha$-fractional derivative satisfies the following properties.** [30]
Let $\alpha \in (0,1]$ and $g, h$ be $\alpha$- differentiable for $t > 0$. Then

$(p_1)$ $D^\alpha(t^n) = nt^{n-\alpha}$ for all $n \in \mathbb{R}$.
$(p_2)$ $D^\alpha(C) = 0$ for all constant functions, $g(t) = C$.
$(p_3)$ $D^\alpha(gh) = gD^\alpha(h) + hD^\alpha(g)$.
$(p_4)$ $D^\alpha(\frac{g}{h}) = \frac{hD^\alpha(g) - gD^\alpha(h)}{h^2}$.
$(p_5)$ $D^\alpha(g \circ h)(t) = g'(h(t))D^\alpha h(t)$, for $g$ is differentiable at $h(t)$.
$(p_6)$ If $g$ is differentiable, then $D^\alpha(g)(t) = t^{1-\alpha}\frac{dg}{dt}(t)$.

**Definition 2.** [30] *Let $a \geq 0$, $t \geq a$ and a function $y$ defined on $(a,t]$ with $\alpha \in \mathbb{R}$. Then, $\alpha$-fractional integral as follows*

$$I_a^\alpha(y)(t) := \int_a^t \frac{y(x)}{x^{1-\alpha}} dx \tag{3}$$

*provided improper integral exists.*

## 3. Main Results

In this section, the oscillatory behavior of solutions for the system (1) under certain conditions are established. Next we give the following lemmas that will be used in our further discussion.

**Lemma 1.** *If $U(t) \in \mathbb{P}$ is a nonoscillatory solution for (1), then the component function $x(t)$ is always nonoscillatory.*

**Lemma 2.** *Suppose that $(A_1)$ and $(A_4)$ holds. Then there exists a $t_1 \geq t_0$ such that either*
*(I) $u(t) > 0, D^\alpha u(t) > 0, D^\alpha(a(t)D^\alpha u(t)) > 0$ for $t \geq t_1$.*
*or*
*(II) $u(t) > 0, D^\alpha u(t) < 0, D^\alpha(a(t)D^\alpha u(t)) > 0$ for $t \geq t_1$ holds.*

**Proof.** Let $u(t)$ be an eventually positive solution for (1) on $[t_0, \infty)$. Now, system (1) will be reduced to the following inequality

$$D^\alpha\left(\frac{1}{q(t)}D^\alpha\left(\frac{1}{p(t)}D^\alpha u(t)\right)\right) + l^2 l' m' r(t) f(u(\delta(\sigma(t)))) \leq 0, \; t \geq t_1, \tag{4}$$

which implies,

$$D^\alpha\left(b(t)D^\alpha\left(a(t)D^\alpha u(t)\right)\right) + c(t)f(u(\delta(\sigma(t)))) \leq 0, \; t \geq t_1. \tag{5}$$

From (5), we get $D^\alpha(b(t)D^\alpha(a(t)D^\alpha u(t))) \leq 0$ for $t \geq t_0$. Then $b(t)D^\alpha(a(t)D^\alpha u(t))$ is decreasing on $(t_0, \infty)$. Thus the proof completes on using the Lemma 3.2 in [34]. □

The following notations are employed in the sequel.

$$(A_\alpha)_* := \liminf_{t\to\infty} t \int_t^\infty s^{\alpha-1} A^\alpha(s) ds \text{ and } (B_\alpha)_* := \liminf_{t\to\infty} \frac{1}{t} \int_{t_0}^t s^{\alpha+1} A_\alpha(s) ds, \tag{6}$$

where $A_\alpha(t) = \frac{k}{2}\frac{c(t)}{a(t)}\frac{\delta(\sigma(t)) - T}{t}(\delta(\sigma(t)))^\alpha$.

$$d := \liminf_{t\to\infty} tw(t) \text{ and } D := \limsup_{t\to\infty} tw(t). \tag{7}$$

**Theorem 1.** *Suppose that* $(A_1) - (A_4)$ *hold. Assume also that*

$$\int_{t_2}^{\infty} c(s)(s-T)\delta(\sigma(s))ds = \infty, \tag{8}$$

*there exists a positive function* $\rho \in C^{\alpha}([0,\infty);\mathbb{R}_+)$ *such that*

$$\limsup_{t\to\infty} \int_{t_0}^{t} \left( s^{\alpha-1}\rho(s)A_{\alpha}(s) - \frac{1}{4}\frac{(\rho'(s))^2}{\rho(s)} s^{1-\alpha} b(s) \right) ds = \infty. \tag{9}$$

*Then every solution of system* (1) *is oscillatory.*

**Proof.** Suppose that (1) has a nonoscillatory solution $(u(t), v(t), w(t))$ on $[t_0, \infty)$. From Lemma 1, $u(t)$ is always nonoscillatory. Without loss of generality, we shall assume that $u(t) > 0$, $u(\delta(t)) > 0$ and $u(\delta(\sigma(t))) > 0$ for $t \geq T \geq t_0$, since similar arguments can be made for $u(t) < 0$ eventually. Suppose that Case (I) of Lemma 2 holds for $t \geq t_1$. Define

$$w(t) = \rho(t)\frac{b(t)D^{\alpha}(a(t)D^{\alpha}u(t))}{a(t)D^{\alpha}u(t)}, \ t \geq t_1. \tag{10}$$

Thus $w(t) > 0$, differentiating $\alpha$ times with respect to $t$, using (5) and $(A_2)$, we have

$$D^{\alpha}w(t) \leq \frac{D^{\alpha}\rho(t)}{\rho(t)}w(t) - \frac{k\rho(t)c(t)}{a(t)}\frac{u(\delta(\sigma(t)))}{D^{\alpha}u(t)} - \frac{1}{\rho(t)b(t)}w^2(t). \tag{11}$$

Now, let $z_1(t,T) = (t-T)$, $z_2(t,T) = \frac{(t-T)^2}{2}$ and define $U(t) := (t-T)t^{1-\alpha}u(t) - z_2(t,T)D^{\alpha}u(t)$. Then $U(T) = 0$ and differentiating the above, we get

$$D^{\alpha}U(t) = t^{1-\alpha}\Big( t^{1-\alpha}u(t) + (t-T)(1-\alpha)t^{-\alpha}u(t) + (t-T)t^{1-\alpha}u'(t) - z_2'(t,T)D^{\alpha}u(t) - z_2(t,T)(D^{\alpha}u(t))' \Big),$$

which implies

$$U'(t) \geq t^{1-\alpha}u(t) - z_2(t,T)(D^{\alpha}u(t))'. \tag{12}$$

By Taylor's Theorem, we have

$$\int_T^t s^{1-\alpha}u'(s)ds = z_1(t,T)D^{\alpha}u(T) + \int_T^t z_1(t,s)(D^{\alpha}u(s))'ds,$$

since $D^{\alpha}(a(t)D^{\alpha}u(t))$ is decreasing, we get

$$t^{1-\alpha}u(t) \geq t^{1-\alpha}u(T) + z_1(t,T)D^{\alpha}u(T) + (D^{\alpha}u(t))'\int_T^t z_1(t,s)ds.$$

Thus $U'(t) > 0$ on $[T,\infty)$. From this we get $U(t) > 0$ on $[T,\infty)$, which implies that

$$\frac{u(t)}{D^{\alpha}u(t)} > \frac{z_2(t,T)}{(t-T)t^{1-\alpha}} = \frac{t-T}{2}t^{1-\alpha}, \ t \in [T,\infty). \tag{13}$$

Next, define $V(t) := D^{\alpha}u(t) - t(D^{\alpha}u(t))'$. In view of the fact that $D^{\alpha}V(t) = -t^{2-\alpha}(D^{\alpha}u(t))''$, which implies $V'(t) = -t^{2-\alpha}(D^{\alpha}u(t))'' > 0$ for $t \in [T,\infty)$, therefore $V(t)$ is strictly increasing on $[T,\infty)$.

We claim that there is a $t_1 \in [T,\infty)$ such that $V(t) > 0$ on $[t_1,\infty)$. Suppose not, $V(t) < 0$ on $[t_1,\infty)$. Hence,

$$D^{\alpha}\left(\frac{D^{\alpha}u(t)}{t}\right) = -\frac{t^{1-\alpha}}{t^2}(t(D^{\alpha}u(t))' - D^{\alpha}u(t)),$$

which gives

$$\left(\frac{D^{\alpha}u(t)}{t}\right)' = -\frac{1}{t^2}V(t) > 0,\ t \in [t_1,\infty).$$

Choose $t_2 \in (t_1,\infty)$, for $t \geq t_2$, $\delta(\sigma(t)) \geq \delta(\sigma(t_2))$. Since, $\frac{D^{\alpha}u(t)}{t}$ is strictly increasing,

$$\frac{D^{\alpha}u(\delta(\sigma(t)))}{\delta(\sigma(t))} \geq \frac{D^{\alpha}u(\delta(\sigma(t_2)))}{\delta(\sigma(t_2))} := m > 0,$$

the Equation (13) implies that

$$u(\delta(\sigma(t))) \geq \frac{t-T}{2}t^{1-\alpha}m\delta(\sigma(t)). \tag{14}$$

Now, integrating (5) from $t_2$ to $t$, using $(A_2)$ and inequality in (14), we have

$$\int_{t_2}^{t}\left((b(s)D^{\alpha}(a(s)D^{\alpha}u(s)))' + \frac{km}{2}c(s)(s-T)\delta(\sigma(s))\right)ds \leq 0.$$

Then

$$b(t_2)D^{\alpha}(a(t_2)D^{\alpha}u(t_2)) \geq \frac{km}{2}\int_{t_2}^{t} c(s)(s-T)\delta(\sigma(s))ds,$$

which contradicts to (8). Hence $V(t) > 0$ on $[t_1,\infty)$. Accordingly,

$$t^{1-\alpha}\left(\frac{D^{\alpha}u(t)}{t}\right)' = -\frac{t^{1-\alpha}}{t^2}(t(D^{\alpha}u(t))' - D^{\alpha}u(t)) = -\frac{t^{1-\alpha}}{t^2}V(t) < 0,\ t \in (t_1,\infty),$$

which gives $t(D^{\alpha}u(t))' < D^{\alpha}u(t)$. Then $\delta(\sigma(t)) \leq \delta(t) \leq t$,

$$\frac{D^{\alpha}u(\delta(\sigma(t)))}{\delta(\sigma(t))} \geq \frac{D^{\alpha}u(t)}{t}, \tag{15}$$

since $\frac{D^{\alpha}u(t)}{t}$ is strictly increasing. Using (13) and (15) in (11), we get

$$D^{\alpha}w(t) \leq \frac{D^{\alpha}\rho(t)}{\rho(t)}w(t) - \frac{k\rho(t)c(t)}{ta(t)}\frac{(\delta(\sigma(t)))^{\alpha}(\delta(\sigma(t))-T)}{2} - \frac{1}{\rho(t)b(t)}w^2(t). \tag{16}$$

Therefore

$$D^{\alpha}w(t) \leq -\frac{k\rho(t)c(t)}{ta(t)}\frac{(\delta(\sigma(t)))^{\alpha}(\delta(\sigma(t))-T)}{2} + \frac{1}{4}b(t)\frac{(D^{\alpha}\rho(t))^2}{\rho(t)},$$

using (6) and $(p_6)$, we get

$$w'(t) \leq -t^{\alpha-1}\rho(t)A_\alpha(t) + \frac{1}{4}\frac{(\rho'(t))^2}{\rho(t)}t^{1-\alpha}b(t). \tag{17}$$

Integrating,

$$\int_{t_1}^{t}\left(s^{\alpha-1}\rho(s)A_\alpha(s) - \frac{1}{4}\frac{(\rho'(s))^2}{\rho(s)}s^{1-\alpha}b(s)\right)ds \leq w(t_1),$$

which contradicts the hypothesis (9). □

We now derive various oscillatory criteria on using the earlier results and we can generalize the Philos type kernel. Let us define a class of functions $\Omega$. Consider

$$\mathbb{D}_0 = \{(t,s) : t > s \geq t_0\}, \text{and } \mathbb{D} = \{(t,s) : t \geq s \geq t_0\}.$$

The function $H \in C(\mathbb{D}, \mathbb{R})$ belongs to the class $\Omega$, if

$(T_1)$ $H(t,t) = 0$ for $t \geq t_0$, and $H(t,s) > 0$ for $(t,s) \in \mathbb{D}_0$.

$(T_2)$ The nonpositive partial derivative $\frac{\partial H}{\partial s}$ exist on $\mathbb{D}_0$ such that $h(t,s) = H(t,s)\frac{\rho'(s)}{\rho(s)} + \frac{\partial H}{\partial s}(t,s)$.

**Theorem 2.** *Assume that $(A_1) - (A_4)$ hold. Further there exists $\rho \in C^\alpha([0,\infty);\mathbb{R}_+)$ such that*

$$\limsup_{t\to\infty}\frac{1}{H(t,t_1)}\int_{t_1}^{t}\left(H(t,s)s^{\alpha-1}\rho(s)A_\alpha(s) - \frac{1}{4}\frac{\rho(s)b(s)}{H(t,s)}s^{1-\alpha}h^2(t,s)\right)ds = \infty. \tag{18}$$

*Then each solution of system (1) is oscillatory.*

**Proof.** As we proceed in the proof of Theorem 1 and from (16), we have the inequality

$$w'(t) \leq \frac{\rho'(t)}{\rho(t)}w(t) - t^{\alpha-1}\rho(t)A_\alpha(t) - \frac{t^{\alpha-1}}{\rho(t)b(t)}w^2(t). \tag{19}$$

Integrating,

$$\int_{t_1}^{t} H(t,s)s^{\alpha-1}\rho(s)A_\alpha(s)ds$$

$$\leq \int_{t_1}^{t} H(t,s)\frac{\rho'(s)}{\rho(s)}w(s)ds - \int_{t_1}^{t} H(t,s)w'(s)ds - \int_{t_1}^{t} H(t,s)\frac{s^{\alpha-1}}{\rho(s)b(s)}w^2(s)ds,$$

$$\leq H(t,t_1)w(t_1) + \int_{t_1}^{t}\left(H(t,s)\frac{\rho'(s)}{\rho(s)} + \frac{\partial H}{\partial s}(t,s)\right)w(s)ds - \int_{t_1}^{t} H(t,s)\frac{s^{\alpha-1}}{b(s)\rho(s)}w^2(s)ds,$$

$$\leq H(t,t_1)w(t_1) + \int_{t_1}^{t}\left(w(s)h(t,s) - H(t,s)\frac{s^{\alpha-1}}{\rho(s)b(s)}w^2(s)\right)ds,$$

$$\leq H(t,t_1)w(t_1) + \int_{t_1}^{t} \frac{1}{4}\frac{\rho(s)b(s)}{H(t,s)} s^{1-\alpha} h^2(t,s)ds.$$

From this we conclude that

$$\int_{t_1}^{t} \left( H(t,s)s^{\alpha-1}\rho(s)A_\alpha(s) - \frac{1}{4}\frac{\rho(s)b(s)}{H(t,s)} s^{1-\alpha} h^2(t,s) \right) ds \leq H(t,t_1)w(t_1).$$

Since $0 < H(t,s) \leq H(t,t_1)$ for $t > s > t_1$, we have $0 < \frac{H(t,s)}{H(t,t_1)} \leq 1$, hence

$$\frac{1}{H(t,t_1)} \int_{t_1}^{t} \left( H(t,s)s^{\alpha-1}\rho(s)A_\alpha(s) - \frac{1}{4}\frac{\rho(s)b(s)}{H(t,s)} s^{1-\alpha} h^2(t,s) \right) ds \leq w(t_1).$$

Letting $t \to \infty$,

$$\limsup_{t\to\infty} \frac{1}{H(t,t_1)} \int_{t_1}^{t} \left( H(t,s)s^{\alpha-1}\rho(s)A_\alpha(s) - \frac{1}{4}\frac{\rho(s)b(s)}{H(t,s)} s^{1-\alpha} h^2(t,s) \right) ds \leq w(t_1).$$

Therefore assumption (18) is contradicted. Thus every solution of (1) oscillates. □

We immediately obtain the following oscillation result for (1).

**Theorem 3.** *Assume that $(A_1)$–$(A_4)$ hold. Also assume that there exists a function $\rho \in C^\alpha([0,\infty);\mathbb{R}_+)$ such that*

$$\limsup_{t\to\infty} \frac{1}{H(t,t_1)} \int_{t_1}^{t} \left( H(t,s)s^{\alpha-1}\rho(s)A_\alpha(s) - \frac{1}{4}\frac{H(t,s)(\rho'(s))^2}{\rho(s)} s^{1-\alpha} b(s) \right) ds = \infty. \tag{20}$$

*Then every solution of system $(1)$ is oscillatory.*

**Proof.** Proceeding as in the proof of Theorem 1, multiplying inequality (17) by $H(t,s)$ and integrating, we get

$$\int_{t_1}^{t} \left( H(t,s)s^{\alpha-1}\rho(s)A_\alpha(s) - \frac{1}{4}\frac{H(t,s)(\rho'(s))^2}{\rho(s)} s^{1-\alpha} b(s) \right) ds \leq -\int_{t_1}^{t} H(t,s)w'(s)ds \leq H(t,t_1)w(t_1).$$

Taking limsup as $t \to \infty$, and hence

$$\limsup_{t\to\infty} \frac{1}{H(t,t_1)} \int_{t_1}^{t} \left( s^{\alpha-1}\rho(s)A_\alpha(s)H(t,s) - \frac{1}{4}\frac{H(t,s)(\rho'(s))^2}{\rho(s)} s^{1-\alpha} b(s) \right) ds \leq w(t_1),$$

which contradicts the hypothesis in (20). □

The following theorem is to be proved using the techniques employed in the previous theorems.

**Theorem 4.** *Suppose that the assumptions $(A_1)$–$(A_4)$ and (8) hold. Further assume also that Case (I) of Lemma 2 holds, then*

$$(A_\alpha)_* \leq d - t^{\alpha-1}\frac{1}{b(s)}d^2, \tag{21}$$

*and*

$$(B_\alpha)_* \leq D - D^2. \tag{22}$$

**Proof.** Let $u(t)$ be a nonoscillatory solution of (5) such that $u(t) > 0$, $u(\delta(t)) > 0$ and $u(\delta(\sigma(t))) > 0$ for $t \geq T > t_0$, consider the case (I) of Lemma 2 holds, $u(t)$ satisfies the inequality $D^\alpha\left(b(t)D^\alpha\left(a(t)D^\alpha u(t)\right)\right) \leq 0,\ t \in [T, \infty)$. Define Riccati transformation

$$w(t) = \frac{b(t)D^\alpha(a(t)D^\alpha u(t))}{a(t)D^\alpha u(t)}.$$

Thus $w(t) > 0$, differentiating $\alpha$ times with respect to $t$, using (5) and $(A_2)$, we have

$$D^\alpha w(t) \leq -\frac{kc(t)}{a(t)}\frac{u(\delta(\sigma(t)))}{D^\alpha u(t)} - \frac{1}{b(t)}w^2(t).$$

By using (15), (13) and (6), we obtain the above inequality

$$w'(t) + t^{\alpha-1}A_\alpha(t) + t^{\alpha-1}\frac{1}{b(t)}w^2(t) \leq 0. \tag{23}$$

Given that $A_\alpha(t) > 0$ and $w(t) > 0$, which gives $w'(t) \leq 0$ and

$$-b(t)(w'(t)t^{1-\alpha}/w^2(t)) > 1.$$

which yields that

$$\left(\frac{1}{w(t)}\right)' > t^{\alpha-1}\frac{1}{b(t)}.$$

Integrating the above inequality and denote $t_1^{\alpha-1}\frac{1}{b(t_1)} = M$, we have

$$M(t - t_1)w(t) < 1. \tag{24}$$

which implies that

$$\lim_{t\to\infty} w(t) = 0,\ \lim_{t\to\infty} tw(t) = 0. \tag{25}$$

From (9) and (24), $0 < d < 1$ and $0 < D < 1$. Even though if $d = 0$ and $D = 0$, there is nothing to prove. Now, to claim (21). Integrating (23) from $t$ to $\infty$ and use (25), we get

$$w(t) \geq \int_t^\infty s^{\alpha-1}A_\alpha(s)ds + \int_t^\infty s^{\alpha-1}\frac{1}{b(s)}w^2(s)ds.$$

Multiplying by $t$ and taking liminf as $t \to \infty$, by (25), $d \geq (A_\alpha)_*$. For given $\epsilon > 0$, there exists a $t_2 \geq t_1$ as

$$d - \epsilon < tw(t) < d + \epsilon \; and \; t \int_t^\infty s^{\alpha-1} A_\alpha(s) ds \geq (A_\alpha)_* - \epsilon, \; t \geq t_2. \tag{26}$$

Again from (26),

$$\begin{aligned} tw(t) &\geq t \int_t^\infty s^{\alpha-1} A_\alpha(s) ds + t \int_t^\infty s^{\alpha-1} \tfrac{1}{b(s)} w^2(s) ds \\ &\geq t \int_t^\infty s^{\alpha-1} A_\alpha(s) ds + t^\alpha \tfrac{1}{b(t)} \int_t^\infty \tfrac{(sw(s))^2}{s^2} ds \\ &\geq t \int_t^\infty s^{\alpha-1} A_\alpha(s) ds + t^\alpha \tfrac{1}{b(t)} (d-\epsilon)^2 \int_t^\infty \tfrac{1}{s^2} ds \\ &= t \int_t^\infty s^{\alpha-1} A_\alpha(s) ds + t^{\alpha-1} \tfrac{1}{b(t)} (d-\epsilon)^2. \end{aligned} \tag{27}$$

Therefore from (26) and (27), $d \geq (A_\alpha)_* - \epsilon + (d-\epsilon)^2$. Then

$$d \geq (A_\alpha)_* + t^{\alpha-1} \frac{1}{b(t)} d^2,$$

since $\epsilon$ is arbitrarily small. Next to prove that (22). Multiply (23) by $s^2$, integrating from $t_1$ to $t$, and integration by parts follows that

$$\begin{aligned} \int_{t_1}^t s^{\alpha+1} A_\alpha(s) ds &\leq -\int_{t_1}^t s^2 w'(s) ds - \int_{t_1}^t s^{\alpha+1} \frac{1}{b(s)} w^2(s) ds \\ &\leq -t^2 w(t) + t_1^2 w(t_1) + 2 \int_{t_1}^t sw(s) ds - \int_{t_1}^t s^{\alpha+1} \frac{1}{b(s)} w^2(s) ds, \end{aligned}$$

implies

$$t^2 w(t) \leq t_1^2 w(t_1) - \int_{t_1}^t s^{\alpha+1} A_\alpha(s) ds + \int_{t_1}^t \left( 2sw(s) - s^{\alpha+1} \frac{1}{b(s)} w^2(s) \right) ds. \tag{28}$$

Thus, we obtain

$$\begin{aligned} tw(t) &\leq \tfrac{t_1^2 w(t_1)}{t} - \tfrac{1}{t} \int_{t_1}^t s^{\alpha+1} A_\alpha(s) ds + \tfrac{1}{t} \int_{t_1}^t s^{1-\alpha} b(s) ds, \\ &\leq \tfrac{t_1^2 w(t_1)}{t} - \tfrac{1}{t} \int_{t_1}^t s^{\alpha+1} A_\alpha(s) ds + \tfrac{1}{t} t^{1-\alpha} b(t) \int_{t_1}^t ds. \end{aligned} \tag{29}$$

By $(A_4)$, (29) imply that

$$tw(t) \leq \frac{t_1^2 w(t_1)}{t} - \frac{1}{t} \int_{t_1}^t s^{\alpha+1} A_\alpha(s) ds + \frac{1}{t}(t - t_1).$$

Thus

$$\limsup_{t\to\infty} tw(t) \leq 1 - \liminf_{t\to\infty} \frac{1}{t} \int_{t_1}^{t} s^{\alpha+1} A_\alpha(s) ds.$$

Hence from (6), (7), $D \leq 1 - (B_\alpha)_*$. For any $\epsilon > 0$, there exists a $t_2 \geq t_1$ such that

$$D - \epsilon < tw(t) < D + \epsilon \text{ and } \frac{1}{t} \int_{t_0}^{t} s^{\alpha+1} A_\alpha(s) ds > (B_\alpha)_* - \epsilon,\ t \geq t_2. \tag{30}$$

Now, from (28) and (30) we get

$$D \leq -(B_\alpha)_* + \epsilon(D+\epsilon)(2 - D + \epsilon),\ t \geq t_2,$$

since $\epsilon$ is arbitrarily small, we have

$$(B_\alpha)_* \leq D - D^2,$$

which proves (22). □

**Lemma 3.** *Suppose that $(A_1)$–$(A_4)$ and (8) hold. Also assume that Case (II) of Lemma 2 holds. If*

$$\int_{t_2}^{\infty} \eta^{\alpha-1} \frac{1}{a(\eta)} \left( \int_{\eta}^{\infty} \int_{\mu}^{\infty} s^{\alpha-1} c(s) ds d\mu \right) d\eta = \infty. \tag{31}$$

*Then* $\lim_{t\to\infty} u(t) = 0$.

**Proof.** We consider the Case (II) of Lemma 2, $D^\alpha u(t) < 0, D^\alpha(a(t)D^\alpha u(t)) > 0$ for $t \geq t_1$. Since $u(t)$ is positive and decreasing, we have $\lim_{t\to\infty} u(t) = d' \geq 0$. Suppose not, $d' > 0$. Given that $u(\delta(\sigma(t))) \leq \delta(t) \leq t$, then $u(\delta(\sigma(t))) \geq u(t) > d'$ for $t \geq t_2 \geq t_1$ sufficiently large, $u(t)$ is decreasing. Integrating (5) from $t$ to $\infty$ and using $u(\delta(\sigma(t))) \geq d'$, we get

$$\int_t^\infty \left( b(s) D^\alpha \left(a(s) D^\alpha u(s)\right) \right)' ds \leq - \int_t^\infty k s^{\alpha-1} c(s) u(\delta(\sigma(s))) ds \leq -kd' \int_t^\infty s^{\alpha-1} c(s) ds,$$

then,

$$b(t) D^\alpha \left(a(t) D^\alpha u(t)\right) \geq kd' \int_t^\infty s^{\alpha-1} c(s) ds.$$

By $(A_4)$, we get

$$\left(a(t) D^\alpha u(t)\right)' \geq kd' \frac{1}{b(t) t^{1-\alpha}} \int_t^\infty s^{\alpha-1} c(s) ds \geq kd' \int_t^\infty s^{\alpha-1} c(s) ds.$$

Again integrating, we obtain

$$-a(t) D^\alpha u(t) \geq kd' \int_t^\infty \int_\mu^\infty s^{\alpha-1} c(s) ds d\mu,$$

this implies that,

$$-u^{'}(t) \geq kd^{'}t^{\alpha-1}\frac{1}{a(t)}\int_{t}^{\infty}\int_{\mu}^{\infty}s^{\alpha-1}c(s)dsd\mu.$$

By integrating, once again it is get as

$$u(t_2) \geq kd^{'}\int_{t_2}^{\infty}\left(\eta^{\alpha-1}\frac{1}{a(\eta)}\int_{\eta}^{\infty}\int_{\mu}^{\infty}s^{\alpha-1}c(s)dsd\mu\right)d\eta,$$

which contradicts to (31). Thus $d^{'} = 0$ and hence $\lim_{t\to\infty} u(t) = 0$. □

From Theorem 4, Nehari type oscillation criteria for (1).

**Theorem 5.** *Assume that $(A_1) - (A_4)$, (8) and (31) hold. If*

$$\liminf_{t\to\infty}\frac{1}{t}\int_{t_0}^{t}\left(ks^{\alpha+1}\frac{c(s)}{a(s)}\frac{u(\delta(\sigma(s))) - T}{s}(u(\delta(\sigma(s))))^{\alpha}\right)ds > \frac{1}{2}, \tag{32}$$

*then u(t) is oscillatory or satisfies $u(t) = 0$ as $t \to \infty$.*

## 4. Examples

In this section, we provide some examples in order to see the effect of the main results.

**Example 1.** *Consider $\frac{1}{2}$-fractional delay differential system*

$$\begin{aligned} D^{\frac{1}{2}}(u(t)) &= \frac{1}{\sqrt{t}}g(v(\frac{t}{2})) \\ D^{\frac{1}{2}}(v(t)) &= -\frac{1}{\sqrt{t}}h(w(t)), \\ D^{\frac{1}{2}}(w(t)) &= \frac{1}{\sqrt{t}}f(u(\frac{t}{2})), \quad t \geq t_0, \end{aligned} \tag{33}$$

*where $C_1 = \cos(\ln 2)$, $C_2 = \sin(\ln 2)$, $A_1 = \cos(\ln 4)$, $A_2 = \sin(\ln 4)$.*
*Here $\alpha = \frac{1}{2}$, $p(t) = \frac{1}{a(t)} = \frac{1}{\sqrt{t}}$, $q(t) = \frac{1}{b(t)} = \frac{1}{\sqrt{t}}$, $r(t) = \frac{1}{\sqrt{t}}$, $f(u) = A_1\sqrt{(1-u^2)} - A_2 u$, $g(v) = v$ and $h(w) = w$.*

*It is easy to see that*

$$\begin{aligned} D^{\alpha}g(v) &= \frac{1}{\sqrt{t}}(C_1 + C_2) \geq l^{'} > 0, \\ D^{\alpha}h(w) &= \frac{1}{\sqrt{t}}(C_1 - C_2) \geq m^{'} > 0, \\ f(u)/u &= A_1\sqrt{\frac{1}{u^2} - 1} - A_2 \geq 0.2579 = k > 0, \end{aligned}$$

*since $u^2 < 1$, $\sigma(t) = \delta(t) = \frac{t}{2}$ and $D^{\alpha}\sigma(t) = \frac{\sqrt{t}}{2} \geq l > 0$, $c(t) = \frac{C_1^2 - C_2^2}{4\sqrt{t}}$, $A_{\alpha}(t) = \frac{0.2579}{16}\frac{\frac{t}{4} - T}{\sqrt{t}}$. Now it is considered as,*

$$\int_{t_2}^{\infty} c(s)(s-T)\delta(\sigma(s))ds = \frac{C_1^2 - C_2^2}{4}\int_{t_2}^{\infty}\frac{(s-T)}{\sqrt{s}}\frac{s}{4}ds \to \infty \text{ as } t \to \infty.$$

*By taking* $\rho(t) = 16/k$ *then* $\rho'(t) = 0$. *Consider*

$$\begin{aligned}
&\limsup_{t\to\infty} \int_{t_1}^{t} \left( s^{\alpha-1}\rho(s)A_\alpha(s) - \frac{1}{4}\frac{(\rho'(s))^2}{\rho(s)} s^{1-\alpha} b(s) \right) ds \\
&= \limsup_{t\to\infty} \int_{t_1}^{t} \left( s^{-\frac{1}{2}} \frac{16}{k} \frac{k(\frac{s}{4} - T)}{16} \frac{1}{\sqrt{s}} \right) ds \\
&= \limsup_{t\to\infty} \frac{1}{4} \int_{t_1}^{t} \left( \frac{s - 4T}{s} \right) ds \to \infty \text{ as } t \to \infty.
\end{aligned}$$

*Since, each of the conditions are verified in Theorem 1, all solutions of* (33) *are oscillatory. Thus* $(u(t), v(t), w(t)) = (\sin(\ln t), C_1 \cos(\ln t) - C_2 \sin(\ln t), C_1 \sin(\ln t) + C_2 \cos(\ln t))$ *is one such solution.*

**Note:** The decreasing condition imposed on $q(t)$ and $r(t)$ is only a sufficient condition, however it is not a necessary one. The following example ensures the oscillatory behavior of the system (34) even though $q(t)$ and $r(t)$ is nondecreasing.

**Example 2.** *Consider* $\frac{1}{3}$*-fractional following differential system*

$$\begin{aligned}
D^{\frac{1}{3}}(u(t)) &= \frac{t^{\frac{2}{3}}}{1 + \frac{3}{4}\cos^{\frac{5}{3}}(t)} g(v(t - 2\pi)), \\
D^{\frac{1}{3}}(v(t)) &= -t^{\frac{2}{3}} w(t), \\
D^{\frac{1}{3}}(w(t)) &= \frac{t^{\frac{2}{3}}}{1 + \cos^2(t)} f(u(t - \frac{3\pi}{2})), \quad t \geq t_0.
\end{aligned} \tag{34}$$

*Here* $\alpha = \frac{1}{3}$, $p(t) = \frac{1}{a(t)} = \frac{t^{\frac{2}{3}}}{1+\frac{3}{4}\cos^{\frac{5}{3}}(t)}$, $q(t) = \frac{1}{b(t)} = t^{\frac{2}{3}}$, $r(t) = \frac{t^{\frac{2}{3}}}{1+\cos^2(t)}$, $f(u) = u(1 + u^2)$, $g(v) = v(1 + \frac{3}{4}v^{\frac{5}{3}})$ *and* $h(w) = w$. *It is easy to see that* $D^\alpha g(v) = v^{\frac{1}{3}} + 2v^2 \geq 1 = l' > 0$ *such that* $y^2 > 1$, $y' > \frac{1}{3}$, $D^\alpha h(w) \geq 1 = m' > 0$, $f(u)/u = 1 + u^2 \geq 1 = k > 0$, $\sigma(t) = t - 2\pi$, $\delta(t) = t - \frac{3\pi}{2}$ *and* $D^\alpha \sigma(t) = t^{\frac{2}{3}} \geq l$ *such that* $t_1 = l^{\frac{2}{3}}$, $t \geq t_1$, $c(t) = l^2 \frac{t^{\frac{2}{3}}}{1+\cos^2(t)}$, $A_\alpha(t) = \frac{l^2}{2} \frac{1+\frac{3}{4}\cos^{\frac{5}{3}}(t)}{1+\cos^2(t)} \frac{t-\frac{3\pi}{2}-T}{t} (t - \frac{3\pi}{2})^{\frac{1}{3}}$. *Now consider,*

$$\begin{aligned}
\int_{t_2}^{\infty} c(s)(s - T)\delta(\sigma(s))ds &= \int_{t_2}^{\infty} l^2 \frac{s^{\frac{2}{3}}}{1 + \cos^2(s)} (s - T)(s - \frac{3\pi}{2})ds \\
&\geq \frac{l^2}{2} \int_{t_2}^{\infty} s^{\frac{2}{3}} (s - T)(s - \frac{3\pi}{2})ds \to \infty \text{ as } t \to \infty.
\end{aligned}$$

*If we take* $\rho(t) = 1$ *then* $\rho'(t) = 0$. *Consider*

$$\limsup_{t\to\infty} \int_{t_1}^{t} \left( s^{\alpha-1}\rho(s)A_\alpha(s) - \frac{1}{4}\frac{(\rho'(s))^2}{\rho(s)} s^{1-\alpha} b(s) \right) ds$$

$$= \limsup_{t\to\infty} \int_{t_1}^{t} \left( s^{-\frac{2}{3}} l^2 \frac{s^{\frac{2}{3}}}{1+\cos^2(s)} \frac{1}{2} \frac{1+\frac{3}{4}\cos^{\frac{5}{3}}(s)}{1+\cos^2(s)} \frac{s-\frac{3\pi}{2}-T}{s} (s-\frac{3\pi}{2})^{\frac{1}{3}} \right) ds$$

$$\geq \limsup_{t\to\infty} \frac{7l^2}{16} \int_{t_1}^{t} \left( (1-\frac{\frac{3\pi}{2}-T}{s})(s-\frac{3\pi}{2})^{\frac{1}{3}} \right) ds \to \infty \; as \; t \to \infty.$$

*Theorem 1 are satisfying the new conditions arriving at the solution for* (34) *is oscillatory and it is given as* $(u(t), v(t), w(t)) = (\sin t, \cos t, \sin t)$.

**Example 3.** *Consider the* $\frac{1}{2}$*-fractional differential system*

$$\begin{aligned} D^{\frac{1}{2}}(u(t)) &= e^{2t} t^{\frac{1}{2}} g(v(t-1)), \\ D^{\frac{1}{2}}(v(t)) &= -e^{-2t} t^{\frac{1}{2}} w(t), \\ D^{\frac{1}{2}}(w(t)) &= (et)^{\frac{1}{2}} f(u(t-\frac{1}{2})), \; t \geq t_0. \end{aligned} \tag{35}$$

*Here* $\alpha = \frac{1}{2}$, $\frac{1}{a(t)} = p(t) = e^{2t}t^{\frac{1}{2}}$, $\frac{1}{b(t)} = q(t) = e^{-2t}t^{\frac{1}{2}}$, $r(t) = (et)^{\frac{1}{2}}$, $g(v) = v$, $h(w) = w$ *and* $f(u) = u$. *Now it is easy to check that* $D^\alpha g(v) = v^{\frac{1}{2}} = e^{-\frac{t}{2}} = l' > 0$, $D^\alpha h(w) = w^{\frac{1}{2}} = e^{\frac{t}{2}} = m' > 0$, $f(u)/u = 1 = k > 0$, $\sigma(t) = t-1$, $\delta(t) = t - \frac{1}{2}$ *and* $D^\alpha \sigma(t) = t^{\frac{1}{2}} \geq l$ *such that* $t_1 = l^{\frac{1}{2}}$ *for* $t \geq t_1$, $c(t) = l^2(et)^{\frac{1}{2}}$, $A_\alpha(t) = \frac{l^2}{2} e^{\frac{1}{2}} e^{2t}(t-\frac{1}{2}-T)(t-\frac{1}{2})^{\frac{1}{2}}$. *Now,*

$$\int_{t_2}^{\infty} c(s)(s-T)\delta(\sigma(s))ds = \int_{t_2}^{\infty} l^2(es)^{\frac{1}{2}}(s-T)(s-\frac{1}{2})ds = l^2 e^{\frac{1}{2}} \int_{t_2}^{\infty} s^{\frac{1}{2}}(s-T)(s-\frac{1}{2})ds \to \infty.$$

*Taking* $\rho(t) = \frac{1}{t^{\frac{7}{2}}e^{2t}}$ *then* $\rho'(t) = -\frac{t^{\frac{7}{2}}}{2t^7 e^{2t}}(4t+7)$. *Consider*

$$\limsup_{t\to\infty} \int_{t_1}^{t} \left( s^{\alpha-1}\rho(s)A_\alpha(s) - \frac{1}{4}\frac{(\rho'(s))^2}{\rho(s)} s^{1-\alpha} b(s) \right) ds$$

$$= \limsup_{t\to\infty} \int_{t_1}^{t} \left( s^{-\frac{1}{2}} \frac{l^2 e^{\frac{1}{2}}}{s^{\frac{7}{2}} e^{2s}} \frac{1}{2} e^{2s} (s-\frac{1}{2}-T)(s-\frac{1}{2})^{\frac{1}{2}} - \frac{(4s+7)^2}{4s^9 e^{4s}} e^{4s} s^{\frac{7}{2}} \right) ds$$

$$\leq \limsup_{t\to\infty} \int_{t_1}^{t} \left( \frac{l^2 e^{\frac{1}{2}}}{2s^{\frac{7}{2}}} s - \frac{s^2}{4s^{\frac{11}{2}}} \right) ds \leq \limsup_{t\to\infty} \int_{t_1}^{t} \left( \frac{l^2 e^{\frac{1}{2}}}{2s^{\frac{5}{2}}} - \frac{1}{4s^{\frac{7}{2}}} \right) ds < \infty.$$

*Here further the condition (9) of the above Theorem 1 seems to be not satisfied, in view of the fact that* $(A_4)$ *fails to hold, and hence the system (35) is not oscillatory. In fact,* $(u(t), v(t), w(t)) = (e^t, e^{-t}, e^t)$ *it is a solution for (35), and nonoscillatory.*

**Remark 1.** *The results obtained in this article further can be extended to a neutral system with forced term*

$$\begin{aligned} D^{\alpha}\left(u(t)+p(t)u(\delta(t))\right) &= a(t)h_1\left(v(\tau(t))\right), \\ D^{\alpha}\left(v(t)\right) &= -b(t)h_2\left(w(t)\right), \\ D^{\alpha}\left(w(t)\right) &= c(t)h_3\left(u(\sigma(t))\right)+e(t),\ t \geq t_0, \end{aligned}$$

*for the cases*

$$\int_{t_0}^{\infty} a(s)d_{\alpha}s < \infty, \int_{t_0}^{\infty} b(s)d_{\alpha}s = \infty,$$

*and*

$$\int_{t_0}^{\infty} a(s)d_{\alpha}s < \infty, \int_{t_0}^{\infty} b(s)d_{\alpha}s < \infty.$$

## 5. Conclusions

Through this article, we have derived some new oscillation results for a certain class of nonlinear three-dimensional $\alpha$-fractional differential systems by using the Riccati transformation and inequality technique. This work extends and also improves some classical results in the literature [16,18,32] to the $\alpha$-fractional systems and studied the oscillation criteria. Further, the present results are essentially new and, in order to illustrate the validity of the obtained results, we have provided three examples.

**Author Contributions:** The all authors contributed equally and all authors read the manuscript and approved the final submission.

**Acknowledgments:** The authors are very grateful for the comments of the reviewers which helped to improve the present manuscript.

**Conflicts of Interest:** It is hereby the authors declare that there is no conflict of interest.

## References

1. Abbas, S.; Benchohra, M.; N'Guerekata, G.M. *Topics in Fractional Differential Equations*; Springer: New York, NY, USA, 2012.
2. Daftardar-Gejji, V. *Fractional Calculus Theory and Applications*; Narosa Publishing House Pvt. Ltd.: Kolkata, India, 2014.
3. Hilfer, R. *Applications of Fractional Calculus in Physics*; World Scientific Pub. Co.: Singapore, 2000.
4. Kilbas, A.A.; Srivastava, H.M.; Trujillo, J.J. *Theory and Applications of Fractional Differential Equations*; Elsevier Science: Amsterdam, The Netherlands, 2006.
5. Milici, C.; Draganescu, G. *Introduction to Fractional Calculus*; Lambert Academic Publishing: Saarbrucken, Germany, 2016.
6. Samko, S.G.; Kilbas, A.A.; Mirchev, O.I. *Fractional Integrals and Derivatives, Theory and Applications*; Gordon and Breach Science Publishers: Singapore, 1993.
7. Zhou, Y. *Basic Theory of Fractional Differential Equations*; World Scientific: Singapore, 2014.
8. Akin, E.; Dosla, Z.; Lawrence, B. Oscillatory properties for three-dimensional dynamic systems. *Nonlinear Anal.* **2008**, *69*, 483–494. [CrossRef]
9. Akin, E.; Dosla, Z.; Lawrence, B. Almost oscillatory three-dimensional dynamical systems. *Adv. Differ. Equ.* **2012**, *2012*, 46. [CrossRef]
10. Dosoudilova, M.; Lomtatidze, A.; Sremr, J. *Oscillatory Properties of Solutions to Two-Dimensional Emden-Fowler Type Equations*; International Workshop Qualitde; 2014. Available online: http://www.rmi.ge/eng/QUALITDE-2014/Dosoudilova_et_al_workshop_2014.pdf (accessed on 1 December 2018)
11. Huo, H.F.; Li, W.T. Oscillation criteria for certain two-dimensional differential systems. *Int. J. Appl. Math.* **2001**, *6*, 253–261.

12. Kordonis, I.-G.E.; Philos, C.G. On the oscillation of nonlinear two-dimensional differential systems. *Proc. Am. Math. Soc.* **1998**, *126*, 1661–1667. [CrossRef]
13. Li, W.T.; Cheng, S.S. Limiting behaviours of non-oscillatory solutions of a pair of coupled nonlinear differential equations. *Proc. Edinb. Math. Soc.* **2000**, *43*, 457–473. [CrossRef]
14. Lomtatidze, A.; Sremr, J. On oscillation and nonoscillation of two-dimensional linear differential systems. *Georgian Math. J.* **2013**, *20*, 573–600. [CrossRef]
15. Sadhasivam, V.; Deepa, M.; Nagajothi, N. On the oscillation of nonlinear fractional differential systems. *Int. J. Adv. Res. Sci. Eng. Technol.* **2017**, *4*, 4861–4867.
16. Thandapani, E.; Selvaraj, B. Oscillatory behavior of solutions of three-dimensional delay difference systems. *Rodov. Mater.* **2004**, *13*, 39–52.
17. Troy, W.C. Oscillations in a third order differential equation modeling a nuclear reactor. *SIAM J. Appl. Math.* **1977**, *32*, 146–153. [CrossRef]
18. Schmeidel, E. Oscillation of nonlinear three-dimensional difference systems with delays. *Math. Biochem.* **2010**, *135*, 163–170.
19. Vreeke, S.D.; Sandquist, G.M. Phase plane analysis of reactor kinetics. *Nuclear Sci. Eng.* **1970**, *42*, 259–305. [CrossRef]
20. Bainov, D.; Mishev, D. *Oscillatory Theory of Operator-Differential Equations*; World Scientific: Singapore, 1987.
21. Erbe, L.H.; Kong, Q.K.; Zhang, B.G. *Oscillation Theory for Functional Differential Equations*; Marcel Dekker: New York, NY, USA, 1995.
22. Padhi, S.; Pati, S. *Theory of Third Order Differential Equations*; Electronic Journal of Differential Equations; Springer: New Delhi, India, 2014.
23. Saker, S.H. *Oscillation Theory of Delay Differential and Difference Equations*; VDM Verlag Dr. Muller Aktiengesellschaft and Co.: Mansoura, Egypt, 2010.
24. Kiguradze, I.T.; Chaturia, T.A. *Asymptotic Properties of Solutions of Nonautonomous Ordinary Differential Equations*; Kluwer Acad. Publ.: Dordrecht, The Netherlands, 1993.
25. Abdeljawad, T. On conformable fractional calculus. *J. Comput. Appl. Math.* **2015**, *279*, 57–66. [CrossRef]
26. Atangana, A.; Baleanu, D.; Alsaedi, A. New properties of conformable derivative. *Open Math.* **2015**, *13*, 889–898. [CrossRef]
27. Khalil, R.; Al Horani, M.; Yousef, A.; Sababheh, M. A new definition of fractional derivatives. *J. Comput. Appl. Math.* **2014**, *264*, 65–70. [CrossRef]
28. Lazo, M.J.; Torres, D.F. Variational calculus with conformable fractional derivatives. *IEEE/CAA J. Autom. Sin.* **2017**, *4*, 340–352. [CrossRef]
29. Anderson, D.R.; Ulnessn, D.J. Properties of the Katugampola fractional derivative with potential application in quantum mechanics. *J. Math. Phys.* **2015**, *56*, 063502. [CrossRef]
30. Katugampola, U.N. A new fractional derivative with classical properties. *arXiv* **2014**, arXiv:1410.6535.
31. Katugampola, U.N. A new approach to generalized fractional derivatives. *Bull. Math. Anal. Appl.* **2014**, *6*, 1–15.
32. Spanikova, E. Oscillatory properties of solutions of three-dimensional differential systems of neutral type. *Czechoslov. Math. J.* **2000**, *50*, 879–887. [CrossRef]
33. Sadhasivam, V.; Kavitha, J.; Deepa, M. Existence of solutions of three-dimensional fractional differential systems. *Appl. Math.* **2017**, *8*, 193–208. [CrossRef]
34. Chatzarakis, G.E.; Deepa, M.; Nagajothi, N.; Sadhasivam, V. On the oscillation of three-dimensional $\alpha$-fractional differential systems. *Dyn. Syst. Appl.* **2018**, *27*, 873–893.

symmetry

MDPI

Article

# Generalized Preinvex Functions and Their Applications

**Adem Kiliçman [1] and Wedad Saleh [2,*]**

[1] Department of Mathematics, Putra University of Malaysia (UPM), Serdang 43400, Malaysia; akilic@upm.edu.my
[2] Department of Mathematics, Taibah University, Al-Medina 20012, Saudi Arabia
* Correspondence: wed_10_777@hotmail.com

Received: 6 September 2018; Accepted: 11 October 2018; Published: 13 October 2018

**Abstract:** A class of function called sub-b-s-preinvex function is defined as a generalization of s-convex and b-preinvex functions, and some of its basic properties are presented here. The sufficient conditions of optimality for unconstrainded and inquality constrained programming are discussed under the sub-b-s-preinvexity. Moreover, some new inequalities of the Hermite—Hadamard type for differentiable sub-b-s-preinvex functions are presented. Examples of applications of these inequalities are shown.

**Keywords:** generalized convexity; b-vex functions; sub-b-s-convex functions

## 1. Introduction

Convex functions play an important role in economics, management science, engineering, finance, and optimization theory. Many interesting generalizations and extensions of classical convexity have been used in optimization and mathematical inequalities. Generalized convex functions called b-vex functions were introduced by Bector and Singh [1], and some of their basic properties have been discussed. Choa et al. [2] investigated a new class of functions called sub-b-convex functions and proved the sufficient conditions of optimality for both unconstrained and inequality-constrained sub-b-convex programming. Hudzik and Maligranda [3] studied certain classes of functions introduced by Orlicz [4], namely, the classes of s-convex functions. Meftah [5] introduced a new class of non-negative functions called s-preinvex functions in the second sense with respect to $\eta$, for some $s \in (0,1]$. Jiagen and Tingsong Du [6] presented a class of generalized convex function has some similar properties of sub-b-convex function and s-convex functions.

Ben-Isreal and Mond [7] defined preinvex functions, and, in [8], Weir and Mond studied how and where preinvex functions could replace convex functions. Mohan and Neogy [9] presented certain properties of preinvex functions. Suneja et al. [10] considered a class of function called b-preinvex functions that are generalizations of preinvex and b-vex functions. A generalization of the b-vex function, called semi-b-preinvex, was given by Long et al. [11]. Refinements of the mathematical inequalities on convex and generalized convex functions have been investigated [12–20].

Motivated by earlier research works [6,12,21–23], the purpose of this article is to present a new class of functions, called sub-b-s-preinvex functions, that can be reduced to sub-b-preinvex when s = 1. Some of their properties are studied. Furthermore, a new class of sets, called sub-b-s-preinvex sets, is defined. A new sub-b-s-preinvex programming is introduced, and the sufficient conditions of optimality under this type of function is established. Moreover, some examples of applications are given.

## 2. Preliminaries

Throughout the paper, the convention bellow will be followed:

Let $\mathbb{R}^n$ denote the n-dimensional Euclidean space, and let $K$ be a non-empty convex subset in $\mathbb{R}^n$. In addition, let $b(u_1, u_2, t) : K \times K \times [0,1] \longrightarrow \mathbb{R}$ and $\eta : K \times K \longrightarrow \mathbb{R}^n$ be two fixed mappings.

The following definitions about b-vex, sub-b-convex, s-convex, sub-b-s-convex, and preinvex functions that will be used throughout the paper are given:

**Definition 1** ([1]). *The function $h : K \longrightarrow \mathbb{R}^n$ is called*

1. *a b-vex function on K with respect to (w.r.t. in short) b if*

$$h\left(tu_1 + (1-t)u_2\right) \leq tbh(u_1) + (1-tb)h(u_2), \forall u_1, u_1 \in K, t \in [0,1]$$

2. *and a b-linear function on K w.r.t. b if*

$$h\left(tu_1 + (1-t)u_2\right) = tbh(u_1) + (1-tb)h(u_2), \forall u_1, u_1 \in K, t \in [0,1].$$

**Definition 2** ([2]). *The function $h : K \longrightarrow \mathbb{R}^n$ is called a sub-b-convex function on K w.r.t. b if*

$$h\left(tu_1 + (1-t)u_2\right) = th(u_1) + (1-t)h(u_2) + b(u_1, u_2, t), \forall u_1, u_1 \in K, t \in [0,1].$$

**Definition 3** ([3]). *The function $h : K \longrightarrow \mathbb{R}^n$ is called an s-convex function in the second sense if*

$$h\left(tu_1 + (1-t)u_2\right) = t^s h(u_1) + (1-t)^s h(u_2), \forall u_1, u_1 \in K, t \in [0,1], s \in (0,1]\ .$$

**Definition 4** ([6]). *A function $h : K \longrightarrow \mathbb{R}$ is called a sub-b-s-convex function on a non-empty convex set $K \subset \mathbb{R}^n$ w.r.t. b: $K \times K \times [0,1] \longrightarrow \mathbb{R}$ if*

$$h\left(tu_1 + (1-\delta)u_2\right) \leq t^s h(u_1) + (1-t)^s h(u_2) + b(u_1, u_2, t)$$

*and $\forall u_1, u_2 \in K, t \in [0,1], s \in (0,1]$ .*

Recall [9] that, by definition, a set $K \subset \mathbb{R}^n$ is called an invex set w.r.t $\eta$ if $u_2 + t\eta(u_1, u_2) \in K$, $\forall u_1, u_2 \in K$ and $t \in [0,1]$.

Ben-Israel and Mond [7] defined a class of functions called preinvex in the non-empty invex set $K \subset \mathbb{R}^n$ *w.r.t.* $\eta$, as follows:

**Definition 5.** *A function $h : K \longrightarrow \mathbb{R}$ is preinvex on K w.r.t. $\eta$ if there exists an n-dimensional vector function $\eta$ such that*

$$h\left(u_2 + \delta\eta(u_1, u_2)\right) \leq th(u_1) + (1-t)h(u_2),$$

$\forall u_1, u_2 \in K, t \in [0,1]$.

## 3. Sub-b-s-Preinvex Function and Their Properties

In this section, the concepts of sub-b-s-preinvex function and sub-b-s-preinvex set are given. Furthermore, some of their properties are studied.

**Definition 6.** *A function $h : K \longrightarrow \mathbb{R}$ is called a sub-b-s-preinvex function on a non-empty invex set $K \subset \mathbb{R}^n$ w.r.t. $\eta, b$ if*

$$h\left(u_2 + t\eta(u_1, u_2)\right) \leq t^s h(u_1) + (1-t)^s h(u_2) + b(u_1, u_2, t) \tag{1}$$

*where $b : K \times K \times [0,1] \longrightarrow \mathbb{R}, \forall u_1, u_2 \in Kt \in [0,1], s \in (0,1]$ .*

**Remark 7.**

1. *If $\eta(u_1, u_2) = u_1 - u_2$ in Equation (1), then sub-b-s-preinvex w.r.t. $\eta$, b becomes a sub-b-s-convex function. Moreover, if s = 1, then Equation (1) becomes a sub-b-convex function.*
2. *When $\eta(u_1, u_2) = u_1 - u_2$ and $b(u, u, t) \leq 0$ in Equation (1), then the sub-b-s-preinvex function becomes a convex function.*

**Theorem 8.** *If $h_1, h_2 : K \longrightarrow \mathbb{R}$ are sub-b-s-preinvex functions w.r.t. $\eta$,b, then $h_1 + h_2$ and $\beta h_1 (\beta \geq 0)$ are also sub-b-s-preinvex functions w.r.t. $\eta$,b.*

**Corollary 9.** *If $h_k : K \longrightarrow \mathbb{R}$, where $k = 1, 2, \cdots, n$ are sub-b-s-preinvex functions w.r.t. $\eta$,$b_k$, then the function which is $H = \sum_{k=1}^{n} a_k h_k$,$a_k \geq 0$ $(k = 1, 2, \cdots, n)$ is also sub-b-s-preinvex function w.r.t. $\eta$, b where $b = \sum_{k=1}^{n} a_k b_k$.*

**Proposition 10.** *If $h_k : K \longrightarrow \mathbb{R}$, where $k = 1, 2, \cdots, n$ are sub-b-s-preinvex functions w.r.t. $\eta$,$b_k$, then the function which is $H = \max h_k$,$k = 1, 2, \cdots, n$ is also a sub-b-s-preinvex function w.r.t. $\eta$,b, where $b = \max b_k$.*

**Theorem 11.** *Let $h_1 : K \longrightarrow \mathbb{R}$ be a sub-b-s-preinvex function w.r.t. $\eta$,$b_1$ and $h_2 : \mathbb{R} \longrightarrow \mathbb{R}$ be an increasing function. Then $h_1 o h_2$ is a sub-b-s-preinvex function w.r.t.$\eta$, b where $b = h_2 o b_1$ if $h_2$ satisfies the following conditions:*

1. $h_2 (\beta u_1) = \beta h_2(u_1), \forall u_1 \in \mathbb{R}, \beta \geq 0;$
2. $h_2 (u_1 + u_2) = h_2(u_1) + h_2(u_2),, \forall u_1, u_2 \in \mathbb{R}, \beta \geq 0.$

**Proof.**

$$\begin{aligned} (h_2 o h_1) (u_2 + \delta\eta(u_1, u_2)) &= h_2 (h_1 (u_2 + t\eta(u_1, u_2))) \\ &\leq h_2 (t^s h_1(u_1) + (1-t)^s h_1(u_2) + b_1(u_1, u_2, t)) \\ &= t^s h_2 (h_1(u_1)) + (1-t)^s h_2 (h_1(u_2)) + h_2 (b_1(u_1, u_2, t)) \\ &= t^s (h_2 o h_1)(u_1) + (1-t)^s (h_2 o h_1)(u_2) + b(u_1, u_2, t), \end{aligned}$$

which means that $h_2 o h_1$ is a sub-b-s-preinvex function *w.r.t.*$\eta$, *b*. □

We introduce a definition of a sub-b-s-preinvex set *w.r.t.*$\eta$, *b* as follows.

**Definition 12.** *A set $K \subseteq \mathbb{R}^{n+1}$ is called a sub-b-s-preinvex set w.r.t.$\eta$, b if*

$$(u_2 + t\eta(u_1, u_2), t^s \beta_1 + (1-t)^s \beta_2 + b(u_1, u_2, t)) \in K.$$

$\forall (u_1, \beta_1), (u_2, \beta_2) \in K$, $u_1, u_2 \in \mathbb{R}^n$, $t \in [0, 1]$, $s \in (0, 1]$ *and* $b : \mathbb{R}^n \times \mathbb{R}^n \times [0, 1] \longrightarrow \mathbb{R}$.

The epigraph of the sub-b-s-preinvex function $h : k \longrightarrow \mathbb{R}$ can be given as

$$G(h) = \{(u, \beta) : u \in K, \beta \in \mathbb{R}, h(u) \leq \beta\}.$$

Now, we are going to investigate characterizations of the sub-b-s-preinvex function in terms of their epigraph G(h), and we start with sufficient and necessary conditions for h to be a sub-b-s-preinvex function *w.r.t.*$\eta$,*b*.

**Theorem 13.** *$h : k \longrightarrow \mathbb{R}$ is a sub-b-s-preinvex function w.r.t.$\eta$, b iff its epigraph is also a sub-b-s-preinvex set w.r.t.$\eta$, b.*

**Proof.** Let $h$ be a sub-b-s-preinvex and let $(u_1, \beta_1), (u_2, \beta_2) \in G(h)$. Then, by using the hypothesis, we have $h(u_1) \leq \beta_1$ and $h(u_2) \leq \beta_2$.

Moreover,

$$\begin{aligned} h(u_2 + t\eta(u_1, u_2)) &\leq t^s h(u_1) + (1-t)^s h(u_2) + b(u_1, u_2, t) \\ &\leq t^s \beta_1 + (1-t)^s \beta_2 + b(u_1, u_2, t). \end{aligned} \tag{2}$$

Hence,

$$(u_2 + t\eta(u_1, u_2), \beta_1 + (1-t)^s \beta_2 + b(u_1, u_2, t)) \in G(h).$$

Therefore, $G(h)$ is sub-b-s-preinvex set $w.r.t.\eta, b$.

Now, assume that $G(h)$ is a sub-b-s-preinvex set $w.r.t.\eta, b$. Then

$$(u_1, h(u_1)), (u_2, h(u_2)) \in G(h),$$

where $u_1, u_2 \in K$.

$(u_2 + t\eta(u_1, u_2), t^s h(u_1) + (1-t)^s h(u_2) + b(u_1, u_2, \delta)) \in G(h)$, which means that

$$h(u_2 + t\eta(u_1, u_2)) \leq t^s h(u_1) + (1-t)^s h(u_2) + b(u_1, u_2, t).$$

Then $h$ is sub-b-s-preinvex function $w.r.t.\eta, b$. □

**Proposition 14.** *Assume that $K_i$ is a family of is sub-b-s-preinvex sets $w.r.t.\eta, b$. Then $\cap_{i \in I} K_i$ is also a sub-b-s-preinvex set $w.r.t.\eta, b$.*

**Proof.** Consider $(u_1, \beta_1), (u_2, \beta_2) \in \cap_{i \in I} K_i$. Then we have $(u_1, \beta_1), (u_2, \beta_2) \in K_i, \forall i \in I$

$$(u_2 + t\eta(u_1, u_2), \beta_1 + (1-t)^s \beta_2 + b(u_1, u_2, t)) \in K_i, \forall i \in I.$$

$\Rightarrow$

$$(u_2 + t\eta(u_1, u_2), \beta_1 + (1-t)^s \beta_2 + b(u_1, u_2, t)) \in \cap_{i \in I} K_i.$$

Hence, $\cap_{i \in I} K_i$ is a sub-b-s-preinvex set. □

According to Theorem 13 and Proposition 14, the following proposition holds:

**Proposition 15.** *Let $h_i$ be a sub-b-s-preinvex function $w.r.t.\eta, b$. Then a function $H = \sup_{i \in I} h_i$ is also a sub-b-s-preinvex function $w.r.t.\eta, b$.*

**Theorem 16.** *Let $h : k \longrightarrow \mathbb{R}$ be a non-negative differentiable sub-b-s-preinvex function $w.r.t.\eta, b$. Then*

1. $dh_{u_2}\eta(u_1, u_2) \leq t^{s-1}(h(u_1) + h(u_2)) + \lim_{t \longrightarrow 0_+} \frac{b(u_1, u_2, t)}{t}$;
2. $dh_{u_2}\eta(u_1, u_2) \leq t^{s-1}(h(u_1) - h(u_2)) + \frac{h(u_2)}{t} + \lim_{t \longrightarrow 0_+} \frac{b(u_1, u_2, t)}{t}$.

**Proof.**

1. By using the hypothesis, we can write

$$h(u_2 + t\eta(u_1, u_2)) = h(u_2) + tdh_{u_2}\eta(u_1, u_2) + O(t).$$

Additionally,

$$h(u_2 + t\eta(u_1, u_2)) \leq t^s h(u_1) + (1-t)^s h(u_2) + b(u_1, u_2, t).$$

Furthermore,

$$\begin{aligned} h\left(u_2+t\eta(u_1,u_2)\right) &\leq t^s h(u_1)+(1-t)^s h(u_2)+b(u_1,u_2,t) \\ &\leq t^s h(u_1)+(1+t^s)h(u_2)+b(u_1,u_2,t). \end{aligned}$$

Then

$$h(u_2)+tdh_{u_2}\eta(u_1,u_2)+O(t)\leq t^s h(u_1)+(1+t^s)h(u_2)+b(u_1,u_2,t)$$

by taking $\lim_{\delta\longrightarrow 0_+}\frac{b(u_1,u_2,t)}{t}$, which is the maximum of $\frac{b(u_1,u_2,t)}{t}-\frac{O(t)}{t}$. The first result is thus obtained.

2. Similarly,

$$\begin{aligned} &h(u_2)+tdh_{u_2}\eta(u_1,u_2)+O(t) \\ &\quad\leq t^s h(u_1)+(1+t^s)h(u_2)+b(u_1,u_2,t) \\ &\quad= t^s h(u_1)+(1+t^s)h(u_2)-t^s h(u_2)+t^s h(u_2)+b(u_1,u_2,t) \\ &\quad= t^s\left(h(u_1)-h(u_2)\right)+b(u_1,u_2,t)+\left((1-t)^s+t^s\right)h(u_2). \end{aligned}$$

However, we know that $(1-t)^s+\delta^s, \forall t\in[0,1]$, and $s\in(0,1]$ and since $h$ is non-negative function; hence,

$$h(u_2)+\delta h_{u_2}\eta(u_1,u_2)+O(t)\leq t^s\left(h(u_1)-h(u_2)\right)+2h(u_2)+b(u_1,u_2,t).$$

Then, by dividing the last inequality by $t$ and taking $\lim_{t\longrightarrow 0_+}$, we obtain the second part of the theorem.

□

**Theorem 17.** *Let $h:k\longrightarrow\mathbb{R}$ be a negative differentiable sub-b-s-preinvex function w.r.t.$\eta, b$. Then*

$$dh_{u_2}\eta(u_1,u_2)\leq t^{s-1}\left(h(u_1)-h(u_2)\right)+\lim_{t\longrightarrow 0_+}\frac{b(u_1,u_2,t)}{t}.$$

**Proof.** We obtain the result by using the hypotheses, since

$$dh_{u_2}\eta(u_1,u_2)\leq t^{s-1}\left(h(u_1)-h(u_2)\right)+\frac{b(u_1,u_2,t)}{t}-\frac{O(t)}{t}.$$

Then, by taking $\lim_{t\longrightarrow 0_+}\frac{b(u_1,u_2,t)}{t}$, which is the maximum of $\frac{b(u_1,u_2,t)}{t}-\frac{O(t)}{t}$, we obtain the result. □

**Corollary 18.** *Assume that $h:k\longrightarrow\mathbb{R}$ is a differentiable sub-b-s-preinvex function w.r.t.$\eta, b$, and*

1. *$h$ is a non-negative function, then*

$$d\left(h_{u_2}-h_{u_1}\right)\eta(u_1,u_2)\leq\frac{h(u_1)+h(u_2)}{t}+\lim_{t\longrightarrow 0_+}\frac{b(u_1,u_2,t)+b(u_2,u_1,t)}{t},$$

2. *$h$ is a negative function, then*

$$d\left(h_{u_2}-h_{u_1}\right)\eta(u_1,u_2)\leq\lim_{t\longrightarrow 0_+}\frac{b(u_1,u_2,t)-b(u_2,u_1,t)}{t}.$$

**Proof.**

1. Since $h$ is a non-negative function and by using Theorem 16,

$$dh_{u_2}\eta(u_1,u_2) \leq t^{s-1}\left(h(u_1)-h(u_2)\right)+\frac{h(u_2)}{t}+\lim_{t\longrightarrow 0_+}\frac{b(u_1,u_2,t)}{t}.$$

Additionally,

$$dh_{u_1}\eta(u_1,u_2) \leq t^{s-1}\left(h(u_2)-h(u_1)\right)+\frac{h(u_1)}{t}+\lim_{t\longrightarrow 0_+}\frac{b(u_2,u_1,t)}{t}.$$

Thus,

$$d\left(h_{u_2}-h_{u_1}\right)\eta(u_1,u_2) \leq \frac{h(u_1)+h(u_2)}{t}+\lim_{t\longrightarrow 0_+}\frac{b(u_1,u_2,t)+b(u_2,u_1,t)}{t}.$$

2. Since $h$ is a negative function, and according to Theorem 17, the second result can be obtained directly.

□

## 4. Hermite–Hadamard-Type Integral Inequalities for Differentiable Sub-B-S-Preinvex Functions

There are a great deal of inequalities related to the class of convex functions. For example, Hermite–Hadamard's inequality is one of the well-known results in the literature, which can be stated as follows.

**Theorem 19.** *(Hermite–Hadamard's inequality) Let $h$ be a convex function on $[u_1,u_2]$ with $u_1 < u_2$. If $h$ is an integral on $[u_1,u_2]$, then*

$$h\left(\frac{u_1+u_2}{2}\right) \leq \frac{1}{u_2-u_1}\int_{u_1}^{u_2} h(x)dx \leq \frac{h(u_1)+h(u_2)}{2}. \tag{3}$$

For more properties about the above inequality, we refer the interested readers to [24,25]. Dragomir and Fitzpatrick [26] demonstrated a variation of Hadamard's inequality, which holds for $s$-convex functions in the second sense.

**Theorem 20.** *Theorem Let $h:\mathbf{R}_+ \longrightarrow \mathbf{R}_+$ be an s-convex function in the second sense $s \in (0,1)$ and $u_1,u_2 \in \mathbf{R}_+$, $u_1 < u_2$. If $h \in L^1([u_1,u_2])$, then*

$$2^{s-1}h\left(\frac{u_1+u_2}{2}\right) \leq \frac{1}{u_2-u_1}\int_{u_1}^{u_2} h(x)dx \leq \frac{h(u_1)+h(u_2)}{s+1}.$$

Now, we will present new inequalities of Hermite–Hadamard for functions whose derivatives in absolute value are sub-b-s-preinvex functions. Our results generalize those results presented in [27] concerning Hermite–Hadamard type inequalities for preinvex functions.

**Lemma 21** ([27]). *Assume that $K \subset \mathbf{R}$ is an open invex subset w.r.t$\eta$ and $u_1,u_2 \in K$ with $u_1 < u_1+\eta(u_2,u_1)$. Let $h: K \longrightarrow \mathbf{R}$ be a differentiable mapping on $K$ such that $h' \in L\left([u_1,u_1+\eta(u_2,u_1)]\right)$. Then the following equality holds:*

$$-\frac{h(u_1)+h(u_1+\eta(u_2,u_1))}{2}+\frac{1}{\eta(u_2,u_1)}\int_{u_1}^{u_1+\eta(u_2,u_1)} h(x)dx$$
$$=\frac{\eta(u_2,u_1)}{2}\int_0^1 (1-2t)h'\left(u_1+t\eta(u_2,u_1)\right)dt.$$

**Theorem 22.** *Assume that* $K \subset [0,c], c > 0$ *is an open invex subset w.r.t*$\eta$ *and* $u_1, u_2 \in K$ *with* $u_1 < u_1 + \eta(u_2,u_1)$. *Let* $h : K \longrightarrow \mathbf{R}$ *be a differentiable mapping on K such that* $h' \in L([u_1, u_1 + \eta(u_2,u_1)])$. *If* $|h'|$ *is a sub-b-s-preinvex function on K, then we have the following inequality:*

$$\left|\frac{h(u_1) + h(u_1 + \eta(u_2,u_1))}{2} - \frac{1}{\eta(u_2,u_1)}\int_{u_1}^{u_1+\eta(u_2,u_1)} h(x)dx\right|$$
$$\leq \frac{\eta(u_2,u_1)}{2}\left[\frac{(2^{s+1}-1)(s+1)+(1-2^s)(s+2)}{2^s(s+1)(s+2)}\left[|h'(u_2)| + |h'(u_1)|\right] + \frac{1}{2}|b(u_1,u_2,t)|\right]. \quad (4)$$

**Proof.** From Lemma 21, we have

$$\left|\frac{h(u_1) + h(u_1 + \eta(u_2,u_1))}{2} - \frac{1}{\eta(u_2,u_1)}\int_{a}^{u_1+\eta(u_2,u_1)} h(x)dx\right|$$
$$\leq \frac{\eta(u_2,u_1)}{2}\int_0^1 |1-2t|\,|h'(u_1 + t\eta(u_2,u_1))|\,dt. \quad (5)$$

Since $|h'|$ is a sub-b-s-preinvex on K, for every $u_1, u_2 \in K, \in (0,1]$ and $s \in (0,1)$, we obtain

$$|h'(u_1 + t\eta(u_2,u_1))| \leq t^s|h'(u_2)| + (1-t)^s|h'(u_1)| + |b(u_1,u_2,t)|.$$

Hence, we have

$$\left|\frac{h(u_1) + h(u_1 + \eta(u_1,u_2))}{2} - \frac{1}{\eta(u_1,u_2)}\int_{a}^{u_1+\eta(u_1,u_2)} h(x)dx\right|$$
$$\leq \frac{\eta(u_1,u_2)}{2}\left[|h'(u_2)|\int_0^1 |1-2t|t^s dt + |h'(u_1)|\int_0^1 |1-2t|(1-t)^s dt + |b(u_1,u_2,t)|\int_0^1 |1-2t|dt\right]. (6)$$

since

$$\int_0^1 1-2t|(1-t)^s dt = \int_0^1 1-2t|t^s dt$$
$$= \int_0^{\frac{1}{2}} (1-2t)t^s dt - \int_{\frac{1}{2}}^1 (1-2t)t^s dt$$
$$= \frac{(2^{s+1}-1)(s+1)+(1-2^s)(s+2)}{2^s(s+1)(s+2)}.$$

Additionally,

$$\int_0^1 |1-2t|dt = \frac{1}{2}.$$

Therefore, the proof of Theorem 22 is complete. □

**Corollary 23.** *If* $\eta(u_2,u_1) = u_2 - u_1$ *in Theorem 22, then Inequality 4 reduces to the following inequality:*

$$\left|\frac{h(u_1) + h(u_2)}{2} - \frac{1}{u_2 - u_1}\int_{u_1}^{u_2} h(x)dx\right|$$
$$\leq \frac{u_2 - u_1}{2}\left[\frac{(2^{s+1}-1)(s+1)+(1-2^s)(s+2)}{2^s(s+1)(s+2)}\left[|h'(u_2)| + |h'(u_1)|\right] + \frac{1}{2}|b(u_1,u_2,t)|\right]. \quad (7)$$

**Theorem 24.** *Assume that $K \subset [0,c], c > 0$ is an open invex subset w.r.t$\eta$ and $u_1, u_2 \in K$ with $u_1 < u_1 + \eta(u_2,u_1)$. Let $h : K \longrightarrow \mathbf{R}$ be a differentiable mapping on K such that $h' \in L([u_1, u_1 + \eta(u_2,u_1)])$. If $|h'|^q$ is a sub-b-s-preinvex function on K for $q > 1$, then we have the following inequality:*

$$\left|\frac{h(u_1) + h(u_1 + \eta(u_2,u_1))}{2} - \frac{1}{\eta(u_2,u_1)} \int_{u_1}^{u_1+\eta(u_2,u_1)} h(x)dx\right| \leq \frac{\eta(u_2,u_1)}{2(p+1)^{\frac{1}{p}}} \left[\frac{|h'(u_2)|^q + |h'(u_1)|^q}{s+1} + |b(u_1,u_2,t)|\right]^{\frac{1}{q}} \tag{8}$$

*where $\frac{1}{p} + \frac{1}{q} = 1$.*

**Proof.** From Lemma 21 and using the Hölder's integral inequality, we have

$$\left|\frac{h(u_1) + h(u_1 + \eta(u_2,u_1))}{2} - \frac{1}{\eta(u_2,u_1)} \int_{a}^{u_1+\eta(u_2,u_1)} h(x)dx\right| \leq \frac{\eta(u_2,u_1)}{2} \left(\int_0^1 |1-2t|^p \, dt\right)^{\frac{1}{p}} \left(\int_0^1 \left|\acute{h}(u_1 + t\eta(u_2,u_1))\right|^q dt\right)^{\frac{1}{q}}. \tag{9}$$

Since $|h'|^q$ is a sub-b-s-preinvex on $K$, for every $u_1, u_2 \in K$, $\in (0,1]$ and s $\in (0,1)$, we obtain

$$\left|h'(u_1 + t\eta(u_2,u_1))\right|^q \leq t^s|h'(u_2)|^q + (1-t)^s|h'(u_1)|^q + |b(u_1,u_2,t)|.$$

Hence,

$$\begin{aligned}\int_0^1 |h'(u_1 + t\eta(u_2,u_1))|^q dt &\leq [|h'(u_2)|^q + |h'(u_1)|^q] \int_0^1 t^s dt + |b(u_1,u_2,t)| \\ &= \frac{1}{s+1} [|h'(u_2)|^q + |h'(u_1)|^q] + |b(u_1,u_2,t)|.\end{aligned}$$

Moreover, via basic calculus, we obtain $\int_0^1 |1-2t|^p dt = \frac{1}{p+1}$. Thus, the proof of Theorem 24 is complete. □

**Corollary 25.** *If $\eta(u_2,u_1) = u_2 - u_1$ in Theorem 24, then Inequality 8 reduces to the following inequality:*

$$\left|\frac{h(u_1) + h(u_2)}{2} - \frac{1}{u_2 - u_1} \int_{u_1}^{u_2} h(x)dx\right| \leq \frac{u_2 - u_1}{2(p+1)^{\frac{1}{p}}} \left[\frac{\left|\acute{h}(u_2)\right|^q + \left|\acute{h}(u_1)\right|^q}{s+1} + |b(u_1,u_2,t)|\right]^{\frac{1}{q}} \tag{10}$$

*where $\frac{1}{p} + \frac{1}{q} = 1$.*

## 5. Application

In this section, we apply our results to the non-linear programming problem and to special means. Let us consider the unconstraint problem $(P)$

$$(P) : \min\{h(u), u \in K\}. \tag{11}$$

**Theorem 26.** *Consider that $h : k \longrightarrow \mathbb{R}$ is a non-negative differentiable sub-b-s-preinvex function w.r.t.$\eta, b$. If $u^* \in K$ and*

$$dh_{u^*}\eta(u,u^*) \geq \frac{h(u^*)}{t} + \lim_{t \longrightarrow 0_+} \frac{b(u,u^*,t)}{t}, \ \forall u \in K, t \in [0,1], s \in (0,1], \tag{12}$$

*then $u^*$ is the optimal solution to $(P)$ with respect to $h$ on $K$.*

**Proof.** By using the hypothesis and the second pair of Theorem 16, we obtain

$$dh_{u^*}\eta(u,u^*) - \frac{h(u^*)}{t} - \lim_{t \longrightarrow 0_+} \frac{b(u,u^*,t)}{t} \leq t^{s-1}(h(u) - h(u^*)),$$

$\forall t \in [0,1], s \in (0,1]$, and since

$$dh_{u^*}\eta(u,u^*) \geq \frac{h(u^*)}{t} + \lim_{t \longrightarrow 0_+} \frac{b(u,u^*,t)}{t}.$$

That is $h(u) - h(u^*) \geq 0$, which means that $u^*$ is the optimal solution. □

**Example 27.** *Let us take the following function $h : \mathbb{R}^+ \longrightarrow \mathbb{R}$ such that $h(u) = 2u^s$, where $s \in (0,1]$. Additionally, let $b(u_1, u_2, t) = tu_1^2 + 4tu_2^2$ and*

$$\eta(u_1, u_2) = \begin{cases} -u_2; u_1 = u_2 \\ 1 - u_2; u_1 \neq 2 \end{cases}.$$

*Since $b(u_1, u_2, t) \geq 0, \forall t \in (0,1]$, it is easy to say that $h$ is a sub-b-s-preinvex function. Additionally, $h(u)$ is a non-negative differentiable, and $\lim_{t \longrightarrow 0_+} \frac{b(u_1,u_2,t)}{t}$ exists for every $u_1, u_2 \in \mathbb{R}^+$ and $t \in (0,1]$. Thus, the following unconstraint sub-b-s-preinvex programming can be given as*

$$(P) : \min \{h(u), u \in \mathbb{R}^+\}.$$

$$dh_{u^*}\eta(u,u^*) = 2s(u^*)^{s-1}\eta(u,u^*), \frac{h(u^*)}{t} = \frac{2(u^*)^s}{t}$$

*and*

$$\lim_{t \longrightarrow 0_+} \frac{b(u,u^*,t)}{t} = u^2 + 4(u^*)^2.$$

*Thus, we see that $u^* = 0$ and*

$$dh_{u^*}\eta(u,u^*) \geq \frac{h(u^*)}{t} + \lim_{t \longrightarrow 0_+} \frac{b(u,u^*,t)}{t}$$

*holds $\forall u \in K, t \in (0,1], s \in (0,1)$. Hence, according to Theorem 26, the minimum value of $h(u)$ at zero.*

**Corollary 28.** *Assume that $h : K \longrightarrow \mathbb{R}$ is a strictly non-negative differentiable sub-b-s-preinvex function w.r.t. $\eta, b$. If $u^* \in K$ and satisfies the condition of Equation (12), then $u^*$ is the unique optimal solution of $h$ on $K$.*

**Proof.** Since $h$ is a strictly non-negative differentiable sub-b-s-preinvex function *w.r.t.*$\eta, b$ and by using Theorem 16, we obtain

$$dh_{u_2}\eta(u_1,u_2) < t^{s-1}(h(u_1) - h(u_2)) + \frac{h(u_2)}{t} + \lim_{t \longrightarrow 0_+} \frac{b(u_1,u_2,t)}{t}. \tag{13}$$

Let $v_1, v_2 \in K$ where $v_1 \neq v_2$ be optimal solutions of $(P)$. Then $h(v_1) = h(v_2)$, and Equation (13) yields that

$$dh_{v_2}\eta(v_1, v_2) - \frac{h(v_2)}{t} - \lim_{t \longrightarrow 0_+} \frac{b(v_1, v_2, t)}{t} < t^{s-1}\left(h(v_1) - h(v_2)\right).$$

By using Equation (12), we have $t^{s-1}\left(h(v_1) - h(v_2)\right) > 0$, but $h(v_1) = h(v_2)$. Thus, $v_1 = v_2 = u^*$. It follows that $u^*$ is the unique optimal optimal solution of $h$ on $K$. □

Now, grant nonlinear programming as follows:

$$(P_*) : \min\left\{h(u) : u \in \mathbb{R}^n, f_i(u) \leq 0, i \in I, where I = 1, 2, \cdots, m\right\}.$$

$F_e$ is the feasible set of $(P_*)$, which is given as

$$S_f = \{u \in \mathbb{R}^n : f_i(u) \leq 0, i \in I\}.$$

In addition, for $u^* \in S_f$, we define $N(u^*) = \{i : f_i(u^*) = 0, i \in I\}$.

**Theorem 29 (Karush-Kuhn-Tucker Sufficient Conditions).** *Assume that $h : \mathbb{R}^n \longrightarrow \mathbb{R}$ is a non-negative differentiable sub-b-s-preinvex function w.r.t.$\eta, b$ and $f_i : \mathbb{R}^n \longrightarrow \mathbb{R}$ are differentiable sub-b-s-preinvex functions w.r.t.$\eta, b_i, i \in I$. Additionally, let*

$$dh_{u^*}\eta(u, u^*) + \sum_{i \in I} v_i df_{iu^*}\eta(u, u^*) = 0,\ u^* \in S_f,\ v_i \geq 0,\ i \in I. \tag{14}$$

*If*

$$\frac{h(u^*)}{t} + \lim_{t \longrightarrow 0_+} \frac{b(u, u^*, t)}{t} \leq -\sum_{i=1} v_i lim_{t \longrightarrow 0_+} \frac{b(u, u^*, t)}{t}, \tag{15}$$

*then $u^*$ is an optimal solution of $(P_*)$.*

**Proof.** For any $u \in S_f$, then we obtain $f_i(u) \leq 0 = f_i(u^*), \forall u \in S_f$. Therefore, from the sub-b-s-preinvexity of $f_i$ and Theorem 17, we get

$$df_{iu^*}\eta(u, u^*) - \lim_{t \longrightarrow 0_+} \frac{b(u, u^*, t)}{t} \leq t^{s-1}\left(f_i(u) - f_i(u^*)\right) \leq 0. \tag{16}$$

From Equation (14), we obtain

$$\begin{aligned} dh_{u^*}\eta(u, u^*) &= -\sum_{i \in I} v_i df_{iu^*}\eta(u, u^*) \\ &= -\sum_{i \in N(u^*)} v_i df_{iu^*}\eta(u, u^*). \end{aligned} \tag{17}$$

Equations (15) and (17) yields that

$$\begin{aligned} &dh_{u^*}\eta(u, u^*) - \frac{h(u^*)}{t} - \lim_{t \longrightarrow 0_+} \frac{b(u, u^*, t)}{t} \\ &\quad \geq -\sum_{i \in N(u^*)} v_i \left(df_{iu^*}\eta(u, u^*) - \lim_{t \longrightarrow 0_+} \frac{b(u, u^*, t)}{t}\right). \end{aligned} \tag{18}$$

Here, we use Equations (16) and (18) to obtain

$$dh_{u^*}\eta(u, u^*) \geq \frac{h(u^*)}{t} + \lim_{t \longrightarrow 0_+} \frac{b(u, u^*, t)}{t}$$

and according to Theorem 26, one has

$$h(u) \geq h(u^*), \forall u \in S_f.$$

Hence, $u^*$ is an optimal solution of $(P_*)$. □

Now, some applications to special means are given. The following result is established in [28].

Assume that $H : I_1 \longrightarrow I_2 \subset [0, \infty)$ is a non-negative convex function on $I_1$. Then $H^s(x)$ is s-convex on $I_1$, where $s \in (0,1)$. For arbitrary positive real numbers $u_1, u_2 (u_1 \neq u_2)$, the following special means are given:

1. The arithmetic mean:
$$A = A(u_1, u_2) = \frac{u_1 + u_2}{2}, \quad u_1, u_2 \geq 0.$$

2. The logarithmic mean:
$$L = L(u_1, u_2) = \begin{cases} u_1 & if u_1 = u_2, \\ \frac{u_2 - u_1}{\ln u_1, \ln u_2} & if u_1 \neq u_2, \end{cases} .u_1, u_2 > 0$$

3. The P-logarithmic mean:
$$L_p = L_p(u_1, u_2) = \begin{cases} u_1 & if u_1 = u_2, \\ \left[\frac{u_2^{p+1} - u_1^{p+1}}{(p+1)(u_2 - u_1)}\right]^{\frac{1}{p}} & if u_1 \neq u_2, \end{cases} p \in \mathbf{R}\{-1, 0\}, u_1, u_2 > 0.$$

It is well known that $L_p$ is monotonic non-decreasing over $p \in \mathbf{R}$ with $L_{-1} = L$ and $L_0 = 1$. In particular, we have the following inequality

$$L \leq A.$$

Now, some new inequalities are derived for the above means

Let $h : [u_1, u_2] \longrightarrow \mathbf{R}, 0 < u_1 < u_2, h(x) = x^s$ and $s \in (0, 1]$, Then

$$\frac{1}{u_2 - u_1} \int_{u_1}^{u_2} h(x) dx = L_s^s(u_1, u_2)$$

$$\frac{h(u_1) + h(u_2)}{2} = A(u_1^s, u_2^s).$$

1. From Corollary 23,
$$|A(u_1^s, u_2^s) - L_s^s(u_1, u_2)| \\ \leq \frac{u_2 - u_1}{2}\left[\frac{(2^{s+1} - 1)(s+1) + (1 - 2^s)(s+2)}{2^s(s+1)(s+2)}\left[\left|\acute{h}(u_2)\right| + \left|\acute{h}(u_1)\right|\right] + \frac{1}{2}|b(u_1, u_2, t)|\right]. \quad (19)$$

2. From Corollary 25,
$$|A(u_1^s, u_2^s) - L_s^s(u_1, u_2)| \leq \frac{u_2 - u_1}{2(p+1)^{\frac{1}{p}}}\left[\frac{\left|\acute{h}(u_2)\right|^q + \left|\acute{h}(u_1)\right|^q}{s+1} + |b(u_1, u_2, t)|\right]^{\frac{1}{q}}. \quad (20)$$

If $s = 1$, then

$$|A(u_1,u_2)-L(u_1,u_2)| \le \frac{u_2-u_1}{4}\left[\frac{1}{2}[|h'(u_2)|+|h'(u_1)|]+|b(u_1,u_2,t)|\right]$$

$$|A(u_1,u_2)-L(u_1,u_2)| \le \frac{u_2-u_1}{2(p+1)^{\frac{1}{p}}}\left[\frac{1}{2}[|h'(u_2)|^q+|h'(u_1)|^q]+|b(u_1,u_2,t)|\right]$$

where $q>1$ and $\frac{1}{p}+\frac{1}{q}=1$.

## 6. Conclusions

In this paper, we introduce a new class of functions and sets called sub-b-s-preinvex functions and sub-b-s-preinvex sets and discuss some of their properties. In addition, the optimality conditions for a non-linear programming problem are also established. Hermite–Hadamard-type integral inequalities for differentiable sub-b-s-preinvex functions have been studied. Relationships between these inequalities and the classical inequalities have been established. The ideas and techniques of this paper may motivate further research, for example, in manifolds.

**Author Contributions:** Both authors have contributed equally to this paper. The idea of this whole paper was brought about by W.S., she also prepared the whole article; A.K. looked into the final manuscript.

**Funding:** This research received no external funding.

**Conflicts of Interest:** The authors declare no conflict of interest.

## References

1. Bector, C.R.; Singh, C. B-vex functions. *J. Optim. Theory Appl.* **1991**, *71*, 439–453. [CrossRef]
2. Chao, M.T.; Jian, J.B.; Liang, D.Y. Sub-b-convex functions and sub-b-convex programming. *Oper. Res. Trans.* **2012**, *16*, 1–8.
3. Hudzik, H.; Maligranda, L. Some remarks on s-convex functions. *Aequ. Math.* **1994**, *48*, 100–111. [CrossRef]
4. Orlicz, W. A note on modular spaces I. *Bull. Acad. Polon. Sci. Ser. Math. Astronom. Phys.* **1961**, *9*, 157–162.
5. Meftah, B. New integral inequalities for s-preinvex functions. *Int. J. Nonlinear Anal.* **2017**, *8*, 331–336.
6. Liao, J.; Tingsong, D. On Some Characterizations of Sub-b-s-Convex Functions. *Filomat* **2016**, *30*, 3885–3895. [CrossRef]
7. Ben-Israel, A.; Mond, B. What is invexity? *ANZIAM J.* **1986**, *28*, 1–9. [CrossRef]
8. Weir, T.; Mond, B. Pre-invex functions in multiple objective optimization. *J. Math. Anal. Appl.* **1988**, *136*, 29–38. [CrossRef]
9. Mohan, S.R.; Neogy, S.K. On invex sets and preinvex functions. *J. Math. Anal. Appl.* **1995**, *189*, 901–908. [CrossRef]
10. Suneja, S.K.; Singh, C.; Bector, C.R. Generalization of preinvex and B-vex functions. *J. Optim. Theory Appl.* **1993**, *76*, 577–587. [CrossRef]
11. Long,X.J.; Peng, J.W. Semi-B-preinvex functions. *J. Optim. Theory Appl.* **2006**, *131*, 301–305. [CrossRef]
12. Ahmad, I.; Jayswalb, A.; Kumarib, B. Characterizations of geodesic sub-b-s-convex functions on Riemannian manifolds. *J. Nonlinear Sci. Appl.* **2018**, *11*, 189–197. [CrossRef]
13. Boltyanski, V.; Martini, H.; Soltan, P.S. *Excursions into Combinatorial Geometry*; Springer Science & Business Media: Berlin/Heidelberg, Germany, 1997.
14. Chen, X. Some properties of semi-E-convex functions. *J. Math. Anal. Appl.* **2002**, *275*, 251–262. [CrossRef]
15. Duca, D.I.; Lupsa, L. On the E-epigraph of an E-convex functions. *J. Optim. Theory Appl.* **2006**, *129*, 341–348. [CrossRef]
16. Dwilewicz, R.J. A short History of Convexity. *Differ. Geom. Dyn. Syst.* **2009**, *11*, 112–129.
17. Kiliçman, A.; Saleh, W. A note on starshaped sets in2-dimensional manifolds without conjugate points. *J. Funct. Spaces* **2014**, *2014*, 3.
18. Liu, W. New integral inequalities involving beta function via P-convexity. *Miskolc Math. Notes* **2014**, *15*, 585–591.

19. Martini, H.; Swanepoel, K.J. Generalized convexity notions and combinatorial geometry. *Gongr. Numer.* **2003**, *164*, 65–93.
20. Syau, Y.R.; Jia, L.; Lee, E.S. Generalizations of E-convex and B-vex functions. *Comput. Math. Appl.* **2009**, *58*, 711–716. [CrossRef]
21. Liao, J.; Tingsong, D. Optimality Conditions in Sub-(b, m)-Convex Programming. *Univ. Politeh. Buchar. Sci. Bull.-Ser. A-Appl. Math. Phys.* **2017**, *79*, 95–106.
22. Fok, H.; Von, S. Generalizations of some Hermite—Hadamard-type inequalities. *Indian J. Pure Appl. Math.* **2015**, *46*, 359–370 [CrossRef]
23. Noor, M.A.; Noor, K.I.; Iftikhar, S.; Safdar, F. Some Properties of Generalized Strongly Harmonic Convex Functions. *Int. J. Anal. Appl.* **2018**, *16*, 427–436.
24. Dragomir, S.S. On Hadamards inequality for convex functions on the co-ordinates in a rectangle from the plane. *Taiwan. J. Math.* **2001**, *5*, 775–788. [CrossRef]
25. Hua, J.; Xi, B.; Feng, Q. Inequalities of Hermite–Hadamard type involving an s-convex function with applications. *Appl. Math. Comput.* **2014**, *246*, 752–760. [CrossRef]
26. Dragomir, S.S.; Fitzpatrick, S. The Hadamard inequalities for s-convex functions in the second sense. *Demonstr. Math.* **1999**, *32*, 687–696. [CrossRef]
27. Barani, A.; Ghazanfari, A.G.; Dragomir, S.S. Hermite–Hadamard inequality for functions whose derivatives absolute values are preinvex. *J. Inequal. Appl.* **2012**, *2012*, 247. [CrossRef]
28. Pearce, C.E.M.; Pecaric, J. Inequalities for differentiable mappings with application to specialmeans and quadrature formulae. *Appl. Math. Lett.* **2000**, *13*, 51–55. [CrossRef]

symmetry

MDPI

*Article*

# Multiplicity of Small or Large Energy Solutions for Kirchhoff–Schrödinger-Type Equations Involving the Fractional $p$-Laplacian in $\mathbb{R}^N$

**Jae-Myoung Kim [1], Yun-Ho Kim [2,*] and Jongrak Lee [3]**

1 Center for Mathematical Analysis and Computation, Yonsei University, Seoul 03722, Korea; cauchy02@naver.com
2 Department of Mathematics Education, Sangmyung University, Seoul 110-743, Korea
3 Department of Mathematics, Ewha Womans University, Seoul 120-750, Korea; jrlee0124@ewha.ac.kr
* Correspondence: kyh1213@smu.ac.kr; Tel.: +82-10-3272-8655

Received: 06 September 2018 ; Accepted: 24 September 2018; Published: 26 September 2018

**Abstract:** We herein discuss the following elliptic equations: $\mathcal{M}\left(\int_{\mathbb{R}^N}\int_{\mathbb{R}^N}\frac{|u(x)-u(y)|^p}{|x-y|^{N+ps}}dx\,dy\right)(-\Delta)_p^s u + V(x)|u|^{p-2}u = \lambda f(x,u)$ in $\mathbb{R}^N$, where $(-\Delta)_p^s$ is the fractional $p$-Laplacian defined by $(-\Delta)_p^s u(x) = 2\lim_{\varepsilon\searrow 0}\int_{\mathbb{R}^N\setminus B_\varepsilon(x)}\frac{|u(x)-u(y)|^{p-2}(u(x)-u(y))}{|x-y|^{N+ps}}dy,\quad x\in\mathbb{R}^N$. Here, $B_\varepsilon(x) := \{y\in\mathbb{R}^N : |x-y|<\varepsilon\}$, $V:\mathbb{R}^N\to(0,\infty)$ is a continuous function and $f:\mathbb{R}^N\times\mathbb{R}\to\mathbb{R}$ is the Carathéodory function. Furthermore, $\mathcal{M}:\mathbb{R}_0^+\to\mathbb{R}^+$ is a Kirchhoff-type function. This study has two aims. One is to study the existence of infinitely many large energy solutions for the above problem via the variational methods. In addition, a major point is to obtain the multiplicity results of the weak solutions for our problem under various assumptions on the Kirchhoff function $\mathcal{M}$ and the nonlinear term $f$. The other is to prove the existence of small energy solutions for our problem, in that the sequence of solutions converges to 0 in the $L^\infty$-norm.

**Keywords:** fractional $p$-Laplacian; Kirchhoff-type equations; fountain theorem; modified functional methods; Moser iteration method

## 1. Introduction

Significant attention has been focused on the study of fractional-type operators in view of the mathematical theory to some phenomena: the social sciences, quantum mechanics, continuum mechanics, phase transition phenomena, game theory, and Levy processes [1–5].

Herein, we discuss the results regarding the existence and multiplicity of nontrivial weak solutions for Kirchhoff-type equations

$$\mathcal{M}\left(\int_{\mathbb{R}^N}\int_{\mathbb{R}^N}\frac{|u(x)-u(y)|^p}{|x-y|^{N+ps}}dx\,dy\right)(-\Delta)_p^s u + V(x)|u|^{p-2}u = \lambda f(x,u) \quad \text{in } \mathbb{R}^N, \tag{1}$$

where $(-\Delta)_p^s$ is the fractional $p$-Laplacian operator defined by

$$(-\Delta)_p^s u(x) = 2\lim_{\varepsilon\searrow 0}\int_{\mathbb{R}^N\setminus B_\varepsilon(x)}\frac{|u(x)-u(y)|^{p-2}(u(x)-u(y))}{|x-y|^{N+ps}}dy$$

for $x\in\mathbb{R}^N$, with $0<s<1<p<\infty$, $ps<N$, $B_\varepsilon(x) := \{y\in\mathbb{R}^N : |x-y|<\varepsilon\}$, $V:\mathbb{R}^N\to(0,\infty)$ is a continuous function and $f:\mathbb{R}^N\times\mathbb{R}\to\mathbb{R}$ is the Carathéodory function. Furthermore, $\mathcal{M}\in C(\mathbb{R}_0^+,\mathbb{R}^+)$ is a Kirchhoff-type function.

Considering the effects of the change in the length of the stings that occurred by transverse vibrations, Kirchhoff in [6] originally proposed the following equation:

$$\rho\frac{\partial^2 u}{\partial t^2}-\Big(\frac{\rho_0}{h}+\frac{E}{2L}\int_0^L\Big|\frac{\partial u}{\partial x}\Big|dx\Big)\frac{\partial^2 u}{\partial x^2}=0,$$

which is the generalization of the classical D'Alembert's wave equation.

Subsequently, most researchers have extensively studied Kirchhoff-type equations associated with the fractional $p$-Laplacian problems in various ways; see [7–14] and the references therein. The critical point theory, originally introduced in [15] is critical in obtaining the solutions to elliptic equations of the variational type. It is considered that one of the crucial aspects for assuring the boundedness of the Palais–Smale sequence of the Euler–Lagrange functional, which is important to apply the critical point theory, is the Ambrosetti and Rabinowitz condition ((AR)-condition, briefly) in [15].

(AR) There exist positive constants C and $\zeta$ such that $\zeta > p$ and

$$0<\zeta\mathcal{F}(x,t)\le f(x,t)t \quad \text{for} \quad x\in\Omega \quad \text{and} \quad |t|\ge C,$$

where $\mathcal{F}(x,t)=\int_0^t f(x,s)\,ds$ and $\Omega$ is a bounded domain in $\mathbb{R}^N$.

Most results for our problem (1) are to establish the existence of nontrivial solutions under the (AR)-condition; see [7,10,14,16] for bounded domains and [11] for a whole space $\mathbb{R}^N$. The (AR)-condition is natural and important to guarantee the boundedness of the Palais–Smale sequence; this condition, however, is too restrictive and gets rid of many nonlinearities. Many authors have attempted to eliminate the (AR)-condition for elliptic equations associated with the $p$-Laplacian; see [17–20] and also see [21–25] for the superlinear problems of the fractional Laplacian type.

In this regard, we show that problem (1) permits the existence of multiple solutions under various conditions weaker than the (AR)-condition. In particular, following ([17], Remark 1.8), there exist many examples that do not fulfill the condition of the nonlinear term $f$ in [18,19,21,22,24–26]. Thus, motivated by these examples, the first aim of this paper is to demonstrate the existence of infinitely many large solutions for the problem above using the fountain theorem. One of novelties of this study is to obtain the multiplicity results for problem (1) when $f$ contains mild assumptions different from those of [18,19,21,22,24–26] (see Theorem 1). The other is to demonstrate this result with sufficient conditions for the modified Kirchhoff function $\mathcal{M}$, and the assumption on $f$ similar to that in [18,26] (see Theorem 2). As far as we are aware, none have reported such multiplicity results for our problem under the assumptions given in Theorem 2 of Section 2.

The second aim is to investigate that the existence of small energy solutions for problem (1), whose $L^\infty$-norms converge to zero, depends only on the local behavior and assumptions on $f(x,t)$, and only sufficiently small $t$ are required. Wang [27] initially investigated that nonlinear boundary value problems

$$\begin{cases}-\Delta u=\lambda\,|u|^{q-1}u+f(x,u), & \text{in } \Omega,\\ u=0, & \text{on } \mathbb{R}^N\backslash\Omega,\end{cases}$$

admit a sequence of infinitely many small solutions where $0<q<1$, and the nonlinear term $f$ was considered as a perturbation term. He employed global variational formulations and cut off techniques to obtain this existence result that is a local phenomenon and is forced by the sublinear term. Utilizing the argument in [27], Guo [28] showed that the $p$-Laplacian equations with indefinite concave nonlinearities have infinitely many solutions. In this regard, lots of authors have considered the results for the elliptic equations with nonlinear terms on a bounded domain in $\mathbb{R}^N$; see [29–31]. It is well known that the studies in [14,17,19,21,22,26,29,32,33] as well as our first primary result essentially demand some global conditions on $f(x,t)$ for $t$, such as oddness and behavior at infinity, for applying the fountain theorem to allow an infinite number of solutions. In contrast to these studies that yield large solutions in that they form an unbounded sequence, by modifying and extending the function $f(x,t)$ to a adequate function $\tilde{f}(x,t)$, the authors in [27–29] investigated the existence of small energy solutions to equations of the elliptic type. A natural question is whether the results in [27–31] may be extended to Equation (1). As is known, such a result for Kirchhoff–Schrödinger-type equations involving the non-local fractional $p$-Laplacian on the whole space $\mathbb{R}^N$ has not been much studied, although a given domain is bounded. In particular, no results are available even though the fractional $p$-Laplacian problems without Kirchhoff function $\mathcal{M}$ are considered, and we are only aware of paper [34] in this direction. In comparison with the papers [27–29], the main difficulty to obtain our second aim is to show the $L^\infty$-bound of weak solutions for problem (1). We remark that the strategy for obtaining this multiplicity is to assign a regularity-type result based on the work of Drábek, Kufner, and Nicolosi in [35]. Furthermore, it is noteworthy that the conditions on $f(x,t)$ are imposed near zero; in particular, $f(x,t)$ is odd in $t$ for a small $t$, and no conditions on $f(x,t)$ exist at infinity.

This paper is structured as follows. In Section 2, we state the basic results to solve the Kirchhoff-type equation, and review the well-known facts for the fractional Sobolev spaces. Moreover, under certain conditions on $f$, our problem admits a sequence of infinitely many large energy solutions of our problem (1) via the fountain theorem. Moreover, we assign the existence of nontrivial weak solutions for our problem with new conditions for the modified Kirchhoff function $\mathcal{M}$ and the nonlinear term $f$. In Section 3, we present the existence of small energy solutions for our problem in that the sequence of solutions converges to 0 in the $L^\infty$-norm. Hence, we employ the regularity result on the $L^\infty$-bound of a weak solution and the modified functional method.

## 2. Existence of Infinitely Many Large Energy Solutions

In this section, we recall some elementary concepts and properties of the fractional Sobolev spaces. We refer the reader to [4,36–38] for the detailed descriptions.

Suppose that

($\mathcal{V}1$) $V \in C(\mathbb{R}^N)$, $\inf_{x\in\mathbb{R}^N} V(x) > 0$.
($\mathcal{V}2$) $\text{meas}\{x \in \mathbb{R}^N : V(x) \le V_0\} < +\infty$ for all $V_0 \in \mathbb{R}$.

Let $0 < s < 1$ and $1 < p < +\infty$. We define the fractional Sobolev space $W^{s,p}(\mathbb{R}^N)$ by

$$W^{s,p}(\mathbb{R}^N) := \left\{ u \in L^p(\mathbb{R}^N) : \int_{\mathbb{R}^N}\int_{\mathbb{R}^N} \frac{|u(x)-u(y)|^p}{|x-y|^{N+ps}}\,dxdy < +\infty \right\},$$

endowed with the norm

$$\|u\|^p_{W^{s,p}(\mathbb{R}^N)} := |u|^p_{W^{s,p}(\mathbb{R}^N)} + \|u\|^p_{L^p(\mathbb{R}^N)} \quad \text{with} \quad |u|^p_{W^{s,p}(\mathbb{R}^N)} := \int_{\mathbb{R}^N}\int_{\mathbb{R}^N} \frac{|u(x)-u(y)|^p}{|x-y|^{N+ps}}\,dxdy.$$

Furthermore, we denote the basic function space $W(\mathbb{R}^N)$ by the completion of $C_0^\infty(\mathbb{R}^N)$ in $W^{s,p}(\mathbb{R}^N)$, equipped with the norm

$$\|u\|^p_{W(\mathbb{R}^N)} := |u|^p_{W^{s,p}(\mathbb{R}^N)} + \|u\|^p_{p,V} \quad \text{with} \quad \|u\|^p_{p,V} := \int_{\mathbb{R}^N} V(x)|u|^p\,dx.$$

Following a similar argument in [11,12], we can easily show that the space $W(\mathbb{R}^N)$ is a separable and reflexive Banach space.

We recall the continuous or compact embedding theorem in ([11], Lemma 1) and ([24], Lemma 2.1).

**Lemma 1.** *Let $0 < s < 1 < p < +\infty$ with $ps < N$. Then, there exists a positive constant $C = C(N,p,s)$ such that, for all $u \in W^{s,p}(\mathbb{R}^N)$,*

$$\|u\|_{L^{p_s^*}(\mathbb{R}^N)} \le C\,|u|_{W^{s,p}(\mathbb{R}^N)},$$

*where* $p_s^* = \frac{Np}{N-sp}$ *is the fractional critical exponent. Consequently, the space* $W^{s,p}(\mathbb{R}^N)$ *is continuously embedded in* $L^q(\mathbb{R}^N)$ *for any* $q \in [p, p_s^*]$. *Moreover, the space* $W^{s,p}(\mathbb{R}^N)$ *is compactly embedded in* $L^q_{loc}(\mathbb{R}^N)$ *for any* $q \in [p, p_s^*)$.

**Lemma 2.** *Let* $0 < s < 1 < p < +\infty$ *with* $ps < N$. *Suppose that the assumptions* $(\mathcal{V}1)$ *and* $(\mathcal{V}2)$ *hold. If* $r \in [p, p_s^*]$, *then the embeddings*

$$W(\mathbb{R}^N) \hookrightarrow W^{s,p}(\mathbb{R}^N) \hookrightarrow L^r(\mathbb{R}^N)$$

*are continuous with* $\|u\|^p_{W^{s,p}(\mathbb{R}^N)} \le C\|u\|^p_{W(\mathbb{R}^N)}$ *for all* $u \in W(\mathbb{R}^N)$. *In particular, there exists a constant* $K_r > 0$ *such that* $\|u\|_{L^r(\mathbb{R}^N)} \le K_r\|u\|_{W(\mathbb{R}^N)}$ *for all* $u \in W(\mathbb{R}^N)$. *If* $r \in [p, p_s^*)$, *then the embedding*

$$W(\mathbb{R}^N) \hookrightarrow L^r(\mathbb{R}^N)$$

*is compact.*

**Definition 1.** *Let* $0 < s < 1 < p < +\infty$. *We say that* $u \in W(\mathbb{R}^N)$ *is a weak solution of problem* (1) *if*

$$\mathcal{M}(|u|^p_{W^{s,p}(\mathbb{R}^N)}) \int_{\mathbb{R}^N}\int_{\mathbb{R}^N} \frac{|u(x)-u(y)|^{p-2}(u(x)-u(y))(v(x)-v(y))}{|x-y|^{N+ps}}\,dxdy \tag{2}$$

$$+\int_{\mathbb{R}^N} V(x)\,|u|^{p-2}\,uv\,dx = \lambda\int_{\mathbb{R}^N} f(x,u)v\,dx$$

*for any* $v$ *in* $W(\mathbb{R}^N)$.

We assume that the Kirchhoff function $\mathcal{M} : \mathbb{R}_0^+ \to \mathbb{R}^+$ satisfies the following conditions:

($\mathcal{M}1$) $\mathcal{M} \in C(\mathbb{R}_0^+, \mathbb{R}^+)$ satisfies $\inf_{t\in\mathbb{R}_0^+} \mathcal{M}(t) \ge m_0 > 0$, where $m_0$ is a constant.

($\mathcal{M}2$) There exists $\theta \in [1, \frac{N}{N-ps})$ such that $\theta\mathfrak{M}(t) \ge \mathcal{M}(t)t$ for any $t \ge 0$, where $\mathfrak{M}(t) := \int_0^t \mathcal{M}(\tau)d\tau$.

A typical example for $\mathcal{M}$ is given by $\mathcal{M}(t) = b_0 + b_1t^n$ with $n > 0, b_0 > 0$ and $b_1 \ge 0$.

Next, we consider the appropriate assumptions for the nonlinear term $f$. Let us denote $\mathcal{F}(x,t) = \int_0^t f(x,s)\,ds$ and let $\theta \in \mathbb{R}$ be given in ($\mathcal{M}2$).

($\mathcal{F}1$) $f : \mathbb{R}^N \times \mathbb{R} \to \mathbb{R}$ satisfies the Carathéodory condition.

($\mathcal{F}2$) There exist nonnegative functions $\rho \in L^{p'}(\mathbb{R}^N) \cap L^\infty(\mathbb{R}^N)$ and $\sigma \in L^{\frac{p_s^*}{p_s^*-q}}(\mathbb{R}^N) \cap L^\infty(\mathbb{R}^N)$ such that

$$|f(x,t)| \le \rho(x) + \sigma(x)\,|t|^{q-1}, \quad q \in (\theta p, p_s^*)$$

for all $(x,t) \in \mathbb{R}^N \times \mathbb{R}$.

($\mathcal{F}3$) $\lim_{|t|\to\infty} \frac{\mathcal{F}(x,t)}{|t|^{\theta p}} = \infty$ uniformly for almost all $x \in \mathbb{R}^N$.

($\mathcal{F}4$) There exist real numbers $c_0 > 0$, $r_0 \ge 0$, and $\kappa > \frac{N}{ps}$ such that

$$|\mathcal{F}(x,t)|^\kappa \le c_0\,|t|^{\kappa p}\,\mathfrak{F}(x,t)$$

for all $(x,t) \in \mathbb{R}^N \times \mathbb{R}$ and $|t| \ge r_0$, where $\mathfrak{F}(x,t) = \frac{1}{\theta p}f(x,t)t - \mathcal{F}(x,t) \ge 0$.

($\mathcal{F}5$) There exist $\mu > \theta p$ and $\varrho > 0$ such that

$$\mu\mathcal{F}(x,t) \le tf(x,t) + \varrho t^p$$

for all $(x,t) \in \mathbb{R}^N \times \mathbb{R}$.

For $u \in W(\mathbb{R}^N)$, the Euler–Lagrange functional $\mathcal{E}_\lambda : W(\mathbb{R}^N) \to \mathbb{R}$ is defined by

$$\mathcal{E}_\lambda(u) = A_{s,p}(u) - \lambda\Psi(u),$$

where

$$A_{s,p}(u) := \frac{1}{p}(\mathfrak{M}(|u|^p_{W^{s,p}(\mathbb{R}^N)}) + \|u\|^p_{p,V}) \quad \text{and} \quad \Psi(u) := \int_{\mathbb{R}^N} \mathcal{F}(x,u)\,dx.$$

Then, it is easily verifiable that $A_{s,p} \in C^1(W(\mathbb{R}^N),\mathbb{R})$ and $\Psi \in C^1(W(\mathbb{R}^N),\mathbb{R})$. Therefore, the functional $\mathcal{E}_\lambda$ is Fréchet differentiable on $W(\mathbb{R}^N)$ and its (Fréchet) derivative is as follows:

$$\begin{aligned}\langle \mathcal{E}'_\lambda(u), v\rangle &= \langle A'_{s,p}(u) - \lambda\Psi'(u), v\rangle \\ &= \mathcal{M}(|u|^p_{W^{s,p}(\mathbb{R}^N)}) \int_{\mathbb{R}^N}\int_{\mathbb{R}^N} \frac{|u(x)-u(y)|^{p-2}(u(x)-u(y))(v(x)-v(y))}{|x-y|^{N+ps}}\,dxdy \\ &\quad + \int_{\mathbb{R}^N} V(x)\,|u|^{p-2}\,uv\,dx - \lambda\int_{\mathbb{R}^N} f(x,u)v\,dx\end{aligned} \tag{3}$$

for any $u, v \in W(\mathbb{R}^N)$. Following Lemmas 2 and 3 in [11], the functional $A_{s,p}$ is weakly lower semi-continuous in $W(\mathbb{R}^N)$ and $\Psi$ is weakly continuous in $W(\mathbb{R}^N)$.

We now show that the functional $\mathcal{E}_\lambda$ satisfies the Cerami condition ($(C)_c$-condition, briefly), i.e., for $c \in \mathbb{R}$, any sequence $\{u_n\} \subset W(\mathbb{R}^N)$ such that $\mathcal{E}_\lambda(u_n) \to c$ and $\|\mathcal{E}'_\lambda(u_n)\|_{W^*(\mathbb{R}^N)}(1 + \|u_n\|_{W(\mathbb{R}^N)}) \to 0$ as $n \to \infty$ has a convergent subsequence. Here, $W^*(\mathbb{R}^N)$ is a dual space of $W(\mathbb{R}^N)$. This plays a decisive role in establishing the existence of nontrivial weak solutions.

**Lemma 3.** *Let $0 < s < 1 < p < +\infty$ and $ps < N$. Assume that $(\mathcal{V}1)$, $(\mathcal{V}2)$, $(\mathcal{M}1)$, $(\mathcal{M}2)$, and $(\mathcal{F}1)$–$(\mathcal{F}4)$ hold. Then, the functional $\mathcal{E}_\lambda$ satisfies the $(C)_c$-condition for any $\lambda > 0$.*

**Proof.** For $c \in \mathbb{R}$, let $\{u_n\}$ be a $(C)_c$-sequence in $W(\mathbb{R}^N)$, that is,

$$\mathcal{E}_\lambda(u_n) \to c \quad \text{and} \quad \|\mathcal{E}'_\lambda(u_n)\|_{W^*(\mathbb{R}^N)}(1 + \|u_n\|_{W(\mathbb{R}^N)}) \to 0 \qquad \text{as} \quad n \to \infty, \tag{4}$$

which implies that

$$c = \mathcal{E}_\lambda(u_n) + o(1) \quad \text{and} \quad \langle \mathcal{E}'_\lambda(u_n), u_n\rangle = o(1), \tag{5}$$

where $o(1) \to 0$ is $n \to \infty$. If $\{u_n\}$ is bounded in $W(\mathbb{R}^N)$, it follows from the proceeding as in the proof of Lemma 6 in [11] that $\{u_n\}$ converges strongly to $u$ in $W(\mathbb{R}^N)$. Hence, it suffices to verify that the sequence $\{u_n\}$ is bounded in $W(\mathbb{R}^N)$. However, we argue by contradiction and suppose that the conclusion is false, i.e., $\{u_n\}$ is a unbounded sequence in $W(\mathbb{R}^N)$. Therefore, we may assume that

$$\|u_n\|_{W(\mathbb{R}^N)} > 1 \quad \text{and} \quad \|u_n\|_{W(\mathbb{R}^N)} \to \infty, \quad \text{as} \quad n \to \infty. \tag{6}$$

Define a sequence $\{w_n\}$ by $w_n = u_n/\|u_n\|_{W(\mathbb{R}^N)}$. Then, it is clear that $\{w_n\} \subset W(\mathbb{R}^N)$ and $\|w_n\|_{W(\mathbb{R}^N)} = 1$. Hence, up to a subsequence (still denoted as the sequence $\{w_n\}$), we obtain $w_n \rightharpoonup w$ in $W(\mathbb{R}^N)$ as $n \to \infty$. Furthermore, by Lemma 2, we have

$$w_n(x) \to w(x) \text{ a.e. in } \mathbb{R}^N \text{ and } \quad w_n \to w \text{ in } L^r(\mathbb{R}^N) \quad \text{as} \quad n \to \infty \tag{7}$$

for $p \le r < p_s^*$. Owing to the condition (5), we have

$$c = \mathcal{E}_\lambda(u_n) + o(1) = \frac{1}{p}(\mathfrak{M}(|u_n|^p_{W^{s,p}(\mathbb{R}^N)}) + \|u_n\|^p_{p,V}) - \lambda\int_{\mathbb{R}^N} \mathcal{F}(x,u_n)\,dx + o(1).$$

Since $\|u_n\|_{W(\mathbb{R}^N)} \to \infty$ as $n \to \infty$, we assert that

$$\begin{aligned}\int_{\mathbb{R}^N} \mathcal{F}(x,u_n)\,dx &= \frac{1}{\lambda p}(\mathfrak{M}(|u_n|^p_{W^{s,p}(\mathbb{R}^N)}) + \|u_n\|^p_{p,V}) - \frac{c}{\lambda} + \frac{o(1)}{\lambda} \\ &\geq \frac{1}{\lambda p}(\frac{1}{\theta}\mathcal{M}(|u_n|^p_{W^{s,p}(\mathbb{R}^N)})|u_n|^p_{W^{s,p}(\mathbb{R}^N)} + \|u_n\|^p_{p,V}) - \frac{c}{\lambda} + \frac{o(1)}{\lambda} \\ &\geq \frac{\min\{m_0/\theta, 1\}}{\lambda p}\|u_n\|^p_{W(\mathbb{R}^N)} - \frac{c}{\lambda} + \frac{o(1)}{\lambda} \to \infty \quad \text{as} \quad n \to \infty. \end{aligned} \tag{8}$$

The assumption ($\mathcal{F}3$) implies that there exists $t_0 > 1$ such that $\mathcal{F}(x,t) > |t|^{\theta p}$ for all $x \in \mathbb{R}^N$ and $|t| > t_0$. From the assumptions ($\mathcal{F}1$) and ($\mathcal{F}2$), there is a constant $C > 0$ such that $|\mathcal{F}(x,t)| \leq C$ for all $(x,t) \in \mathbb{R}^N \times [-t_0, t_0]$. Therefore, we can choose $C_0 \in \mathbb{R}$ such that $\mathcal{F}(x,t) \geq C_0$ for all $(x,t) \in \mathbb{R}^N \times \mathbb{R}$; thus,

$$\frac{\mathcal{F}(x,u_n) - C_0}{\mathfrak{M}(|u_n|^p_{W^{s,p}(\mathbb{R}^N)}) + \|u_n\|^p_{p,V}} \geq 0, \tag{9}$$

for all $x \in \mathbb{R}^N$, and for all $n \in \mathbb{N}$. Set $\Omega = \{x \in \mathbb{R}^N : w(x) \neq 0\}$. By the convergence (7), we know that $|u_n(x)| = |w_n(x)|\,\|u_n\|_{W(\mathbb{R}^N)} \to \infty$ as $n \to \infty$ for all $x \in \Omega$. Therefore, it follows from the assumptions ($\mathcal{M}2$), ($\mathcal{F}3$), and the relation (6) that, for all $x \in \Omega$,

$$\begin{aligned}\lim_{n\to\infty} \frac{\mathcal{F}(x,u_n)}{\mathfrak{M}(|u_n|^p_{W^{s,p}(\mathbb{R}^N)}) + \|u_n\|^p_{p,V}} &\geq \lim_{n\to\infty} \frac{\mathcal{F}(x,u_n)}{\mathfrak{M}(1)\,|u_n|^{\theta p}_{W^{s,p}(\mathbb{R}^N)} + \|u_n\|^p_{p,V}} \\ &\geq \lim_{n\to\infty} \frac{\mathcal{F}(x,u_n)}{\mathfrak{M}(1)\|u_n\|^{\theta p}_{W(\mathbb{R}^N)} + \|u_n\|^p_{W(\mathbb{R}^N)}} \\ &\geq \lim_{n\to\infty} \frac{\mathcal{F}(x,u_n)}{\mathfrak{M}(1)\|u_n\|^{\theta p}_{W(\mathbb{R}^N)} + \|u_n\|^{\theta p}_{W(\mathbb{R}^N)}} \\ &= \lim_{n\to\infty} \frac{\mathcal{F}(x,u_n)}{(\mathfrak{M}(1)+1)\|u_n\|^{\theta p}_{W(\mathbb{R}^N)}} \\ &= \lim_{n\to\infty} \frac{\mathcal{F}(x,u_n)}{(\mathfrak{M}(1)+1)\,|u_n(x)|^{\theta p}}|w_n(x)|^{\theta p} \\ &= \infty, \end{aligned} \tag{10}$$

where we use the inequality $\mathfrak{M}(|u_n|^p_{W^{s,p}(\mathbb{R}^N)}) \leq \mathfrak{M}(1)\,|u_n|^{\theta p}_{W^{s,p}(\mathbb{R}^N)}$. Hence, we obtain $\text{meas}(\Omega) = 0$. If $\text{meas}(\Omega) \neq 0$, according to relations (8)–(10) and Fatou's lemma, we deduce that

$$\begin{aligned}\frac{1}{\lambda} &= \liminf_{n\to\infty} \frac{\int_{\mathbb{R}^N} \mathcal{F}(x,u_n)\,dx}{\lambda\int_{\mathbb{R}^N} \mathcal{F}(x,u_n)\,dx + c - o(1)} \\ &= \liminf_{n\to\infty} \int_{\mathbb{R}^N} \frac{p\mathcal{F}(x,u_n)}{\mathfrak{M}(|u_n|^p_{W^{s,p}(\mathbb{R}^N)}) + \|u_n\|^p_{p,V}}\,dx \\ &\geq \liminf_{n\to\infty} \int_{\Omega} \frac{p\mathcal{F}(x,u_n)}{\mathfrak{M}(|u_n|^p_{W^{s,p}(\mathbb{R}^N)}) + \|u_n\|^p_{p,V}}\,dx - \limsup_{n\to\infty} \int_{\Omega} \frac{pC_0}{\mathfrak{M}(|u_n|^p_{W^{s,p}(\mathbb{R}^N)}) + \|u_n\|^p_{p,V}}\,dx \\ &= \liminf_{n\to\infty} \int_{\Omega} \frac{p(\mathcal{F}(x,u_n) - C_0)}{\mathfrak{M}(|u_n|^p_{W^{s,p}(\mathbb{R}^N)}) + \|u_n\|^p_{p,V}}\,dx \\ &\geq \int_{\Omega} \liminf_{n\to\infty} \frac{p(\mathcal{F}(x,u_n) - C_0)}{\mathfrak{M}(|u_n|^p_{W^{s,p}(\mathbb{R}^N)}) + \|u_n\|^p_{p,V}}\,dx \end{aligned}$$

$$= \int_\Omega \liminf_{n\to\infty} \frac{p\mathcal{F}(x,u_n)}{\mathfrak{M}(|u_n|^p_{W^{s,p}(\mathbb{R}^N)}) + \|u_n\|^p_{p,V}}\,dx - \int_\Omega \limsup_{n\to\infty} \frac{pC_0}{\mathfrak{M}(|u_n|^p_{W^{s,p}(\mathbb{R}^N)}) + \|u_n\|^p_{p,V}}\,dx$$
$$= \infty, \tag{11}$$

which yields a contradiction. Thus, $w(x) = 0$ for almost all $x \in \mathbb{R}^N$.

Observe that, for a sufficiently large $n$,

$$\begin{aligned} c+1 &\geq \mathcal{E}_\lambda(u_n) - \frac{1}{\theta p}\langle \mathcal{E}'_\lambda(u_n), u_n\rangle \\ &= \frac{1}{p}(\mathfrak{M}(|u_n|^p_{W^{s,p}(\mathbb{R}^N)}) + \|u_n\|^p_{p,V}) - \lambda\int_{\mathbb{R}^N} \mathcal{F}(x,u_n)\,dx \\ &\quad - \frac{1}{\theta p}(\mathcal{M}(|u_n|^p_{W^{s,p}(\mathbb{R}^N)})|u_n|^p_{W^{s,p}(\mathbb{R}^N)} + \|u_n\|^p_{p,V}) + \frac{\lambda}{\theta p}\int_{\mathbb{R}^N} f(x,u_n)u_n\,dx \\ &\geq \lambda \int_{\mathbb{R}^N} \mathfrak{F}(x,u_n)\,dx, \end{aligned} \tag{12}$$

where $\mathfrak{F}$ is given in ($\mathcal{F}4$). Let us define $\Omega_n(a,b) := \{x \in \mathbb{R}^N : a \leq |u_n(x)| < b\}$ for $a \geq 0$. By the convergence (7),

$$w_n \to 0 \quad \text{in} \quad L^r(\mathbb{R}^N) \quad \text{and} \quad w_n(x) \to 0 \quad \text{a.e.} \quad \text{in} \quad \mathbb{R}^N \quad \text{as} \quad n \to \infty \tag{13}$$

for $p \leq r < p_s^*$. Hence, from the relation (8), we obtain

$$0 < \frac{1}{\lambda p} \leq \limsup_{n\to\infty} \int_{\mathbb{R}^N} \frac{|\mathcal{F}(x,u_n)|}{\mathfrak{M}(|u_n|^p_{W^{s,p}(\mathbb{R}^N)}) + \|u_n\|^p_{p,V}}\,dx. \tag{14}$$

Meanwhile, from the assumptions ($\mathcal{M}2$), ($\mathcal{F}2$), the relation (13), and Lemma 2, we obtain

$$\begin{aligned} &\int_{\Omega_n(0,r_0)} \frac{\mathcal{F}(x,u_n)}{\mathfrak{M}(|u_n|^p_{W^{s,p}(\mathbb{R}^N)}) + \|u_n\|^p_{p,V}}\,dx \\ &\leq \int_{\Omega_n(0,r_0)} \frac{\rho(x)\,|u_n(x)| + \frac{\sigma(x)}{q}\,|u_n(x)|^q}{\mathfrak{M}(|u_n|^p_{W^{s,p}(\mathbb{R}^N)}) + \|u_n\|^p_{p,V}}\,dx \\ &\leq \frac{\|\rho\|_{L^{p'}(\mathbb{R}^N)}\|u_n\|_{L^p(\mathbb{R}^N)}}{\mathfrak{M}(|u_n|^p_{W^{s,p}(\mathbb{R}^N)}) + \|u_n\|^p_{p,V}} + \frac{\|\sigma\|_{L^\infty(\mathbb{R}^N)}}{\min\{1, m_0/\theta\}q}\int_{\Omega_n(0,r_0)} |u_n(x)|^{q-p}\,|w_n(x)|^p\,dx \\ &\leq \frac{\|\rho\|_{L^{p'}(\mathbb{R}^N)}\|u_n\|_{L^p(\mathbb{R}^N)}}{\mathfrak{M}(|u_n|^p_{W^{s,p}(\mathbb{R}^N)}) + \|u_n\|^p_{p,V}} + \frac{\|\sigma\|_{L^\infty(\mathbb{R}^N)}}{\min\{1, m_0/\theta\}q} r_0^{q-p}\int_{\mathbb{R}^N} |w_n(x)|^p\,dx \\ &\leq \frac{K_p\|\rho\|_{L^{p'}(\mathbb{R}^N)}\|u_n\|_{W(\mathbb{R}^N)}}{\min\{1, m_0/\theta\}\|u_n\|^p_{W(\mathbb{R}^N)}} + \frac{\|\sigma\|_{L^\infty(\mathbb{R}^N)}}{\min\{1, m_0/\theta\}q} r_0^{q-p}\int_{\mathbb{R}^N} |w_n(x)|^p\,dx \\ &\leq \frac{C_1}{\min\{1, m_0/\theta\}\|u_n\|^{p-1}_{W(\mathbb{R}^N)}} + \frac{\|\sigma\|_{L^\infty(\mathbb{R}^N)}}{\min\{1, m_0/\theta\}q} r_0^{q-p}\int_{\mathbb{R}^N} |w_n(x)|^p\,dx \to 0, \quad \text{as } n \to \infty, \end{aligned} \tag{15}$$

where $C_1$ is a positive constant, $r_0$ is given in ($\mathcal{F}4$), and we use the following inequality:

$$\mathfrak{M}(|u_n|^p_{W^{s,p}(\mathbb{R}^N)}) + \|u_n\|^p_{p,V} \geq \min\{1, m_0/\theta\}\|u_n\|^p_{W(\mathbb{R}^N)}.$$

We set $\kappa' = \kappa/(\kappa - 1)$. Since $\kappa > N/ps$, we have $p < \kappa' p < p_s^*$. Hence, it follows from $(\mathcal{F}4)$, estimates (12) and (13) that

$$\begin{aligned}
\int_{\Omega_n(r_0,\infty)} \frac{|\mathcal{F}(x,u_n)|}{\mathfrak{M}(|u_n|^p_{W^{s,p}(\mathbb{R}^N)}) + \|u_n\|^p_{p,V}} dx &\leq \int_{\Omega_n(r_0,\infty)} \frac{|\mathcal{F}(x,u_n)|}{\min\{1, m_0/\theta\} |u_n(x)|^p} |w_n(x)|^p dx \\
&\leq \frac{1}{\min\{1, m_0/\theta\}} \left\{ \int_{\Omega_n(r_0,\infty)} \left( \frac{|F(x,u_n)|}{|u_n(x)|^p} \right)^\kappa dx \right\}^{\frac{1}{\kappa}} \left\{ \int_{\Omega_n(r_0,\infty)} |w_n(x)|^{\kappa' p} \right\}^{\frac{1}{\kappa'}} \\
&\leq \frac{c_0^{\frac{1}{\kappa}}}{\min\{1, m_0/\theta\}} \left\{ \int_{\Omega_n(r_0,\infty)} \mathfrak{F}(x,u_n)\, dx \right\}^{\frac{1}{\kappa}} \left\{ \int_{\mathbb{R}^N} |w_n(x)|^{\kappa' p} \right\}^{\frac{1}{\kappa'}} \\
&\leq \frac{c_0^{\frac{1}{\kappa}}}{\min\{1, m_0/\theta\}} \left( \frac{c+1}{\lambda} \right)^{\frac{1}{\kappa}} \left\{ \int_{\mathbb{R}^N} |w_n(x)|^{\kappa' p} \right\}^{\frac{1}{\kappa'}} \to 0, \quad \text{as } n \to \infty. \qquad (16)
\end{aligned}$$

Combining the relation (15) with the convergence (16), we have

$$\begin{aligned}
\int_{\mathbb{R}^N} \frac{|\mathcal{F}(x,u_n)|}{\mathfrak{M}(|u_n|^p_{W^{s,p}(\mathbb{R}^N)}) + \|u_n\|^p_{p,V}} dx &= \int_{\Omega_n(0,r_0)} \frac{|\mathcal{F}(x,u_n)|}{\mathfrak{M}(|u_n|^p_{W^{s,p}(\mathbb{R}^N)}) + \|u_n\|^p_{p,V}} dx \\
&\quad + \int_{\Omega_n(r_0,\infty)} \frac{|\mathcal{F}(x,u_n)|}{\mathfrak{M}(|u_n|^p_{W^{s,p}(\mathbb{R}^N)}) + \|u_n\|^p_{p,V}} dx \to 0
\end{aligned}$$

as $n \to \infty$, which contradicts inequality the convergence (14). The proof is completed. □

**Lemma 4.** *Let $0 < s < 1 < p < +\infty$ and $ps < N$. Assume that $(\mathcal{V}1)$, $(\mathcal{V}2)$, $(\mathcal{M}1)$, $(\mathcal{M}2)$, $(\mathcal{F}1)$–$(\mathcal{F}3)$, and $(\mathcal{F}5)$ hold. Then, the functional $\mathcal{E}_\lambda$ satisfies the $(C)_c$-condition for any $\lambda > 0$.*

**Proof.** For $c \in \mathbb{R}$, let $\{u_n\}$ be a $(C)_c$-sequence in $W(\mathbb{R}^N)$ satisfying (4). Then, relation (5) holds. As in the proof of Lemma 3, we only prove that $\{u_n\}$ is bounded in $W(\mathbb{R}^N)$. However, arguing by contradiction, suppose that $\|u_n\|_{W(\mathbb{R}^N)} \to \infty$ as $n \to \infty$. Let $v_n = u_n/\|u_n\|_{W(\mathbb{R}^N)}$. Then, $\|v_n\|_{W(\mathbb{R}^N)} = 1$ and $\|v_n\|_{L^r(\mathbb{R}^N)} \leq K_r \|v_n\|_{W(\mathbb{R}^N)} = K_r$ for $p \leq r \leq p_s^*$ by the continuous embedding in Lemma 2. Passing to a subsequence, we may assume that $v_n \rightharpoonup v$ in $W(\mathbb{R}^N)$ as $n \to \infty$; then, by compact embedding, $v_n \to v$ in $L^r(\mathbb{R}^N)$ for $p \leq r < p_s^*$, and $v_n(x) \to v(x)$ for almost all $x \in \mathbb{R}^N$ as $n \to \infty$. By the assumption $(\mathcal{F}5)$, one obtains

$$\begin{aligned}
c + 1 &\geq \mathcal{E}_\lambda(u_n) - \frac{1}{\mu} \langle \mathcal{E}'_\lambda(u_n), u_n \rangle \\
&= \frac{1}{p} (\mathfrak{M}(|u_n|^p_{W^{s,p}(\mathbb{R}^N)}) + \|u_n\|^p_{p,V}) - \lambda \int_{\mathbb{R}^N} \mathcal{F}(x,u_n)\, dx \\
&\quad - \frac{1}{\mu} (\mathcal{M}(|u_n|^p_{W^{s,p}(\mathbb{R}^N)}) |u_n|^p_{W^{s,p}(\mathbb{R}^N)} + \|u_n\|^p_{p,V}) + \frac{\lambda}{\mu} \int_{\mathbb{R}^N} f(x,u_n) u_n\, dx \\
&\geq \left( \frac{1}{\theta p} - \frac{1}{\mu} \right) \mathcal{M}(|u_n|^p_{W^{s,p}(\mathbb{R}^N)}) |u_n|^p_{W^{s,p}(\mathbb{R}^N)} + \left( \frac{1}{p} - \frac{1}{\mu} \right) \|u_n\|^p_{p,V} - \frac{\lambda \varrho}{\mu} \int_{\mathbb{R}^N} |u_n(x)|^p\, dx \\
&\geq \min\{1, m_0\} \left( \frac{1}{\theta p} - \frac{1}{\mu} \right) \|u_n\|^p_{W(\mathbb{R}^N)} - \frac{\lambda \varrho}{\mu} \int_{\mathbb{R}^N} |u_n(x)|^p\, dx, \qquad (17)
\end{aligned}$$

which implies

$$1 \leq \frac{\lambda \varrho \theta p}{\min\{1, m_0\}(\mu - \theta p)} \limsup_{n \to \infty} \|v_n\|^p_{L^p(\mathbb{R}^N)} = \frac{\lambda \varrho \theta p}{\min\{1, m_0\}(\mu - \theta p)} \|v\|^p_{L^p(\mathbb{R}^N)}. \qquad (18)$$

Hence, it follows from the inequality (18) that $v \neq 0$. From the same argument as in Lemma 3, we can verify the relations (8)–(10), and hence yield the relation (11). Therefore, we arrive at a contradiction. Thus, $\{u_n\}$ is bounded in $W(\mathbb{R}^N)$. □

Next, based on the fountain theorem in ([39], Theorem 3.6), we demonstrate the infinitely many weak solutions for problem (1). Hence, we let $X$ be a separable and reflexive Banach space. It is well known that there exists $\{e_n\} \subseteq X$ and $\{f_n^*\} \subseteq X^*$ such that

$$X = \overline{\text{span}\{e_n : n = 1, 2, \cdots\}}, \quad X^* = \overline{\text{span}\{f_n^* : n = 1, 2, \cdots\}},$$

and

$$\langle f_i^*, e_j \rangle = \begin{cases} 1, & \text{if } i = j, \\ 0, & \text{if } i \neq j. \end{cases}$$

Let us denote $X_n = \text{span}\{e_n\}$, $Y_k = \bigoplus_{n=1}^k X_n$, and $Z_k = \overline{\bigoplus_{n=k}^\infty X_n}$. Then, we recall the fountain lemma.

**Lemma 5.** *Let $\mathcal{X}$ be a real reflexive Banach space, $\mathcal{E} \in C^1(\mathcal{X}, \mathbb{R})$ satisfies the $(C)_c$-condition for any $c > 0$, and $\mathcal{E}$ is even. If for each sufficiently large $k \in \mathbb{N}$, there exist $\rho_k > \delta_k > 0$ such that the following conditions hold:*

*(1)* $b_k := \inf\{\mathcal{E}(u) : u \in Z_k, \|u\|_{\mathcal{X}} = \delta_k\} \to \infty \quad \text{as} \quad k \to \infty,$
*(2)* $a_k := \max\{\mathcal{E}(u) : u \in Y_k, \|u\|_{\mathcal{X}} = \rho_k\} \leq 0.$

*Then, the functional $\mathcal{E}$ has an unbounded sequence of critical values, i.e., there exists a sequence $\{u_n\} \subset \mathcal{X}$ such that $\mathcal{E}'(u_n) = 0$ and $\mathcal{E}(u_n) \to \infty$ as $n \to \infty$.*

Using Lemma 5, we demonstrate the existence of infinitely many nontrivial weak solutions for our problem.

**Theorem 1.** *Let $0 < s < 1 < p < +\infty$ and $ps < N$. Assume that $(\mathcal{V}1)$, $(\mathcal{V}2)$, $(\mathcal{M}1)$, $(\mathcal{M}2)$, and $(\mathcal{F}1)$–$(\mathcal{F}4)$ hold. If $f(x, -t) = -f(x, t)$ satisfies for all $(x, t) \in \mathbb{R}^N \times \mathbb{R}$, then the functional $\mathcal{E}_\lambda$ has a sequence of nontrivial weak solutions $\{u_n\}$ in $W(\mathbb{R}^N)$ such that $\mathcal{E}_\lambda(u_n) \to \infty$ as $n \to \infty$ for any $\lambda > 0$.*

**Proof.** Clearly, $\mathcal{E}_\lambda$ is an even functional and satisfies the $(C)_c$-condition. Note that $W(\mathbb{R}^N)$ is a separable and reflexive Banach space. According to Lemma 5, it suffices to show that there exists $\rho_k > \delta_k > 0$ such that

(1) $b_k := \inf\{\mathcal{E}_\lambda(u) : u \in Z_k, \|u\|_{W(\mathbb{R}^N)} = \delta_k\} \to \infty \quad \text{as} \quad k \to \infty;$
(2) $a_k := \max\{\mathcal{E}_\lambda(u) : u \in Y_k, \|u\|_{W(\mathbb{R}^N)} = \rho_k\} \leq 0,$

for a sufficiently large $k$. We denote

$$\alpha_k := \sup_{u \in Z_k, \|u\|_{W(\mathbb{R}^N)} = 1} \left( \int_{\mathbb{R}^N} \frac{1}{q} |u(x)|^q \, dx \right), \quad \theta p < q < p_s^*.$$

Then, we know $\alpha_k \to 0$ as $k \to \infty$. Indeed, suppose to the contrary that there exist $\varepsilon_0 > 0$ and a sequence $\{u_k\}$ in $Z_k$ such that

$$\|u_k\|_{W(\mathbb{R}^N)} = 1, \ \int_{\mathbb{R}^N} \frac{1}{q} |u_k(x)|^q \, dx \geq \varepsilon_0$$

for all $k \geq k_0$. Since the sequence $\{u_k\}$ is bounded in $W(\mathbb{R}^N)$, there exists an element $u$ in $W(\mathbb{R}^N)$ such that $u_k \rightharpoonup u$ in $W(\mathbb{R}^N)$ as $k \to \infty$, and

$$\langle f_j^*, u \rangle = \lim_{k \to \infty} \langle f_j^*, u_k \rangle = 0$$

for $j = 1, 2, \cdots$ . Hence, $u = 0$. However, we obtain

$$\varepsilon_0 \le \lim_{k\to\infty} \int_{\mathbb{R}^N} \frac{1}{q}|u_k(x)|^q \, dx = \int_{\mathbb{R}^N} \frac{1}{q}|u(x)|^q \, dx = 0,$$

which yields a contradiction.

For any $u \in Z_k$, it follows from assumptions $(\mathcal{M}2)$, $(\mathcal{F}2)$, and the Hölder inequality that

$$\begin{aligned}
\mathcal{E}_\lambda(u) &= \frac{1}{p}(\mathfrak{M}(|u|^p_{W^{s,p}(\mathbb{R}^N)}) + \|u\|^p_{p,V}) - \lambda \int_{\mathbb{R}^N} \mathcal{F}(x,u)\, dx \\
&\ge \frac{1}{p}(\mathfrak{M}(|u|^p_{W^{s,p}(\mathbb{R}^N)}) + \|u\|^p_{p,V}) - \lambda \int_{\mathbb{R}^N} |\rho(x)|\,|u(x)|\, dx - \lambda \int_{\mathbb{R}^N} \frac{|\sigma(x)|}{q}\,|u(x)|^q \, dx \\
&\ge \frac{1}{p}(\mathfrak{M}(|u|^p_{W^{s,p}(\mathbb{R}^N)}) + \|u\|^p_{p,V}) - \lambda\|\rho\|_{L^{p'}(\mathbb{R}^N)}\|u\|_{L^p(\mathbb{R}^N)} - \frac{\lambda}{q}\|\sigma\|_{L^\infty(\mathbb{R}^N)} \int_{\mathbb{R}^N} |u(x)|^q \, dx \\
&\ge \frac{1}{p}(\mathfrak{M}(|u|^p_{W^{s,p}(\mathbb{R}^N)}) + \|u\|^p_{p,V}) - \lambda C_2\|u\|_{W(\mathbb{R}^N)} - \frac{\lambda}{q} C_3 \|u\|^q_{L^q(\mathbb{R}^N)} \\
&\ge \frac{\min\{1, m_0/\theta\}}{p}\|u\|^p_{W(\mathbb{R}^N)} - \lambda C_2\|u\|_{W(\mathbb{R}^N)} - \frac{\lambda}{q}\alpha_k^q C_4 \|u\|^q_{W(\mathbb{R}^N)},
\end{aligned} \tag{19}$$

where $C_2, C_3$ and $C_4$ are positive constants. We choose $\delta_k = (2\lambda C_4 \alpha_k^q / \min\{1, m_0/\theta\})^{1/(p-q)}$. Since $p < q$ and $\alpha_k \to 0$ as $k \to \infty$, we assert $\delta_k \to \infty$ as $k \to \infty$. Hence, if $u \in Z_k$ and $\|u\|_{W(\mathbb{R}^N)} = \delta_k$, then we deduce that

$$\mathcal{E}_\lambda(u) \ge \left(\frac{1}{p} - \frac{1}{q}\right)\delta_k^p - 2\lambda C_2 \delta_k \to \infty \quad \text{as} \quad k \to \infty,$$

which implies the condition (1).

Next, suppose that condition (2) is not satisfied for some $k$. Then, there exists a sequence $\{u_n\}$ in $Y_k$ such that

$$\|u_n\|_{W(\mathbb{R}^N)} > 1 \quad \text{and} \quad \|u_n\|_{W(\mathbb{R}^N)} \to \infty \text{ as } n \to \infty \quad \text{and} \quad \mathcal{E}_\lambda(u_n) \ge 0. \tag{20}$$

Let $w_n = u_n / \|u_n\|_{W(\mathbb{R}^N)}$. Then, it is obvious that $\|w_n\|_{W(\mathbb{R}^N)} = 1$. Since $\dim Y_k < \infty$, there exists $w \in Y_k \setminus \{0\}$ such that, up to a subsequence,

$$\|w_n - w\|_{W(\mathbb{R}^N)} \to 0 \quad \text{and} \quad w_n(x) \to w(x)$$

for almost all $x \in \mathbb{R}^N$ as $n \to \infty$. For $x \in \Omega := \{x \in \mathbb{R}^N : w(x) \neq 0\}$, we obtain $|u_n(x)| \to \infty$ as $n \to \infty$. Hence, it follows from the assumption $(\mathcal{F}3)$ that

$$\lim_{n\to\infty} \frac{\mathcal{F}(x,u_n)}{\mathfrak{M}(|u_n|^p_{W^{s,p}(\mathbb{R}^N)}) + \|u_n\|^p_{p,V}} \ge \lim_{n\to\infty} \frac{\mathcal{F}(x,u_n)}{(\mathfrak{M}(1)+1)\,|u_n(x)|^{\theta p}}\,|w_n(x)|^{\theta p} = \infty. \tag{21}$$

As shown in the proof of Lemma 3, we can choose $C_1 \in \mathbb{R}$ such that

$$\frac{\mathcal{F}(x,u_n) - C_1}{\mathfrak{M}(|u_n|^p_{W^{s,p}(\mathbb{R}^N)}) + \|u_n\|^p_{p,V}} \ge 0 \tag{22}$$

for $x \in \Omega$. Considering the inequalities (21), (22) and Fatou's lemma, we assert by a similar argument to the inequality (10) that

$$\lim_{n\to\infty} \int_\Omega \frac{\mathcal{F}(x,u_n)}{\mathfrak{M}(|u_n|^p_{W^{s,p}(\mathbb{R}^N)}) + \|u_n\|^p_{p,V}}\, dx \ge \liminf_{n\to\infty} \int_\Omega \frac{\mathcal{F}(x,u_n) - C_1}{\mathfrak{M}(|u_n|^p_{W^{s,p}(\mathbb{R}^N)}) + \|u_n\|^p_{p,V}}\, dx$$

$$\geq \int_{\Omega} \liminf_{n\to\infty} \frac{\mathcal{F}(x,u_n)}{\mathfrak{M}(|u_n|^p_{W^{s,p}(\mathbb{R}^N)}) + \|u_n\|^p_{p,V}}\,dx = \infty. \tag{23}$$

Therefore, using the relation (23), we have

$$\begin{aligned}\mathcal{E}_\lambda(u_n) &\leq \frac{1}{p}(\mathfrak{M}(|u_n|^p_{W^{s,p}(\mathbb{R}^N)}) + \|u_n\|^p_{p,V}) - \lambda \int_\Omega \mathcal{F}(x,u_n)\,dx \\ &= \frac{\mathfrak{M}(|u_n|^p_{W^{s,p}(\mathbb{R}^N)}) + \|u_n\|^p_{p,V}}{p}\left(1 - \lambda p \int_\Omega \frac{\mathcal{F}(x,u_n)}{\mathfrak{M}(|u_n|^p_{W^{s,p}(\mathbb{R}^N)}) + \|u_n\|^p_{p,V}}\,dx\right) \to -\infty\end{aligned}$$

as $n \to \infty$, which yields a contradiction to the relation (20). The proof is complete. □

**Remark 1.** *Although we replaced ($\mathcal{F}4$) with ($\mathcal{F}5$) in the assumption of Theorem 1, we assert that the problem (1) admits a sequence of nontrivial weak solutions $\{u_n\}$ in $W(\mathbb{R}^N)$ such that $\mathcal{E}_\lambda(u_n) \to \infty$ as $n \to \infty$.*

Lastly, we investigate the existence of nontrivial weak solutions for our problem by replacing the assumptions ($\mathcal{F}4$) and ($\mathcal{F}5$) with the following condition, which is from the work of L. Jeanjean [40]:

($\mathcal{F}6$) There exists a constant $\nu \geq 1$ such that

$$\nu\hat{\mathfrak{F}}(x,t) \geq \hat{\mathfrak{F}}(x,st)$$

for $(x,t) \in \mathbb{R}^N \times \mathbb{R}$ and $s \in [0,1]$, where $\hat{\mathfrak{F}}(x,t) = f(x,t)t - \theta p\mathcal{F}(x,t)$.

When the Kirchhoff function $\mathcal{M}$ is constant, and the condition ($\mathcal{F}6$) with $\theta = 1$ holds, the author in [24] obtained the existence of at least one nontrivial weak solution for the superlinear problems of the fractional $p$-Laplacian, which is motivated by the works of [18,26].

To the best of our belief, such existence and multiplicity results are not available for the elliptic equation of the Kirchhoff type under the assumption ($\mathcal{F}6$). Hence, we obtain the following lemma with the sufficient conditions for the modified Kirchhoff function $\mathcal{M}$ and the assumption ($\mathcal{F}6$).

**Lemma 6.** *Let $0 < s < 1 < p < +\infty$ and $ps < N$. Assume that ($\mathcal{V}1$), ($\mathcal{V}2$), ($\mathcal{M}1$), ($\mathcal{M}2$), ($\mathcal{F}1$)–($\mathcal{F}3$), and ($\mathcal{F}6$) hold. Furthermore, we assume that*

($\mathcal{M}3$) *$\mathcal{H}(st) \leq \mathcal{H}(t)$ for $s \in [0,1]$, where $\mathcal{H}(t) = \theta\mathfrak{M}(t) - \mathcal{M}(t)t$ for any $t \geq 0$ and $\theta$ is given in ($\mathcal{M}2$).*

*Then, the functional $\mathcal{E}_\lambda$ satisfies the $(C)_c$-condition for any $\lambda > 0$.*

**Proof.** For $c \in \mathbb{R}$, let $\{u_n\}$ be a $(C)_c$-sequence in $W(\mathbb{R}^N)$ satisfying the convergence (4). Then, the relation (5) holds. By Lemma 3, we only prove that $\{u_n\}$ is bounded in $W(\mathbb{R}^N)$. Therefore, we argue by contradiction and suppose that the conclusion is false, i.e., $\|u_n\|_{W(\mathbb{R}^N)} > 1$ and $\|u_n\|_{W(\mathbb{R}^N)} \to \infty$ as $n \to \infty$. In addition, we define a sequence $\{\omega_n\}$ by $\omega_n = u_n/\|u_n\|_{W(\mathbb{R}^N)}$. Then, up to a subsequence (still denoted as the sequence $\{\omega_n\}$), we obtain $\omega_n \rightharpoonup \omega$ in $W(\mathbb{R}^N)$ as $n \to \infty$,

$$\omega_n(x) \to \omega(x) \text{ a.e. in } \mathbb{R}^N, \quad \omega_n \to \omega \text{ in } L^q(\mathbb{R}^N), \quad \text{and} \quad \omega_n \to \omega \text{ in } L^p(\mathbb{R}^N) \quad \text{as} \quad n \to \infty,$$

where $\theta p < q < p_s^*$.

We set $\Omega = \{x \in \mathbb{R}^N : \omega(x) \neq 0\}$. From the similar manner as in Lemma 3, we obtain meas$(\Omega) = 0$. Therefore, $\omega(x) = 0$ for almost all $x \in \mathbb{R}^N$. Since $\mathcal{E}_\lambda(tu_n)$ is continuous at $t \in [0,1]$, for each $n \in \mathbb{N}$, there exists $t_n \in [0,1]$ such that

$$\mathcal{E}_\lambda(t_nu_n) := \max_{t\in[0,1]} \mathcal{E}_\lambda(tu_n).$$

Let $\{\ell_k\}$ be a positive sequence of real numbers such that $\lim_{k\to\infty} \ell_k = \infty$ and $\ell_k > 1$ for any $k$. Then, it is clear that $\|\ell_k\omega_n\|_{W(\mathbb{R}^N)} = \ell_k > 1$ for any $k$ and $n$. Fix $k$, since $\omega_n \to 0$ strongly in $L^q(\mathbb{R}^N)$ as

$n \to \infty$, the continuity of the Nemytskii operator implies $\mathcal{F}(x, \ell_k \omega_n) \to 0$ in $L^1(\mathbb{R}^N)$ as $n \to \infty$. Hence, we assert

$$\lim_{n\to\infty} \int_{\mathbb{R}^N} \mathcal{F}(x, \ell_k \omega_n)\, dx = 0. \tag{24}$$

Since $\|u_n\|_{W(\mathbb{R}^N)} \to \infty$ as $n \to \infty$, we obtain $\|u_n\|_{W(\mathbb{R}^N)} > \ell_k$ for a sufficiently large $n$. Thus, we know by $(\mathcal{M}2)$ and the convergence (24) that

$$\begin{aligned}
\mathcal{E}_\lambda(t_n u_n) &\geq \mathcal{E}_\lambda\left(\frac{\ell_k}{\|u_n\|_{W(\mathbb{R}^N)}} u_n\right) = \mathcal{E}_\lambda(\ell_k \omega_n) \\
&= \frac{1}{p}(\mathfrak{M}(|\ell_k \omega_n|^p_{W^{s,p}(\mathbb{R}^N)}) + \|\ell_k \omega_n\|^p_{p,V}) - \lambda \int_{\mathbb{R}^N} \mathcal{F}(x, \ell_k \omega_n)\, dx \\
&\geq \frac{1}{p\theta} \mathcal{M}(|\ell_k \omega_n|^p_{W^{s,p}(\mathbb{R}^N)})|\ell_k \omega_n|^p_{W^{s,p}(\mathbb{R}^N)} + \frac{1}{p}\|\ell_k \omega_n\|^p_{p,V} - \lambda \int_{\mathbb{R}^N} \mathcal{F}(x, \ell_k \omega_n)\, dx \\
&\geq \frac{\min\{1, m_0\}}{p\theta} \|\ell_k \omega_n\|^p_{W(\mathbb{R}^N)} - \lambda \int_{\mathbb{R}^N} \mathcal{F}(x, \ell_k \omega_n)\, dx \\
&\geq \frac{\min\{1, m_0\}}{p\theta} \ell_k^p
\end{aligned}$$

for a large enough $n$. Then, letting $n, k \to \infty$, we get

$$\lim_{n\to\infty} \mathcal{E}_\lambda(t_n u_n) = \infty. \tag{25}$$

Since $\mathcal{E}_\lambda(0) = 0$ and $\mathcal{E}_\lambda(u_n) \to c$ as $n \to \infty$, it is obvious that $t_n \in (0,1)$, and $\langle \mathcal{E}'_\lambda(t_n u_n), t_n u_n \rangle = 0$. Therefore, owing to the assumptions $(\mathcal{M}3)$ and $(\mathcal{F}6)$, for all sufficiently large $n$, we deduce that

$$\begin{aligned}
\frac{1}{\nu}\mathcal{E}_\lambda(t_n u_n) &= \frac{1}{\nu}\mathcal{E}_\lambda(t_n u_n) - \frac{1}{p\theta\nu}\langle \mathcal{E}'_\lambda(t_n u_n), t_n u_n \rangle + o(1) \\
&= \frac{1}{p\nu}(\mathfrak{M}(|t_n u_n|^p_{W^{s,p}(\mathbb{R}^N)}) + \|t_n u_n\|^p_{p,V}) - \frac{\lambda}{\nu}\int_{\mathbb{R}^N} \mathcal{F}(x, t_n u_n)\, dx \\
&\quad - \frac{1}{p\theta\nu}(\mathcal{M}(|t_n u_n|^p_{W^{s,p}(\mathbb{R}^N)})|t_n u_n|^p_{W^{s,p}(\mathbb{R}^N)} + \|t_n u_n\|^p_{p,V}) + \frac{\lambda}{p\theta\nu}\int_{\mathbb{R}^N} f(x, t_n u_n) t_n u_n\, dx + o(1) \\
&= \frac{1}{p\theta\nu}\mathcal{H}(t_n u_n) + \frac{1}{p\nu}\|t_n u_n\|^p_{p,V} - \frac{1}{p\theta\nu}\|t_n u_n\|^p_{p,V} + \frac{\lambda}{p\theta\nu}\int_{\mathbb{R}^N} \hat{\mathfrak{F}}(x, t_n u_n)\, dx + o(1) \\
&\leq \frac{1}{p\theta}\mathcal{H}(u_n) + \frac{1}{p}\|t_n u_n\|^p_{p,V} - \frac{1}{p\theta}\|t_n u_n\|^p_{p,V} + \frac{\lambda}{p\theta}\int_{\mathbb{R}^N} \hat{\mathfrak{F}}(x, u_n)\, dx + o(1) \\
&= \frac{1}{p}(\mathfrak{M}(|u_n|^p_{W^{s,p}(\mathbb{R}^N)}) + \|u_n\|^p_{p,V}) - \lambda \int_{\mathbb{R}^N} \mathcal{F}(x, u_n)\, dx \\
&\quad - \frac{1}{p\theta}(\mathcal{M}(|u_n|^p_{W^{s,p}(\mathbb{R}^N)})|u_n|^p_{W^{s,p}(\mathbb{R}^N)} + \|u_n\|^p_{p,V}) + \frac{\lambda}{p\theta}\int_{\mathbb{R}^N} f(x, u_n) u_n\, dx + o(1) \\
&= \mathcal{E}_\lambda(u_n) - \frac{1}{p\theta}\langle \mathcal{E}'_\lambda(u_n), u_n \rangle + o(1) \to c \quad \text{as} \quad n \to \infty,
\end{aligned}$$

which contradicts the convergence (25). This completes the proof. □

We give an example regarding a function $\mathcal{M}$ with the assumptions $(\mathcal{M}1)$–$(\mathcal{M}3)$.

**Example 1.** *Let us see*

$$\mathcal{M}(t) = 1 + \frac{1}{e+t}, \quad t \geq 0.$$

*Then, it is easily checked that this function* $\mathcal{M}$ *complies with the assumptions* $(\mathcal{M}1)$–$(\mathcal{M}3)$.

**Theorem 2.** *Let $0 < s < 1 < p < +\infty$ and $ps < N$. Assume that $(\mathcal{V}1)$, $(\mathcal{V}2)$, $(\mathcal{M}1)$–$(\mathcal{M}3)$, $(\mathcal{F}1)$–$(\mathcal{F}3)$, and $(\mathcal{F}6)$ hold. If $f(x,-t) = -f(x,t)$ holds for all $(x,t) \in \mathbb{R}^N \times \mathbb{R}$, then, for any $\lambda > 0$, the functional $\mathcal{E}_\lambda$ has a sequence of nontrivial weak solutions $\{u_n\}$ in $W(\mathbb{R}^N)$ such that $\mathcal{E}_\lambda(u_n) \to \infty$ as $n \to \infty$.*

**Proof.** The proof is essentially the same as that of Theorem 1. □

## 3. Existence of Infinitely Many Small Energy Solutions

In this section, we prove the existence of a sequence of small energy solutions for the problem (1) converging to zero in $L^\infty$-norm based on the Moser bootstrap iteration technique in ([35], Theorem 4.1) (see also [34]). First, we state the following additional assumptions:

$(\mathcal{F}7)$ There exists a constant $s_0 > 0$ such that $p\mathcal{F}(x,t) - f(x,t)t > 0$ for all $x \in \mathbb{R}^N$ and for $0 < |t| < s_0$.
$(\mathcal{F}8)$ $\lim_{|t|\to 0} \frac{f(x,t)}{|t|^{p-2}t} = +\infty$ uniformly for all $x \in \mathbb{R}^N$.

Because problem (1) includes the potential term and the nonlinear term $f$ is slightly different from that of [35], a more complicated analysis has to be carefully performed when we apply the bootstrap iteration argument.

**Proposition 1.** *Assume that $(\mathcal{V}1)$, $(\mathcal{M}1)$, and $(\mathcal{F}1)$–$(\mathcal{F}2)$ hold. If $u$ is a weak solution of the problem (1), then $u \in L^r(\mathbb{R}^N)$ for all $r \in [p_s^*, \infty]$.*

**Proof.** Suppose that $u$ is non-negative. For $K > 0$, we define

$$v_K(x) = \min\{u(x), K\}$$

and choose $v = v_K^{mp+1}$ $(m \geq 0)$ as a test function in the equality (2). Then, $v \in W(\mathbb{R}^N) \cap L^\infty(\mathbb{R}^N)$, and it follows from the equality (2) that

$$\begin{aligned}\mathcal{M}(|u|^p_{W^{s,p}(\mathbb{R}^N)}) &\int_{\mathbb{R}^N}\int_{\mathbb{R}^N} \frac{|u(x)-u(y)|^{p-2}(u(x)-u(y))(v_K^{mp+1}(x)-v_K^{mp+1}(y))}{|x-y|^{N+ps}}\,dxdy \\ &+ \int_{\mathbb{R}^N} V(x)\,|u|^{p-2}\,uv_K^{mp+1}\,dx = \lambda\int_{\mathbb{R}^N} f(x,u)v_K^{mp+1}\,dx. \end{aligned} \tag{26}$$

The left-hand side of the relation (26) can be estimated as follows:

$$\begin{aligned}&\mathcal{M}(|u|^p_{W^{s,p}(\mathbb{R}^N)}) \int_{\mathbb{R}^N}\int_{\mathbb{R}^N} \frac{|u(x)-u(y)|^{p-2}(u(x)-u(y))(v_K^{mp+1}(x)-v_K^{mp+1}(y))}{|x-y|^{N+ps}}\,dxdy \\ &\quad+ \int_{\mathbb{R}^N} V(x)\,|u|^{p-2}\,uv_K^{mp+1}\,dx \\ &\geq m_0\int_{\mathbb{R}^N}\int_{\mathbb{R}^N} \frac{|u(x)-u(y)|^{p-1}\left|v_K^{mp+1}(x)-v_K^{mp+1}(y)\right|}{|x-y|^{N+ps}}\,dxdy + \int_{\mathbb{R}^N} V(x)v_K^{(m+1)p}\,dx \\ &\geq m_0C_5\int_{\mathbb{R}^N}\int_{\mathbb{R}^N} \frac{|v_K^{m+1}(x)-v_K^{m+1}(y)|^p}{|x-y|^{N+ps}}\,dxdy + \int_{\mathbb{R}^N} V(x)v_K^{(m+1)p}\,dx \\ &\geq \min\{m_0C_5,1\}\|v_K^{m+1}\|^p_{W(\mathbb{R}^N)} \\ &\geq \min\{m_0C_5,1\}C_6\left(\int_{\mathbb{R}^N} |v_K|^{(m+1)p_s^*}\,dx\right)^{\frac{p}{p_s^*}} \end{aligned} \tag{27}$$

for some positive constants $C_5$ and $C_6$. Using the assumption $(\mathcal{F}2)$, the Hölder inequality and the relation (27), the right-hand side of the relation (26) can be estimated:

$$\lambda\int_{\mathbb{R}^N} f(x,u)v_K^{mp+1}\,dx \leq \lambda\int_{\mathbb{R}^N} |f(x,u)||u|^{mp+1}\,dx$$

$$\begin{aligned}
&\le \lambda \int_{\mathbb{R}^N} \rho(x)|u|^{mp+1} + \sigma(x)|u|^{mp+q}\,dx \\
&\le \lambda \int_{\mathbb{R}^N} \rho(x)(|u|^{mp+p} + |u|^{m+1})\,dx \\
&\quad + \lambda \left(\int_{\mathbb{R}^N} \sigma^{\gamma_1}(x)\,dx\right)^{\frac{1}{\gamma_1}} \left(\int_{\mathbb{R}^N} |u|^{(m+1)p\gamma_1'} |u|^{(q-p)\gamma_1'}\,dx\right)^{\frac{1}{\gamma_1'}} \\
&\le \lambda \|\rho\|_{L^\infty(\mathbb{R}^N)} \int_{\mathbb{R}^N} |u|^{(m+1)p}\,dx + \lambda \|\rho\|_{L^{p'}(\mathbb{R}^N)} \left(\int_{\mathbb{R}^N} |u|^{(m+1)p}\,dx\right)^{\frac{1}{p}} \\
&\quad + \lambda \left(\int_{\mathbb{R}^N} \sigma^{\gamma_1}(x)\,dx\right)^{\frac{1}{\gamma_1}} \left(\int_{\mathbb{R}^N} |u|^{(m+1)\beta}\,dx\right)^{\frac{p}{\beta}} \left(\int_{\mathbb{R}^N} |u|^{(q-p)\gamma_1' \frac{\beta}{\beta - p\gamma_1'}}\,dx\right)^{\frac{\beta - p\gamma_1'}{\beta\gamma_1'}},
\end{aligned} \tag{28}$$

where $\gamma_1 = \frac{p_s^*}{p_s^*-q}$, and $\beta = \frac{pp_s^*\gamma_1'}{p_s^*-(q-p)\gamma_1'}$. Obviously $\beta \le p_s^*$, $1 < \frac{\beta}{p\gamma_1'}$, and $\frac{(q-p)\gamma_1'\beta}{\beta - p\gamma_1'} = p_s^*$, and hence the estimate (28) yields

$$\begin{aligned}
\lambda \int_{\mathbb{R}^N} f(x,u) v_K^{mp+1}\,dx &\le \lambda \|\rho\|_{L^\infty(\mathbb{R}^N)} \int_{\mathbb{R}^N} |u|^{(m+1)p}\,dx + \lambda \|\rho\|_{L^{p'}(\mathbb{R}^N)} \left(\int_{\mathbb{R}^N} |u|^{(m+1)p}\,dx\right)^{\frac{1}{p}} \\
&\quad + \lambda \left(\int_{\mathbb{R}^N} \sigma^{\gamma_1}(x)\,dx\right)^{\frac{1}{\gamma_1}} \left(\int_{\mathbb{R}^N} |u|^{p_s^*}\,dx\right)^{\frac{\beta - p\gamma_1'}{\beta\gamma_1'}} \left(\int_{\mathbb{R}^N} |u|^{(m+1)\beta}\,dx\right)^{\frac{p}{\beta}}.
\end{aligned} \tag{29}$$

It follows from relations (26), (27), (29), and the Sobolev inequality that there exists positive constants $C_7, C_8$ and $C_9$ (independent of $K$ and $m > 0$) such that

$$\left(\int_{\mathbb{R}^N} |v_K|^{(m+1)p_s^*}\,dx\right)^{\frac{p}{p_s^*}} \le C_7 \int_{\mathbb{R}^N} |u|^{(m+1)p}\,dx + C_8 \left(\int_{\mathbb{R}^N} |u|^{(m+1)p}\,dx\right)^{\frac{1}{p}} + C_9 \left(\int_{\mathbb{R}^N} |u|^{(m+1)\beta}\,dx\right)^{\frac{p}{\beta}},$$

which implies

$$\|v_K\|_{L^{(m+1)p_s^*}(\mathbb{R}^N)}^{(m+1)p} \le C_7 \|u\|_{L^{(m+1)p}(\mathbb{R}^N)}^{(m+1)p} + C_8 \|u\|_{L^{(m+1)p}(\mathbb{R}^N)}^{m+1} + C_9 \|u\|_{L^{(m+1)\beta}(\mathbb{R}^N)}^{(m+1)p}. \tag{30}$$

To apply the argument that is critical in $L^\infty$-estimates, we first assume that $\|u\|_{L^{(m+1)p}(\mathbb{R}^N)} \ge 1$. From the estimate (30), we have

$$\begin{aligned}
\|v_K\|_{L^{(m+1)p_s^*}(\mathbb{R}^N)}^{(m+1)p} &\le C_7 \|u\|_{L^{(m+1)p}(\mathbb{R}^N)}^{(m+1)p} + C_8 \|u\|_{L^{(m+1)p}(\mathbb{R}^N)}^{m+1} + C_9 \|u\|_{L^{(m+1)\beta}(\mathbb{R}^N)}^{(m+1)p} \\
&\le (C_7 + C_8) \|u\|_{L^{(m+1)p}(\mathbb{R}^N)}^{(m+1)p} + C_9 \|u\|_{L^{(m+1)\beta}(\mathbb{R}^N)}^{(m+1)p},
\end{aligned} \tag{31}$$

which implies

$$\|v_K\|_{L^{(m+1)p_s^*}(\mathbb{R}^N)} \le C_{10}^{\frac{1}{(m+1)p}} \|u\|_{L^{(m+1)t}(\mathbb{R}^N)} \tag{32}$$

for some positive constant $C_{10}$ and for any positive constant $K$, where $t$ is either $p$ or $\beta$. The expression in the estimate (32) is a starting point for a bootstrap technique. Since $u \in W(\mathbb{R}^N)$, hence $u \in L^{p_s^*}(\mathbb{R}^N)$ and we can choose $m := m_1$ in the estimate (32) such that $(m_1+1)t = p_s^*$, *i.e.*, $m_1 = \frac{p_s^*}{t} - 1$. Then, we have

$$\|v_K\|_{L^{(m_1+1)p_s^*}(\mathbb{R}^N)} \le C_{10}^{\frac{1}{(m_1+1)p}} \|u\|_{L^{(m_1+1)t}(\mathbb{R}^N)} \tag{33}$$

for any positive constant $K$. Owing to $u(x) = \lim_{K\to\infty} v_K(x)$ for almost every $x \in \mathbb{R}^N$, Fatou's lemma and the estimate (33) imply

$$\|u\|_{L^{(m_1+1)p_s^*}(\mathbb{R}^N)} \le C_{10}^{\frac{1}{(m_1+1)p}} \|u\|_{L^{(m_1+1)t}(\mathbb{R}^N)}. \tag{34}$$

Thus, we can choose $m = m_2$ in the estimate (32) such that $(m_2+1)t = (m_1+1)p_s^* = \frac{(p_s^*)^2}{t}$. By repeating the similar manner, we obtain

$$\|u\|_{L^{(m_2+1)p_s^*}(\mathbb{R}^N)} \leq C_{10}^{\frac{1}{(m_2+1)p}} \|u\|_{L^{(m_2+1)t}(\mathbb{R}^N)}.$$

By the mathematical induction, we have

$$\|u\|_{L^{(m_n+1)p_s^*}(\mathbb{R}^N)} \leq C_{10}^{\frac{1}{(m_n+1)p}} \|u\|_{L^{(m_n+1)t}(\mathbb{R}^N)} \tag{35}$$

for any $n \in \mathbb{N}$, where $m_n + 1 = \left(\frac{p_s^*}{t}\right)^n$. It follows from relations (34) and (35) that

$$\|u\|_{L^{(m_n+1)p_s^*}(\mathbb{R}^N)} \leq C_{10}^{\frac{1}{p}\sum_{j=1}^{n}\frac{1}{m_j+1}} \|u\|_{L^{p_s^*}(\mathbb{R}^N)}. \tag{36}$$

However, $\sum_{j=1}^{n}\frac{1}{m_j+1} = \sum_{j=1}^{n}\left(\frac{t}{p_s^*}\right)^j$ and $\frac{t}{p_s^*} < 1$. Hence, it follows from the estimate (36) that there exists a constant $C_{11} > 0$ such that

$$\|u\|_{L^{r_n}(\mathbb{R}^N)} \leq C_{11}\|u\|_{L^{p_s^*}(\mathbb{R}^N)} \tag{37}$$

for $r_n = (m_n+1)p_s^* \to \infty$ when $n \to \infty$. An indirect argument concludes that

$$\|u\|_{L^\infty(\mathbb{R}^N)} \leq C_{11}\|u\|_{L^{p_s^*}(\mathbb{R}^N)} \leq C_{12}$$

for some constant $C_{12} > 0$. Meanwhile, we assume that $\|u\|_{L^{(m+1)p}(\mathbb{R}^N)} < 1$. From the relation (30), we have

$$\|v_K\|^{(m+1)p}_{L^{(m+1)p_s^*}(\mathbb{R}^N)} \leq C_7 + C_8 + C_9\|u\|^{(m+1)p}_{L^{(m+1)\beta}(\mathbb{R}^N)} \leq C_{13}\|u\|^{(m+1)p}_{L^{(m+1)\beta}(\mathbb{R}^N)},$$

which implies

$$\|v_K\|_{L^{(m+1)p_s^*}(\mathbb{R}^N)} \leq C_{13}^{\frac{1}{(m+1)p}} \|u\|_{L^{(m+1)\beta}(\mathbb{R}^N)}$$

for some positive constant $C_{13}$. Repeating the iterations as in the arguments above, we derive $\|u\|_{L^\infty(\mathbb{R}^N)} \leq C_{14}$ for some positive constant $C_{14}$.

If $u$ changes sign, we set positive and negative parts as $u^+(x) = \max\{u(x), 0\}$ and $u^-(x) = \min\{u(x), 0\}$. Then, it is obvious that $u^+ \in W(\mathbb{R}^N)$ and $u^- \in W(\mathbb{R}^N)$. For each $K > 0$, we define $v_K(x) = \min\{u^+(x), K\}$. Taking again $v = v_K^{mp+1}$ as a test function in $W(\mathbb{R}^N)$, we obtain

$$\begin{aligned}\mathcal{M}(|u|^p_{W^{s,p}(\mathbb{R}^N)}) &\int_{\mathbb{R}^N}\int_{\mathbb{R}^N} \frac{|u(x)-u(y)|^{p-2}(u(x)-u(y))(v_K^{mp+1}(x) - v_K^{mp+1}(y))}{|x-y|^{N+ps}}\,dxdy \\ &+ \int_{\mathbb{R}^N} V(x)\,|u|^{p-2}\,uv_K^{mp+1}\,dx = \lambda\int_{\mathbb{R}^N} f(x,u)v_K^{mp+1}\,dx,\end{aligned}$$

which implies that

$$\begin{aligned}\mathcal{M}(|u^+|^p_{W^{s,p}(\mathbb{R}^N)}) &\int_{\mathbb{R}^N}\int_{\mathbb{R}^N} \frac{|u^+(x)-u^+(y)|^{p-2}(u^+(x)-u^+(y))(v_K^{mp+1}(x) - v_K^{mp+1}(y))}{|x-y|^{N+ps}}\,dxdy \\ &+ \int_{\mathbb{R}^N} V(x)\,|u^+|^{p-2}\,u^+v_K^{mp+1}\,dx = \lambda\int_{\mathbb{R}^N} f(x,u^+)v_K^{mp+1}\,dx.\end{aligned}$$

Proceeding with the similar way as above, we obtain $u^+ \in L^\infty(\mathbb{R}^N)$. Similarly, we obtain $u^- \in L^\infty(\mathbb{R}^N)$. Therefore, $u = u^+ + u^-$ is in $L^\infty(\mathbb{R}^N)$. The proof is complete. □

The following result can be found in [41].

**Lemma 7.** *Let $\mathcal{E} \in C^1(\mathcal{X}, \mathbb{R})$ where $\mathcal{X}$ is a Banach space. We assume that $\mathcal{E}$ satisfies the $(PS)$-condition, is even and bounded from below, and $\mathcal{E}(0) = 0$. If, for any $n \in \mathbb{N}$, there exist an $n$-dimensional subspace $\mathcal{X}_n$ and $\rho_n > 0$ such that*

$$\sup_{\mathcal{X}_n \cap S_{\rho_n}} \mathcal{E} < 0,$$

*where $S_\rho := \{u \in \mathcal{X} : \|u\|_{\mathcal{X}} = \rho\}$, then $\mathcal{E}$ possesses a sequence of critical values $c_n < 0$ satisfying $c_n \to 0$ as $n \to \infty$.*

Based on the work of [27,29], we provide the following two lemmas.

**Lemma 8.** *Assume that $(\mathcal{V}1)$, $(\mathcal{M}1)$ and $(\mathcal{F}1)$–$(\mathcal{F}2)$ hold. Furthermore, we assume that $\mathfrak{M}(t) \leq \mathcal{M}(t)t$ for any $t \geq 0$, where $\mathfrak{M}$ is given in $(\mathcal{M}2)$. Furthermore, if*

$$p\mathcal{F}(x,t) - f(x,t)t > 0 \tag{38}$$

*for all $x \in \mathbb{R}^N$ and for $t \neq 0$. Then,*

$$\mathcal{E}_\lambda(u) = 0 = \langle \mathcal{E}_\lambda'(u), u\rangle \quad \textit{if and only if} \quad u = 0.$$

**Proof.** Let $\mathcal{E}_\lambda(u) = \langle \mathcal{E}_\lambda'(u), u\rangle = 0$. Then,

$$\begin{aligned} 0 &= -\, p\mathcal{E}_\lambda(u) \\ &= -\,\mathfrak{M}(|u|^p_{W^{s,p}(\mathbb{R}^N)}) - \int_{\mathbb{R}^N} V(x)|u|^p\,dx + \lambda p \int_{\mathbb{R}^N} \mathcal{F}(x,u)\,dx, \end{aligned} \tag{39}$$

and

$$\langle \mathcal{E}_\lambda'(u), u\rangle = \mathcal{M}(|u|^p_{W^{s,p}(\mathbb{R}^N)})|u|^p_{W^{s,p}(\mathbb{R}^N)} + \int_{\mathbb{R}^N} V(x)\,|u|^p\,dx - \lambda \int_{\mathbb{R}^N} f(x,u)u\,dx = 0. \tag{40}$$

It follows from the relations (39) and (40) that

$$\int_{\mathbb{R}^N} \{p\mathcal{F}(x,u) - f(x,u)u\}\,dx \leq 0.$$

Consequently, the assumption (38) implies $u = 0$. □

**Lemma 9.** *Assume that $(\mathcal{F}1)$–$(\mathcal{F}2)$ and $(\mathcal{F}7)$–$(\mathcal{F}8)$ are fulfilled. Then, there exist $0 < t_0 < \min\{s_0, 1\}/2$ and $\tilde{f} \in C^1(\mathbb{R}^N \times \mathbb{R}, \mathbb{R})$ such that $\tilde{f}(x,t)$ is odd in $t$ and satisfies*

$$\tilde{\mathfrak{F}}(x,t) := p\tilde{\mathcal{F}}(x,t) - \tilde{f}(x,t)t \geq 0,$$

$$\tilde{\mathfrak{F}}(x,t) = 0 \quad \textit{iff} \quad t = 0 \quad \textit{or} \quad |t| \geq 2t_0,$$

*where $\frac{\partial}{\partial t}\tilde{\mathcal{F}}(x,t) = \tilde{f}(x,t)$.*

**Proof.** Let us define a cut-off function $\kappa \in C^1(\mathbb{R}, \mathbb{R})$ satisfying $\kappa(t) = 1$ for $|t| \leq t_0$, $\kappa(t) = 0$ for $|t| \geq 2t_0$, $|\kappa'(t)| \leq 2/t_0$, and $\kappa'(t)t \leq 0$. Therefore, we define

$$\tilde{\mathcal{F}}(x,t) = \kappa(t)\mathcal{F}(x,t) + (1-\kappa(t))\xi|t|^p \quad \text{and} \quad \tilde{f}(x,t) = \frac{\partial}{\partial t}\tilde{\mathcal{F}}(x,t), \tag{41}$$

where $\xi > 0$ is a constant. It is straightforward that

$$p\tilde{\mathcal{F}}(x,t) - \tilde{f}(x,t)t = \kappa(t)\mathfrak{F}(x,t) - \kappa'(t)t\mathcal{F}(x,t) + \kappa'(t)t\xi|t|^p,$$

where $\mathfrak{F}(x,t) := p\mathcal{F}(x,t) - f(x,t)t$. For $0 \le |t| \le t_0$ and $|t| \ge 2t_0$, the conclusion is as follows. Owing to $(\mathcal{F}8)$, we choose a sufficiently small $t_0 > 0$ such that $\mathcal{F}(x,t) \ge \xi t^p$ for $t_0 \le |t| \le 2t_0$. By assuming $\kappa'(t)t \le 0$, we obtain the conclusion. □

Now, with the aid of Proposition 1, and Lemmas 7 and 9, we are ready to prove the second primary result.

**Theorem 3.** *Assume that $(\mathcal{V}1)$, $(\mathcal{M}1)$, $(\mathcal{F}1)$–$(\mathcal{F}2)$, and $(\mathcal{F}7)$–$(\mathcal{F}8)$ hold. Moreover, assume that $\mathfrak{M}(t) \le \mathcal{M}(t)t$ for any $t \ge 0$ and $f(x,t)$ is odd in $t$ for a small $t$. Then, there is a positive $\lambda^*$ such that the problem (1) admits a sequence of weak solutions $\{u_n\}$ satisfying $\|u_n\|_{L^\infty(\mathbb{R}^N)} \to 0$ as $n \to \infty$ for every $\lambda \in (0,\lambda^*)$.*

**Proof.** We can modify and extend the given function $f(x,t)$ to $\tilde{f} \in C^1(\mathbb{R}^N \times \mathbb{R}, \mathbb{R})$ satisfying all properties given in Lemma 9. First, we will show that $\tilde{\mathcal{E}}_\lambda := A_{s,p} - \lambda\tilde{\Psi}$ is coercive on $W(\mathbb{R}^N)$. Let $u \in W(\mathbb{R}^N)$ and $\|u\|_{W(\mathbb{R}^N)} > 1$. By Lemma 9, it is easily shown that $\tilde{\mathcal{E}}_\lambda \in C^1(W(\mathbb{R}^N),\mathbb{R})$ and is even on $W(\mathbb{R}^N)$. Moreover, it follows from $(\mathcal{F}2)$ that, for $|u(x)| \le 2t_0$, there exists a positive constant $K_1$ such that $\rho(x)\,|u| + K_1|u|^p \ge |F(x,u)|$.

We set $\Omega_1 := \{x \in \mathbb{R}^N : |u(x)| \le t_0\}$, $\Omega_2 := \{x \in \mathbb{R}^N : t_0 \le |u(x)| \le 2t_0\}$, and $\Omega_3 := \{x \in \mathbb{R}^N : 2t_0 \le |u(x)|\}$, where $t_0$ is given in Lemma 9. From the relation (41) and the conditions of $\kappa$, we have

$$\begin{aligned}
\tilde{\mathcal{E}}_\lambda(u) &:= \frac{1}{p}(\mathfrak{M}(|u|^p_{W^{s,p}(\mathbb{R}^N)}) + \|u\|^p_{p,V}) - \lambda\int_{\mathbb{R}^N}\tilde{\mathcal{F}}(x,u)\,dx \\
&\ge \frac{\min\{1,m_0/\theta\}}{p}\|u\|^p_{W(\mathbb{R}^N)} - \lambda\int_{\Omega_1}\mathcal{F}(x,u)\,dx - \lambda\int_{\Omega_2}\{\kappa(u)\mathcal{F}(x,u) + (1-\kappa(u))\xi|u|^p\}\,dx - \lambda\int_{\Omega_3}\xi|u|^p\,dx \\
&\ge \frac{\min\{1,m_0/\theta\}}{p}\|u\|^p_{W(\mathbb{R}^N)} - \lambda\int_{\Omega_1\cup\Omega_2}\mathcal{F}(x,u)\,dx - \lambda\int_{\Omega_2\cup\Omega_3}\xi|u|^p\,dx \\
&\ge \frac{\min\{1,m_0/\theta\}}{p}\|u\|^p_{W(\mathbb{R}^N)} - \lambda\int_{\Omega_1\cup\Omega_2}\rho(x)|u|\,dx - \lambda\int_{\Omega_1\cup\Omega_2}K_1|u|^p\,dx - \lambda\int_{\Omega_2\cup\Omega_3}\xi|u|^p\,dx \\
&\ge \frac{\min\{1,m_0/\theta\}}{p}\|u\|^p_{W(\mathbb{R}^N)} - 2\lambda\|\rho\|_{L^{p'}(\mathbb{R}^N)}\|u\|_{L^p(\mathbb{R}^N)} - \lambda\,(K_1+\xi)\int_{\mathbb{R}^N}|u|^p\,dx \\
&\ge \frac{\min\{1,m_0/\theta\}}{p}\|u\|^p_{W(\mathbb{R}^N)} - \lambda\left(2C_{15}\|\rho\|_{L^{p'}(\mathbb{R}^N)} + K_1 + \xi\right)\|u\|^p_{W(\mathbb{R}^N)}
\end{aligned}$$

for some positive constant $C_{15}$. If we set

$$\lambda^* := \frac{1}{p(2C_{15}\|\rho\|_{L^{p'}(\mathbb{R}^N)} + K_1 + \xi)},$$

then we deduce that for any $\lambda \in (0,\lambda^*)$, $\tilde{\mathcal{E}}_\lambda$ is coercive, that is, $\tilde{\mathcal{E}}_\lambda(u) \to \infty$ as $\|u\|_{W(\mathbb{R}^N)} \to \infty$.

Next, we claim that the functional $\tilde{\Psi}' : W(\mathbb{R}^N) \to W^*(\mathbb{R}^N)$, defined by

$$\langle\tilde{\Psi}'(u),\varphi\rangle = \int_{\mathbb{R}^N}\tilde{f}(x,u)\varphi\,dx \quad \text{for any} \quad \varphi \in W(\mathbb{R}^N),$$

is compact in $W(\mathbb{R}^N)$. Let us assume that $u_n \rightharpoonup u$ in $W(\mathbb{R}^N)$ as $n \to \infty$. Since the measures of $\Omega_2$ and $\Omega_3$ are finite, we can write $\Omega_2 = \tilde{\Omega}_2 \cup N_2$ and $\Omega_3 = \tilde{\Omega}_3 \cup N_3$, where $\tilde{\Omega}_2$ and $\tilde{\Omega}_3$ are bounded sets and $N_2, N_3$ are of measure zero. Let us denote $B_R(0) := \{x \in \mathbb{R}^N : |x| \le R\}$ contained in the bounded sets $\tilde{\Omega}_2$ and $\tilde{\Omega}_3$ for a sufficiently large $R \in \mathbb{N}$. Then, from the definition of $\tilde{f}(x,u)$, we have $\tilde{f}(x,u) = f(x,u)$ on $\mathbb{R}^N \setminus (\Omega_2 \cup \Omega_3)$. Thus, we deduce that for any $\varphi \in W(\mathbb{R}^N)$

$$\sup_{\|\varphi\|_{W(\mathbb{R}^N)}\le 1}\left|\langle\tilde{\Psi}'(u_n) - \tilde{\Psi}'(u),\varphi\rangle\right| = \sup_{\|\varphi\|_{W(\mathbb{R}^N)}\le 1}\left|\int_{\mathbb{R}^N}(\tilde{f}(x,u_n) - \tilde{f}(x,u))\varphi\,dx\right|$$

$$\leq \sup_{\|\varphi\|_{W(\mathbb{R}^N)}\leq 1} \left| \int_{B_R(0)} (\tilde{f}(x,u_n) - \tilde{f}(x,u))\varphi \, dx \right|$$
$$+ \sup_{\|\varphi\|_{W(\mathbb{R}^N)}\leq 1} \left| \int_{\mathbb{R}^N \setminus (B_R(0)\cup N_4 \cup N_5)} (f(x,u_n) - f(x,u))\varphi \, dx \right|. \tag{42}$$

Owing to Lemma 1, the compact embedding

$$W(\mathbb{R}^N) \hookrightarrow L^p(B_R(0)) \quad \text{implies} \quad u_n \to u \quad \text{in} \quad L^p(B_R(0)) \quad \text{as} \quad n \to \infty.$$

The above, together with the continuity of the Nemytskij operator with $\tilde{f}$ and acting from $L^p(B_R(0))$ into $L^{q'}(B_R(0))$, it is clearly shown that the first term on the right side of the inequality (42) tends to 0 as $n \to \infty$. For the second term in the inequality (42), we have

$$\left| \int_{\mathbb{R}^N \setminus (B_R(0)\cup N_2 \cup N_3)} (f(x,u_n) - f(x,u))\varphi dx \right|$$
$$\leq \int_{\mathbb{R}^N \setminus (B_R(0)\cup N_2 \cup N_3)} \sigma(x)(|u_n(x)|^{q-1} + |u(x)|^{q-1}) \, |\varphi| \, dx$$
$$\leq \|\sigma\|_{L^{\frac{p_s^*}{p_s^*-q}}(\mathbb{R}^N \setminus (B_R(0)\cup N_2 \cup N_3))} \left( \|u_n\|^{q-1}_{L^{p_s^*}(\mathbb{R}^N)} + \|u\|^{q-1}_{L^{p_s^*}(\mathbb{R}^N)} \right) \|\varphi\|_{L^{p_s^*}(\mathbb{R}^N)}.$$

From the assumption $(\mathcal{F}2)$, for $\varepsilon > 0$, there exists $N(R) \in \mathbb{R}$ such that

$$\|\sigma\|_{L^{\frac{p_s^*}{p_s^*-q}}(\mathbb{R}^N \setminus (B_R(0)\cup N_2 \cup N_3))} < \varepsilon$$

for $R > N(R)$. As the sequence $\{u_n\}$ is bounded in $W(\mathbb{R}^N)$, according to Lemma 1, one has $\{u_n\}$ bounded in $L^{p_s^*}(\mathbb{R}^N)$. Thus,

$$\left| \int_{\mathbb{R}^N \setminus (B_R(0)\cup N_2 \cup N_3)} (f(x,u_n) - f(x,u))\varphi \, dx \right| \leq C_{16}\varepsilon \tag{43}$$

for a positive constant $C_{16}$. Owing to the estimate (43), we can deduce that

$$\int_{\mathbb{R}^N} (f(x,u_n) - f(x,u))\varphi \, dx \to 0 \quad \text{as} \quad n \to \infty.$$

This implies that $\tilde{\Psi}'$ is compact in $W(\mathbb{R}^N)$, as claimed.

Since the derivative of $\tilde{\Psi}$ is compact, it follows from the coercivity of $\tilde{\mathcal{E}}_\lambda$ that the functional $\tilde{\mathcal{E}}_\lambda$ satisfies the $(PS)$-condition. The weak lower semicontinuity and the coercivity of $\tilde{\mathcal{E}}_\lambda$ ensure that $\tilde{\mathcal{E}}_\lambda$ is bounded from below. To utilize Lemma 7, we only need to obtain for any $n \in \mathbb{N}$, a subspace $X_n$ and $\rho_n > 0$ such that $\sup_{X_n \cap S_{\rho_n}} \tilde{\mathcal{E}}_\lambda < 0$. For any $n \in \mathbb{N}$, we obtain $n$ independent smooth functions $\phi_i$ for $i = 1, \cdots, n$, and define $X_n := \text{span}\{\phi_1, ..., \phi_n\}$. Owing to Lemma 9, when $\|u\|_{W(\mathbb{R}^N)} < 1$, we have

$$\tilde{\mathcal{E}}_\lambda(u) = \frac{1}{p}(\mathfrak{M}(|u|^p_{W^{s,p}(\mathbb{R}^N)}) + \|u\|^p_{p,V}) - \lambda \int_{\mathbb{R}^N} \tilde{\mathcal{F}}(x,u) \, dx$$
$$\leq \frac{1}{p}\|u\|^p_{W(\mathbb{R}^N)} - \lambda C_{17} \int_{\mathbb{R}^N} \mathcal{F}(x,u) \, dx,$$

for $C_{17} > 0$. Taking the assumption $(\mathcal{F}8)$ into account, it follows that there exists $\delta_0 > 0$ such that $|t| < \delta_0$, which implies

$$\int_{\mathbb{R}^N} \mathcal{F}(x,t) \, dx \geq \frac{K_2}{p} \int_{\mathbb{R}^N} |t|^p \, dx \tag{44}$$

for a sufficiently large $K_2 > 0$. Using the inequality (44) and the fact that all norms on $X_n$ are equivalent, we can choose a appropriate constant $C_{17}$ and a small enough $\rho_n > 0$ to obtain

$$\sup_{X_n \cap S_{\rho_n}} \tilde{\mathcal{E}}_\lambda < 0.$$

According to Lemma 7, we obtain a sequence $c_n < 0$ for $\tilde{\mathcal{E}}_\lambda$ satisfying $c_n \to 0$ when $n$ goes to $\infty$. Then, for any $u_n \in W(\mathbb{R}^N)$ satisfying $\tilde{\mathcal{E}}_\lambda(u_n) = c_n$ and $\tilde{\mathcal{E}}'_\lambda(u_n) = 0$, $\{u_n\}$ is a $(PS)$-sequence of $\tilde{\mathcal{E}}_\lambda(u)$, and $\{u_n\}$ has a convergent subsequence. From Lemmas 8 and 9, we deduce that 0 is the only critical point with 0 energy, and the subsequence of $\{u_n\}$ has to converge to 0. Using an indirect argument, we show that $\{u_n\}$ has to converge to 0. Meanwhile, we obtain $u_n \in L^r(\mathbb{R}^N)$ for all $p_s^* \le r \le \infty$ owing to Proposition 1. Since $\|u_n\|_{L^\infty(\mathbb{R}^N)} \to 0$, by Lemma 9 again, we have $\|u_n\|_{L^\infty(\mathbb{R}^N)} \le t_0$ for a large $n$. Thus, $\{u_n\}$ is a sequence of weak solutions of problem (1). This completes the proof. □

## 4. Conclusions

In summary, this paper is devoted to the study of weak solutions for Kirchhoff–Schrödinger-type equations involving the fractional $p$-Laplacian. In the first part of the present paper, under various assumptions on $\mathcal{M}$ and $f$, we show that our problem admits a sequence of the weak solutions whose energy functional converges to infinity. As we know, a typical example for Kirchhoff function $\mathcal{M}$ is $\mathcal{M}(t) = b_0 + b_1 t^n$ ($n > 0, b_0 > 0, b_1 \ge 0$) and, based on this example, most results for the multiplicity of solutions are presented. From a different point of view, an infinite number of solutions is proved when $\mathcal{M}$ contains new conditions different from those studied in previous related works; see Example 1. The second part is to investigate the existence of small energy solutions for the given problem whose $L^\infty$-norms converge to zero. As mentioned in the Introduction, the main difficulty is to show the $L^\infty$-bound of weak solutions. Our approach is new to the fractional $p$-Laplacian problems even if we utilize the well known Moser bootstrap iteration method to overcome this. To the best of our knowledge, such results have not been studied much in these situations.

**Author Contributions:** All the authors contributed equally to the writing of this paper. Y.-H.K. drafted the manuscript; J.L. and J.-M.K. improved the final version. All authors read and approved the final manuscript.

**Funding:** Jae-Myoung Kim's work is supported by NRF-2015R1A5A1009350 and the National Research Foundation of Korea grant funded by the Korean Government (NRF-2016R1D1A1B03930422). Yun-Ho Kim's work is supported by the Basic Science Research Program through the National Research Foundation of Korea (NRF) funded by the Ministry of Education (NRF-2016R1D1A1B03935866). Jongrak Lee's work is supported by Brain Korea 21 plus Mathematical Science Team for Global Women Leaders at Ewha Womans University and the Basic Science Research Program through the National Research Foundation of Korea (NRF) funded by the Ministry of Education (NRF-2018R1D1A1B07048620).

**Acknowledgments:** The authors gratefully thank the Referee for the constructive comments and recommendations, which definitely help to improve the readability and quality of the paper.

**Conflicts of Interest:** The author declares no conflict of interest.

## References

1. Bertoin, J. *Levy Processes*; Cambridge University Press: Cambridge, UK, 1996.
2. Caffarelli, L. Non-local diffusions, drifts and games. In *Nonlinear Partial Differential Equations*; Springer: Heidelberg, Germany, 2012; pp. 37–52.
3. Chang, X.; Wang, Z.-Q. Nodal and multiple solutions of nonlinear problems involving the fractional Laplacian. *J. Differ. Equ.* **2014**, *256*, 2965–2992. [CrossRef]
4. Gilboa, G.; Osher, S. Nonlocal operators with applications to image processing. *Multiscale Model. Simul.* **2008**, *7*, 1005–1028. [CrossRef]
5. Metzler, R.; Klafter, J. The restaurant at the random walk: Recent developments in the description of anomalous transport by fractional dynamics. *J. Phys. A* **2004**, *37*, 161–208. [CrossRef]
6. Kirchhoff, G. *Mechanik*; Teubner: Leipzig, Germany, 1883.

7. Chen, W. Multiplicity of solutions for a fractional Kirchhoff type problem. *Commun. Pure Appl. Anal.* **2015**, *14*, 2009–2020. [CrossRef]
8. Fiscella, A. Infinitely many solutions for a critical Kirchhoff type problem involving a fractional operator. *Differ. Integral Equ.* **2016**, *29*, 513–530.
9. Lee, J.; Kim, Y.-H.; Kim, J.-M. Existence and multiplicity of solutions for Kirchhoff-Schorödinger type equations involving $p(x)$-Laplacian on the entire space $\mathbb{R}^N$. *Nonlinear Anal. Real World Appl.* **2019**, *45*, 620–649. [CrossRef]
10. Mingqi, X.; Molica, B.; Tian, G.; Zhang, B. Infinitely many solutions for the stationary Kirchhoff problems involving the fractional $p$ -Laplacian. *Nonlinearity* **2016**, *29*, 357–374. [CrossRef]
11. Pucci, P.; Xiang, M.; Zhang, B. Multiple solutions for nonhomogeneous Schrödinger—Kirchhoff type equations involving the fractional $p$-Laplacian in $\mathbb{R}^N$. *Calc. Var. Part. Differ. Equ.* **2015**, *54*, 2785–2806. [CrossRef]
12. Pucci, P.; Xiang, M.; Zhang, B. Existence and multiplicity of entire solutions for fractional $p$-Kirchhoff equations. *Adv. Nonlinear Anal.* **2016**, *5*, 27–55. [CrossRef]
13. Xiang, M.; Zhang, B.; Ferrara, M. Multiplicity results for the non-homogeneous fractional $p$-Kirchhoff equations with concave-convex nonlinearities. *Proc. R. Soc. A* **2015**, *471*, 1–14. [CrossRef]
14. Xiang, M.; Zhang, B.; Guo, X. Infinitely many solutions for a fractional Kirchhoff type problem via fountain theorem. *Nonlinear Anal.* **2015**, *120*, 299–313. [CrossRef]
15. Ambrosetti, A.; Rabinowitz, P. Dual variational methods in critical point theory and applications. *J. Funct. Anal.* **1973**, *14*, 349–381. [CrossRef]
16. Servadei, R. Infinitely many solutions for fractional Laplace equations with subcritical nonlinearity. *Contemp. Math.* **2013**, *595*, 317–340.
17. Lin, X.; Tang, X.H. Existence of infinitely many solutions for $p$-Laplacian equations in $\mathbb{R}^N$. *Nonlinear Anal.* **2013**, *92*, 72–81. [CrossRef]
18. Liu, S.B. On ground states of superlinear $p$-Laplacian equations in $\mathbb{R}^N$. *J. Math. Anal. Appl.* **2010**, *361*, 48–58. [CrossRef]
19. Liu, S.B.; Li, S.J. Infinitely many solutions for a superlinear elliptic equation. *Acta Math. Sinica (Chin. Ser.)* **2003**, *46*, 625–630. (In Chinese)
20. Miyagaki, O.H.; Souto, M.A.S. Superlinear problems without Ambrosetti and Rabinowitz growth condition. *J. Differ. Equ.* **2008**, *245*, 3628–3638. [CrossRef]
21. Binlin, Z.; Bisci, G.M.; Servadei, R. Superlinear nonlocal fractional problems with infinitely many solutions. *Nonlinearity* **2015**, *28*, 2247–2264. [CrossRef]
22. Ge, B. Multiple solutions of nonlinear Schrödinger equation with the fractional Laplacian. *Nonlinear Anal. Real World Appl.* **2016**, *30*, 236–247. [CrossRef]
23. Gou, T.-X.; Sun, H.-R. Solutions of nonlinear Schrödinger equation with fractional Laplacian without the Ambrosetti-Rabinowitz condition. *Appl. Math. Comput.* **2015**, *257*, 409–416. [CrossRef]
24. Torres, C. Existence and symmetry result for fractional $p$-Laplacian in $\mathbb{R}^n$. *Commun. Pure Appl. Anal.* **2017**, *16*, 99–113.
25. Zhang, B.; Ferrara, M. Multiplicity of solutions for a class of superlinear non-local fractional equations. *Complex Var. Elliptic Equ.* **2015**, *60*, 583–595. [CrossRef]
26. Alves, C.O.; Liu, S.B. On superlinear $p(x)$-Laplacian equations in $\mathbb{R}^N$. *Nonlinear Anal.* **2010**, *73*, 2566–2579. [CrossRef]
27. Wang, Z.-Q. Nonlinear boundary value problems with concave nonlinearities near the origin. *NoDEA Nonlinear Differ. Equ. Appl.* **2001**, *8*, 15–33. [CrossRef]
28. Guo, Z. Elliptic equations with indefinite concave nonlinearities near the origin. *J. Math. Anal. Appl.* **2010**, *367*, 273–277. [CrossRef]
29. Choi, E.B.; Kim, J.-M.; Kim, Y.-H. Infinitely many solutions for equations of $p(x)$-Laplace type with the nonlinear Neumann boundary condition. *Proc. Roy. Soc. Edinb.* **2018**, *148*, 1–31. [CrossRef]
30. Jing, Y.; Liu, Z. Infinitely many solutions of $p$-sublinear $p$-Laplacian equations. *J. Math. Anal. Appl.* **2015**, *429*, 1240–1257. [CrossRef]
31. Naimen, D. Existence of infinitely many solutions for nonlinear Neumann problems with indefinite coefficients. *Electron. J. Differ. Equ.* **2014**, *2014*, 1–12.

32. Kim, J.-M.; Kim, Y.-H.; Lee, J. Existence and multiplicity of solutions for equations of $p(x)$-Laplace type in $\mathbb{R}^N$ without AR-condition. *Differ. Integral Equ.* **2018**, *31*, 435–464.
33. Teng, K. Multiple solutions for a class of fractional Schrödinger equations in $\mathbb{R}^N$. *Nonlinear Anal. Real World Appl.* **2015**, *21*, 76–86. [CrossRef]
34. Kim, Y.-H. Infinitely many small energy solutions for equations involving the fractional Laplacian in $\mathbb{R}^N$. *J. Korean Math. Soc.* **2018**, *55*, 1269–1283.
35. Dràbek, P.; Kufner, A.; Nicolosi, F. *Quasilinear Elliptic Equations with Degenerations and Singularities*; Walter de Gruyter & Co.: Berlin, Germany, 1997.
36. Adams, R.A.; Fournier, J.J.F. *Sobolev Spaces*, 2nd ed.; Academic Press: New York, NY, USA; London, UK, 2003.
37. Iannizzotto, A.; Liu, S.; Perera, K.; Squassina, M. Existence results for fractional $p$-Laplacian problems via Morse theory. *Adv. Calc. Var.* **2016**, *9*, 101–125. [CrossRef]
38. Di Nezza, E.; Palatucci, G.; Valdinoci, E. Hitchhiker's guide to the fractional Sobolev spaces. *Bull. Sci. Math.* **2012**, *136*, 521–573. [CrossRef]
39. Willem, M. *Minimax Theorems*; Birkhauser: Basel, Switzerland, 1996.
40. Jeanjean, L. On the existence of bounded Palais-Smale sequences and application to a Landsman-Lazer type problem set on $\mathbb{R}^N$. *Proc. Roy. Soc. Edinb.* **1999**, *129*, 787–809. [CrossRef]
41. Heinz, H.P. Free Ljusternik-Schnirelman theory and the bifurcation diagrams of certain singular nonlinear problems. *J. Differ. Equ.* **1987**, *66*, 263–300. [CrossRef]

symmetry

MDPI

Article

# Laplace Adomian Decomposition Method for Multi Dimensional Time Fractional Model of Navier-Stokes Equation

**Shahid Mahmood [1], Rasool Shah [2], Hassan khan [2,*] and Muhammad Arif [2]**

[1] Department of Mechanical Engineering, Sarhad University of Science and Information Technology, Peshawar, Pakistan; shahidmahmood757@gmail.com
[2] Department of Mathematics, Abdul Wali khan University, Mardan, Pakistan; shahrasool26@gmail.com (R.S.); marifmaths@awkum.edu.pk (M.A.)
* Correspondence: hassanmath@awkum.edu.pk

Received: 8 January 2019; Accepted: 23 January 2019; Published: 29 January 2019

**Abstract:** In this research paper, a hybrid method called Laplace Adomian Decomposition Method (LADM) is used for the analytical solution of the system of time fractional Navier-Stokes equation. The solution of this system can be obtained with the help of Maple software, which provide LADM algorithm for the given problem. Moreover, the results of the proposed method are compared with the exact solution of the problems, which has confirmed, that as the terms of the series increases the approximate solutions are convergent to the exact solution of each problem. The accuracy of the method is examined with help of some examples. The LADM, results have shown that, the proposed method has higher rate of convergence as compare to ADM and HPM.

**Keywords:** Laplace Adomian Decomposition Method (LADM); Navier-Stokes equation; Caputo Operator

## 1. Introduction

In engineering and natural sciences many problems are modeled by linear and non linear parabolic and hyperbolic partial differential equations. For these classical partial differential equations LADM can be used effectively with initial as well as boundary conditions. The present method was initially used by Suheil-A-khuri for the solution of ordinary differential equations [1]. It is slightly difficult to find the exact solutions of non linear differential equations, due to which the combination of two powerful methods, laplace transform and Adomian Decomposition Method called LADM has been used to find the exact solutions of non linear differential equations. The analytical solution of the well known non linear fractional diffusion and wave equations by using LADM are presented in [2,3].

Adomian Decomposition Method (ADM) was first introduced by Gorge Adomian in 1980. It was used very effectively on a wide range of physical models of partial differential equations, such as Burger's equation is a non linear PDE of second order, which have many applications in sciences and technology. The numerical solutions of three dimensional Burger's equation and Riccati differential equations by using LADM have been discussed in [4,5]. LADM is also used for the numerical solution of a special mathematical model for vector born diseases [6]. Delay differential equation have a vital role in the field of biology and economics has been solved by LADM [7,8]. Nonlinear Volterra integral and integro-differential equation solving for Modification LADM [9].

Fractional calculus is a branch of mathematical analysis which can be used in modeling to define derivatives and integrations of fractional order. The fractional calculus is considered an old topic, which is started from some observations of G.W. Leibniz (1695, 1697), and L. Euler (1730). After this, fractional calculus has gained much interest of the researchers towards this subject. This including the contributions of well known mathematicians such as P.S. Laplace (1812), J.B.J. Fourier (1822),

N.H. Abel (1823–1826), J. Liouville (1832–1873). Although it is considered an old topic, but for the last few decade, fractional calculus is launched as an important topic by the scientists and researchers [10,11].

The Navier-Stokes equation is known as Newton second Law for fluid substance, has been derived in 1822 by Claude Louis Navier and Gabriel Stokes. Navier-Stokes equation is an important model to describe many physical phenomena in applied sciences. This model have the capacity of modelling weather, ocean current, water flow in pipes and air flow around a wing. A very special case was considered, which has established the relationship between pressure and external forces acting on the fluid to the responses of fluid flow [12]. The Navier-Stock equation is also used to derive the connection between viscous fluid with rigid bodies and considered a best tool in the field of thermo-hydraulics, meteorology, petroleum industry, plasma physics and technology [13].

Several mathematicians have applied different techniques for the solution of Navier-Stock equation. Among these methods, Kumar et al. have implemented modified Laplace decomposition method for the analytical solution of fractional Navier-Stokes equation [14] coupled method is the combination of He-Laplace transform (HLT) and Fractional Complex Transform (FCT) is used to solve Navier-Stock equation [15]. Fractional Reduced Differential Transformation Method (FRDM) is also implemented for the numerical solution of time fractional Navier-Stock equation [16], see also [17].

## 2. Definitions and Preliminaries Concepts

In this unit, among few definitions of fractional calculus, presented in the article due to Riemann Liouville, Grunwald Letnikov, Caputo, etc., first folks simple descriptions and introductions are reconsidered, which we want to comprehend our education.

**Definition 1.** *The fractional integral of Riemann Liouville $f \in \mathbb{C}_n$ of the direction $\beta \geq 0$ is defined by*

$$I_x^{\beta} g(x) = \begin{cases} g(x) & \text{if } \beta = 0 \\ \frac{1}{\Gamma(\beta)} \int_0^x (x-v)^{\beta-1} g(v) dv & \text{if } \beta > 0, \end{cases}$$

*where $\Gamma$ denote the gamma function define by,*

$$\Gamma(\omega) = \int_0^{\infty} e^{-x} x^{\tau-1} dx \qquad \omega \in \mathbb{C},$$

*In this study, Caputo et al. [18] suggested a revise fractional derivative operator in order to overcome inconsistency measured in Riemann Liouville derivative [19,20]. The above mathematical statement described Caputo fractional derivative operator of initial and boundary condition for fractional as well as integer order derivative.*

**Definition 2.** *The Caputo definition of fractional derivative of order $\beta$ is given by the following mathematical expression*

$$D_x^{\beta} g(x) = \frac{1}{\Gamma(n-\beta)} \int_0^x (x-t)^{n-\beta-1} g^{(n)}(t) dt.$$

*for $n-1 < \beta \leq n$, $n \in \mathbb{N}$, $x > 0$, $g \in \mathbb{C}_t$,$t \geq -1$.*

*Hence, we require the subsequent properties given in next Lemma.*

**Lemma 1.** *If $n-1<\beta\leq n$ with $n\in\mathbb{N}$ and $g\in\mathbb{C}_x$ with $x\geq -1$, then*

$$D_x^{\beta} I_x^{\beta} g(x) = g(x),$$
$$I^{\beta} x^{\lambda} = \frac{\Gamma(\lambda+1)}{\Gamma(\beta+\lambda+1)} x^{\beta+\lambda}, \qquad \beta>0, \lambda>-1, \quad x>0,$$
$$D_x^{\beta} I_x^{\beta} g(x) = g(x) - \sum_{k=0}^{n} g^{(k)}(0^{+})\frac{x^k}{k!}, \qquad for \quad x>0.$$

In this study, Caputo fractional derivative operator is reasonable because other fractional derivative operators have certain disadvantages. Further information about fractional derivatives, are found in [20].

**Definition 3.** *The Laplace transform of $g(x)$, $x>0$ is defined by*

$$G(s) = \mathcal{L}[g(x)] = \int_0^{\infty} e^{-sx} g(x)dx,$$

*where s can be either real or complex.*

**Definition 4.** *The Laplace transform in term of convolution is given by*

$$\mathcal{L}[g_1 \times g_2] = \mathcal{L}[g_1(x)] \times \mathcal{L}[g_2(x)],$$

*where $g_1 \times g_2$, define the convolution between $g_1$ and $g_2$ ,*

$$(g_1 \times g_2)x = \int_0^{x} g_1(t) g_2(x-t)dx.$$

*The Laplace transform of fractional derivative is given by*

$$\mathcal{L}\left[D_x^{\beta} g(x)\right] = s^{\beta}G(s) - \sum_{k=0}^{n-1} s^{\beta-1-k} g^{(k)}(0), \qquad n-1<\beta<n,$$

*where $G(s)$ is the Laplace transform of $g(x)$.*

**Definition 5.** *The Mittag-Leffler function $E_{\beta}(p)$ for $\beta>0$ is defined by the following subsequent series*

$$E_{\beta}(p) = \sum_{n=0}^{\infty} \frac{p^n}{\Gamma(\beta n+1)}, \qquad \beta>0, \quad p\in\mathbb{C}.$$

## 3. Laplace Adomian Decomposition Method

In this unit, we present, Laplace Adomian decomposition method for solving, multi dimensional Naiver-Stokes equation written in an operator form

$$\begin{aligned}
&D_t^{\beta}(f_1) + f_1\frac{\partial f_1}{\partial x_1} + f_2\frac{\partial f_1}{\partial x_2} + f_3\frac{\partial f_1}{\partial x_3} = \rho\left[\frac{\partial^2 f_1}{\partial x_1^2} + \frac{\partial^2 f_1}{\partial x_2^2} + \frac{\partial^2 f_1}{\partial x_3^2}\right] \\
&-\frac{1}{\rho}\frac{\partial p}{\partial x_1}, \\
&D_t^{\beta}(f_2) + f_1\frac{\partial f_2}{\partial x_1} + f_2\frac{\partial f_2}{\partial x_2} + f_3\frac{\partial f_2}{\partial x_3} = \rho\left[\frac{\partial^2 f_2}{\partial x_1^2} + \frac{\partial^2 f_2}{\partial x_2^2} + \frac{\partial^2 f_2}{\partial x_3^2}\right] \\
&-\frac{1}{\rho}\frac{\partial p}{\partial x_2}, \\
&D_t^{\beta}(f_3) + f_1\frac{\partial f_3}{\partial x_1} + f_2\frac{\partial f_3}{\partial x_2} + f_3\frac{\partial f_3}{\partial x_3} = \rho\left[\frac{\partial^2 f_3}{\partial x_1^2} + \frac{\partial^2 f_3}{\partial x_2^2} + \frac{\partial^2 f_3}{\partial x_3^2}\right] \\
&-\frac{1}{\rho}\frac{\partial p}{\partial x_3},
\end{aligned} \tag{1}$$

with initial conditions

$$\begin{cases} f_1(x_1, x_2, x_3, 0) = f(x_1, x_2, x_3), \\ f_2(x_1, x_2, x_3, 0) = h(x_1, x_2, x_3), \\ f_3(x_1, x_2, x_3, 0) = g(x_1, x_2, x_3). \end{cases} \tag{2}$$

Applying the Laplace transform to (1), we have

$$\begin{aligned}
&\mathcal{L}\left[D_t^{\beta}(f_1)\right] + \mathcal{L}\left[f_1\frac{\partial f_1}{\partial x_1} + f_2\frac{\partial f_1}{\partial x_2} + f_3\frac{\partial f_1}{\partial x_3}\right] = \mathcal{L}\rho\left[\frac{\partial^2 f_1}{\partial x_1^2} + \frac{\partial^2 f_1}{\partial x_2^2} + \frac{\partial^2 f_1}{\partial x_3^2}\right] \\
&-\mathcal{L}\left[\frac{1}{\rho}\frac{\partial p}{\partial x_1}\right], \\
&\mathcal{L}\left[D_t^{\beta}(f_2)\right] + \mathcal{L}\left[f_1\frac{\partial f_2}{\partial x_1} + f_2\frac{\partial f_2}{\partial x_2} + f_3\frac{\partial f_2}{\partial x_3}\right] = \mathcal{L}\rho\left[\frac{\partial^2 f_2}{\partial x_1^2} + \frac{\partial^2 f_2}{\partial x_2^2} + \frac{\partial^2 f_2}{\partial x_3^2}\right] \\
&-\mathcal{L}\left[\frac{1}{\rho}\frac{\partial p}{\partial x_2}\right], \\
&\mathcal{L}\left[D_t^{\beta}(f_3)\right] + \mathcal{L}\left[f_1\frac{\partial f_3}{\partial x_1} + f_2\frac{\partial f_3}{\partial x_2} + f_3\frac{\partial f_3}{\partial x_3}\right] = \mathcal{L}\rho\left[\frac{\partial^2 f_3}{\partial x_1^2} + \frac{\partial^2 f_3}{\partial x_2^2} + \frac{\partial^2 f_3}{\partial x_3^2}\right] \\
&-\mathcal{L}\left[\frac{1}{\rho}\frac{\partial p}{\partial x_3}\right],
\end{aligned} \tag{3}$$

and using the differentiation property of Laplace transform, we get

$$
\begin{aligned}
\mathcal{L}(f_1) &= \frac{f(x_1,x_2,x_3)}{s} - \frac{1}{s^{\beta}}\mathcal{L}\left[f_1\frac{\partial f_1}{\partial x_1} + f_2\frac{\partial f_1}{\partial x_2} + f_3\frac{\partial f_1}{\partial x_3}\right] \\
&+ \frac{\rho}{s^{\beta}}\mathcal{L}\left[\frac{\partial^2 f_1}{\partial x_1^2} + \frac{\partial^2 f_1}{\partial x_2^2} + \frac{\partial^2 f_1}{\partial x_3^2}\right] - \frac{1}{s^{\beta}}\mathcal{L}\left[\frac{1}{\rho}\frac{\partial p}{\partial x_1}\right], \\
\mathcal{L}(f_2) &= \frac{h(x_1,x_2,x_3)}{s} - \frac{1}{s^{\beta}}\mathcal{L}\left[f_1\frac{\partial f_2}{\partial x_1} + f_2\frac{\partial f_2}{\partial x_2} + f_3\frac{\partial f_2}{\partial x_3}\right] \\
&+ \frac{\rho}{s^{\beta}}\mathcal{L}\left[\frac{\partial^2 f_2}{\partial x_1^2} + \frac{\partial^2 f_2}{\partial x_2^2} + \frac{\partial^2 f_2}{\partial x_3^2}\right] - \frac{1}{s^{\beta}}\mathcal{L}\left[\frac{1}{\rho}\frac{\partial p}{\partial x_2}\right], \\
\mathcal{L}(f_3) &= \frac{g(x_1,x_2,x_3)}{s} - \frac{1}{s^{\beta}}\mathcal{L}\left[f_1\frac{\partial f_3}{\partial x_1} + f_2\frac{\partial f_3}{\partial x_2} + f_3\frac{\partial f_3}{\partial x_3}\right] \\
&+ \frac{\rho}{s^{\beta}}\mathcal{L}\left[\frac{\partial^2 f_3}{\partial x_1^2} + \frac{\partial^2 f_3}{\partial x_2^2} + \frac{\partial^2 f_3}{\partial x_3^2}\right] - \frac{1}{s^{\beta}}\mathcal{L}\left[\frac{1}{\rho}\frac{\partial p}{\partial x_3}\right],
\end{aligned}
\tag{4}
$$

Adomian solutions are

$$
\begin{cases}
f_1(x_1,x_2,x_3,t) = \sum_{j=0}^{\infty} u_j, \\
f_2(x_1,x_2,x_3,t) = \sum_{j=0}^{\infty} v_j, \\
f_3(x_1,x_2,x_3,t) = \sum_{j=0}^{\infty} w_j,
\end{cases}
\tag{5}
$$

and the nonlinear terms are define by the infinite series of Adomian polynomials,

$$
\begin{cases}
N_1(f_1) = \sum_{j=0}^{\infty} A_j, \\
N_2(f_2) = \sum_{j=0}^{\infty} B_j, \\
N_3(f_3) = \sum_{j=0}^{\infty} C_j.
\end{cases}
\tag{6}
$$

$$
\begin{aligned}
A_j &= \frac{1}{j!}\left[\frac{d^j}{d\lambda^j}\left[N_1\sum_{i=0}^{\infty}(\lambda^j u_j)\right]\right]_{\lambda=0}, \\
B_j &= \frac{1}{j!}\left[\frac{d^j}{d\lambda^j}\left[N_2\sum_{i=0}^{\infty}(\lambda^j v_j)\right]\right]_{\lambda=0}, \\
C_j &= \frac{1}{j!}\left[\frac{d^j}{d\lambda^j}\left[N_3\sum_{i=0}^{\infty}(\lambda^j w_j)\right]\right]_{\lambda=0}.
\end{aligned}
\tag{7}
$$

using LADM solutions in equation (4), we get

$$
\begin{aligned}
\mathcal{L}\left(\sum_{j=0}^{\infty} u_{j+1}\right) &= \frac{f(x_1,x_2,x_3)}{s} - \frac{1}{s^{\beta}}\mathcal{L}\left(\frac{1}{\rho}\frac{\partial p}{\partial x_1}\right) \\
&- \frac{1}{s^{\beta}}\mathcal{L}\left[\left(\sum_{j=0}^{\infty} f_{1_j}\right)\frac{\partial(\sum_{j=0}^{\infty} f_{1_j})}{\partial x_1} + \left(\sum_{j=0}^{\infty} f_{2_j}\right)\frac{\partial(\sum_{j=0}^{\infty} f_{1_j})}{\partial x_2} + \left(\sum_{j=0}^{\infty} f_{3_j}\right)\frac{\partial\left(\sum_{j=0}^{\infty} f_{1_j}\right)}{\partial x_3}\right] \\
&+ \frac{\rho}{s^{\beta}}\mathcal{L}\left[\frac{\partial^2(\sum_{j=0}^{\infty} f_{1_j})}{\partial x_1^2} + \frac{\partial^2(\sum_{j=0}^{\infty} f_{1_j})}{\partial x_2^2} + \frac{\partial^2(\sum_{j=0}^{\infty} f_{1_j})}{\partial x_3^2}\right], \\
\mathcal{L}\left(\sum_{j=0}^{\infty} v_{j+1}\right) &= \frac{h(x_1,x_2,x_3)}{s} - \frac{1}{s^{\beta}}\mathcal{L}\left(\frac{1}{\rho}\frac{\partial p}{\partial x_1}\right) \\
&- \frac{1}{s^{\beta}}\mathcal{L}\left[\left(\sum_{j=0}^{\infty} f_{1_j}\right)\frac{\partial(\sum_{j=0}^{\infty} f_{2_j})}{\partial x_1} + \left(\sum_{j=0}^{\infty} f_{2_j}\right)\frac{\partial(\sum_{j=0}^{\infty} f_{2_j})}{\partial x_2} + \left(\sum_{j=0}^{\infty} f_{3_j}\right)\frac{\partial\left(\sum_{j=0}^{\infty} f_{2_j}\right)}{\partial x_3}\right] \\
&+ \frac{\rho}{s^{\beta}}\mathcal{L}\left[\frac{\partial^2(\sum_{j=0}^{\infty} f_{2_j})}{\partial x_1^2} + \frac{\partial^2(\sum_{j=0}^{\infty} f_{2_j})}{\partial x_2^2} + \frac{\partial^2(\sum_{j=0}^{\infty} f_{2_j})}{\partial x_3^2}\right], \\
\mathcal{L}\left(\sum_{j=0}^{\infty} w_{j+1}\right) &= \frac{g(x_1,x_2,x_3)}{s} - \frac{1}{s^{\beta}}\mathcal{L}\left(\frac{1}{\rho}\frac{\partial p}{\partial x_1}\right) \\
&- \frac{1}{s^{\beta}}\mathcal{L}\left[\left(\sum_{j=0}^{\infty} f_{1_j}\right)\frac{\partial(\sum_{j=0}^{\infty} f_{3_j})}{\partial x_1} + \left(\sum_{j=0}^{\infty} f_{2_j}\right)\frac{\partial(\sum_{j=0}^{\infty} f_{3_j})}{\partial x_2} + \left(\sum_{j=0}^{\infty} f_{3_j}\right)\frac{\partial\left(\sum_{j=0}^{\infty} f_{3_j}\right)}{\partial x_3}\right] \\
&+ \frac{\rho}{s^{\beta}}\mathcal{L}\left[\frac{\partial^2(\sum_{j=0}^{\infty} f_{3_j})}{\partial x_1^2} + \frac{\partial^2(\sum_{j=0}^{\infty} f_{3_j})}{\partial x_2^2} + \frac{\partial^2(\sum_{j=0}^{\infty} f_{3_j})}{\partial x_3^2}\right].
\end{aligned}
\tag{8}
$$

Applying the linearity of the Laplace transform,

$$
\begin{cases}
\mathcal{L}(u_o) = \frac{f(x_1,x_2,x_3)}{s} + \frac{1}{s^{\beta}}\mathcal{L}\left(\frac{1}{\rho}\frac{\partial p}{\partial x_1}\right), \\
\mathcal{L}(v_o) = \frac{h(x_1,x_2,x_3)}{s} + \frac{1}{s^{\beta}}\mathcal{L}\left(\frac{1}{\rho}\frac{\partial p}{\partial x_2}\right), \\
\mathcal{L}(w_o) = \frac{g(x_1,x_2,x_3)}{s} + \frac{1}{s^{\beta}}\mathcal{L}\left(\frac{1}{\rho}\frac{\partial p}{\partial x_3}\right).
\end{cases}
\tag{9}
$$

$$\mathcal{L}\left(\sum_{j=0}^{\infty} u_{j+1}\right) = -\frac{1}{s^{\beta}}\mathcal{L}[\left(\sum_{j=0}^{\infty} f_{1_j}\right)\frac{\partial(\sum_{j=0}^{\infty} f_{1_j})}{\partial x_1} + \left(\sum_{j=0}^{\infty} f_{2_j}\right)\frac{\partial(\sum_{j=0}^{\infty} f_{1_j})}{\partial x_2}$$
$$+\left(\sum_{j=0}^{\infty} f_{3_j}\right)\frac{\partial\left(\sum_{j=0}^{\infty} f_{1_j}\right)}{\partial x_3}] + \frac{\rho}{s^{\beta}}\mathcal{L}\left[\frac{\partial^2(\sum_{j=0}^{\infty} f_{1_j})}{\partial x_1^2} + \frac{\partial^2(\sum_{j=0}^{\infty} f_{1_j})}{\partial x_2^2} + \frac{\partial^2(\sum_{j=0}^{\infty} f_{1_j})}{\partial x_3^2}\right],$$
$$\mathcal{L}\left(\sum_{j=0}^{\infty} v_{j+1}\right) = -\frac{1}{s^{\beta}}\mathcal{L}[\left(\sum_{j=0}^{\infty} f_{1_j}\right)\frac{\partial(\sum_{j=0}^{\infty} f_{2_j})}{\partial x_1} + \left(\sum_{j=0}^{\infty} f_{2_j}\right)\frac{\partial(\sum_{j=0}^{\infty} f_{2_j})}{\partial x_2}$$
$$+\left(\sum_{j=0}^{\infty} f_{3_j}\right)\frac{\partial\left(\sum_{j=0}^{\infty} f_{2_j}\right)}{\partial x_3}] + \frac{\rho}{s^{\beta}}\mathcal{L}\left[\frac{\partial^2(\sum_{j=0}^{\infty} f_{2_j})}{\partial x_1^2} + \frac{\partial^2(\sum_{j=0}^{\infty} f_{2_j})}{\partial x_2^2} + \frac{\partial^2(\sum_{j=0}^{\infty} f_{2_j})}{\partial x_3^2}\right], \tag{10}$$
$$\mathcal{L}\left(\sum_{j=0}^{\infty} w_{j+1}\right) = -\frac{1}{s^{\beta}}\mathcal{L}[\left(\sum_{j=0}^{\infty} f_{1_j}\right)\frac{\partial(\sum_{j=0}^{\infty} f_{3_j})}{\partial x_1} + \left(\sum_{j=0}^{\infty} f_{2_j}\right)\frac{\partial(\sum_{j=0}^{\infty} f_{3_j})}{\partial x_2}$$
$$+\left(\sum_{j=0}^{\infty} f_{3_j}\right)\frac{\partial\left(\sum_{j=0}^{\infty} f_{3_j}\right)}{\partial x_3}] + \frac{\rho}{s^{\beta}}\mathcal{L}\left[\frac{\partial^2(\sum_{j=0}^{\infty} f_{3_j})}{\partial x_1^2} + \frac{\partial^2(\sum_{j=0}^{\infty} f_{3_j})}{\partial x_2^2} + \frac{\partial^2(\sum_{j=0}^{\infty} f_{3_j})}{\partial x_3^2}\right].$$

For $j = 0$, and $j = 1, 2.......\infty$.

$$\begin{aligned}
\mathcal{L}(u_1) &= -\frac{1}{s^{\beta}}\mathcal{L}\left[u_0\frac{\partial(u_0)}{\partial x_1} + v_0\frac{\partial(v_0)}{\partial x_2} + w_0\frac{\partial(u_0)}{\partial x_3}\right] + \frac{\rho}{s^{\beta}}\mathcal{L}\left[\frac{\partial^2(u_0)}{\partial x_1^2} + \frac{\partial^2(u_0)}{\partial x_2^2} + \frac{\partial^2(u_0)}{\partial x_3^2}\right],\\
\mathcal{L}(v_1) &= -\frac{1}{s^{\beta}}\mathcal{L}\left[u_0\frac{\partial(v_0)}{\partial x_1} + v_0\frac{\partial(v_0)}{\partial x_2} + w_0\frac{\partial(v_0)}{\partial x_3}\right] + \frac{\rho}{s^{\beta}}\mathcal{L}\left[\frac{\partial^2(v_0)}{\partial x_1^2} + \frac{\partial^2(v_0)}{\partial x_2^2} + \frac{\partial^2(v_0)}{\partial x_3^2}\right],\\
\mathcal{L}(w_1) &= -\frac{1}{s^{\beta}}\mathcal{L}\left[u_0\frac{\partial(w_0)}{\partial x_1} + v_0\frac{\partial(v_0)}{\partial x_2} + w_0\frac{\partial(w_0)}{\partial x_3}\right] + \frac{\rho}{s^{\beta}}\mathcal{L}\left[\frac{\partial^2(w_0)}{\partial x_1^2} + \frac{\partial^2(w_0)}{\partial x_2^2} + \frac{\partial^2(w_0)}{\partial x_3^2}\right].
\end{aligned} \tag{11}$$

Next applying the inverse Laplace transform, we can calculate $u_j$ , $v_j$ and $w_j$ $(j > 0)$. In specific cases the exact result in the closed form can also be achieve.

**Example 1.** *Consider time-fractional order of two-dimensional Navier-Stock equation with* $q_1 = -q_2 = q$ *as,*

$$\begin{aligned}
D_t^{\beta}(f_1) + f_1\frac{\partial f_1}{\partial x_1} + f_2\frac{\partial f_1}{\partial x_2} &= \rho\left[\frac{\partial^2 f_1}{\partial x_1^2} + \frac{\partial^2 f_1}{\partial x_2^2}\right] + q,\\
D_t^{\beta}(f_2) + f_1\frac{\partial f_2}{\partial x_1} + f_2\frac{\partial f_2}{\partial x_2} &= \rho\left[\frac{\partial^2 f_2}{\partial x_1^2} + \frac{\partial^2 f_2}{\partial x_2^2}\right] - q,
\end{aligned} \tag{12}$$

*with initial conditions*

$$\begin{cases} f_1(x_1, x_2, 0) = -\sin(x_1 + x_2), \\ f_2(x_1, x_2, 0) = \sin(x_1 + x_2). \end{cases} \tag{13}$$

*Applying the Laplace transform to (12), we have*

$$\begin{aligned}
&\mathcal{L}\left(\sum_{j=0}^{\infty} u_{j+1}\right) = \frac{f_{1j}}{s} - \frac{1}{s^{\beta}}\mathcal{L}\left[\left(\sum_{j=0}^{\infty} f_{1j}\right)\frac{\partial(\sum_{j=0}^{\infty} f_{1j})}{\partial x_1} + \left(\sum_{j=0}^{\infty} f_{1j}\right)\frac{\partial(\sum_{j=0}^{\infty} f_{1j})}{\partial x_2}\right] \\
&+ \frac{\rho}{s^{\beta}}\mathcal{L}\left[\frac{\partial^2(\sum_{j=0}^{\infty} f_{1j})}{\partial x_1^2} + \frac{\partial^2(\sum_{j=0}^{\infty} f_{1j})}{\partial x_2^2}\right], \\
&\mathcal{L}\left(\sum_{j=0}^{\infty} v_{j+1}\right) = \frac{f_{2j}}{s} - \frac{1}{s^{\beta}}\mathcal{L}\left[\left(\sum_{j=0}^{\infty} f_{1j}\right)\frac{\partial(\sum_{j=0}^{\infty} f_{2j})}{\partial x_1} + \left(\sum_{j=0}^{\infty} f_{1j}\right)\frac{\partial(\sum_{j=0}^{\infty} f_{2j})}{\partial x_2}\right] \\
&+ \frac{\rho}{s^{\beta}}\mathcal{L}\left[\frac{\partial^2(\sum_{j=0}^{\infty} f_{2j})}{\partial x_1^2} + \frac{\partial^2(\sum_{j=0}^{\infty} f_{2j})}{\partial x_2^2}\right].
\end{aligned} \tag{14}$$

$$\begin{aligned}
u_o &= \mathcal{L}^{-1}\left[\frac{-\sin(x_1+x_2)}{s}\right] = -\sin(x_1+x_2), \\
v_o &= \mathcal{L}^{-1}\left[\frac{\sin(x_1+x_2)}{s}\right] = \sin(x_1+x_2),
\end{aligned} \tag{15}$$

$$\begin{aligned}
\mathcal{L}(u_1) &= -\frac{1}{s^{\beta}}\mathcal{L}\left[-\sin(x_1+x_2)\frac{\partial(-\sin(x_1+x_2))}{\partial x_1} + \sin(x_1+x_2)\frac{\partial(\sin(x_1+x_2))}{\partial x_2}\right] \\
&+ \frac{1}{s^{\beta}}\mathcal{L}\rho\left[\frac{\partial^2(-\sin(x_1+x_2))}{\partial x^2} + \frac{\partial^2(-\sin(x_1+x_2))}{\partial x_2^2}\right] + \frac{1}{s^{\beta}}\mathcal{L}(q), \\
\mathcal{L}(v_1) &= -\frac{1}{s^{\beta}}\mathcal{L}\left[-\sin(x_1+x_2)\frac{\partial(\sin(x_1+x_2))}{\partial x_1} + \sin(x_1+x_2)\frac{\partial(\sin(x_1+x_2))}{\partial x_2}\right] \\
&+ \frac{1}{s^{\beta}}\mathcal{L}\rho\left[\frac{\partial^2(\sin(x_1+x_2))}{\partial x^2} + \frac{\partial^2(\sin(x_1+x_2))}{\partial x_2^2}\right] - \frac{1}{s^{\beta}}\mathcal{L}(q),
\end{aligned} \tag{16}$$

$$\begin{cases}
u_1 = \mathcal{L}^{-1}\left[\frac{2\rho\sin(x_1+x_2)}{s^{\beta+1}} + \frac{q}{s^{\beta+1}}\right], \\
v_1 = \mathcal{L}^{-1}\left[\frac{-2\rho\sin(x_1+x_2)}{s^{\beta+1}} - \frac{q}{s^{\beta+1}}\right],
\end{cases} \tag{17}$$

$$\begin{aligned}
u_1 &= 2\rho\sin(x_1+x_2)\frac{t^{\beta}}{\Gamma(\beta+1)} + \frac{q}{\Gamma(\beta+1)}, \\
v_1 &= -2\rho\sin(x_1+x_2)\frac{t^{\beta}}{\Gamma(\beta+1)} + \frac{q}{\Gamma(\beta+1)},
\end{aligned}$$

$$\begin{cases}
u_2 = -4\rho^2\sin(x_1+x_2)\frac{t^{2\beta}}{\Gamma(2\beta+1)}, \\
v_2 = 4\rho^2\sin(x_1+x_2)\frac{t^{2\beta}}{\Gamma(2\beta+1)}.
\end{cases} \tag{18}$$

*The LADM solution for example (1) is*

$$\begin{aligned}
f_1(x_1,x_2,t) &= u_0(x_1,x_2,t) + u_1(x_1,x_2,t) + u_2(x_1,x_2,t) + u_3(x_1,x_2,t) + \ldots u_n(x_1,x_2,t), \\
f_2(x_1,x_2,t) &= v_0(x_1,x_2,t) + v_1(x_1,x_2,t) + v_2(x_1,x_2,t) + v_3(x_1,x_2,t) + \ldots v_n(x_1,x_2,t),
\end{aligned}$$

$$
\begin{aligned}
&f_1(x_1,x_2,t) = -\sin(x_1+x_2) + 2\rho\sin(x_1+x_2)\frac{t^\beta}{\Gamma(\beta+1)} + \frac{q}{\Gamma(\beta+1)}\\
&-4\rho^2\sin(x_1+x_2)\frac{t^{2\beta}}{\Gamma(2\beta+1)} + \ldots\\
&f_2(x_1,x_2,t) = \sin(x_1+x_2) - 2\rho\sin(x_1+x_2)\frac{t^\beta}{\Gamma(\beta+1)} + \frac{q}{\Gamma(\beta+1)}\\
&+4\rho^2\sin(x_1+x_2)\frac{t^{2\beta}}{\Gamma(2\beta+1)} + \ldots
\end{aligned}
\tag{19}
$$

*when $\beta = 1$, then LADM solution is*

$$
\begin{aligned}
f_1(x_1,x_2,t) &= -e^{2\rho t}(\sin(x_1+x_2)),\\
f_2(x_1,x_2,t) &= e^{2\rho t}(\sin(x_1+x_2)).
\end{aligned}
$$

*For $q = 0$ gave the exact result of classical Navier-Stokes equation for the velocity. The velocity profile of the ordinary Naiver-Stokes equation is shown in Figures, and the velocity profile of Naiver-Stokes equation with $\beta = 1, 0.5$ and $0.8$ is shown in Figures 1–3.*

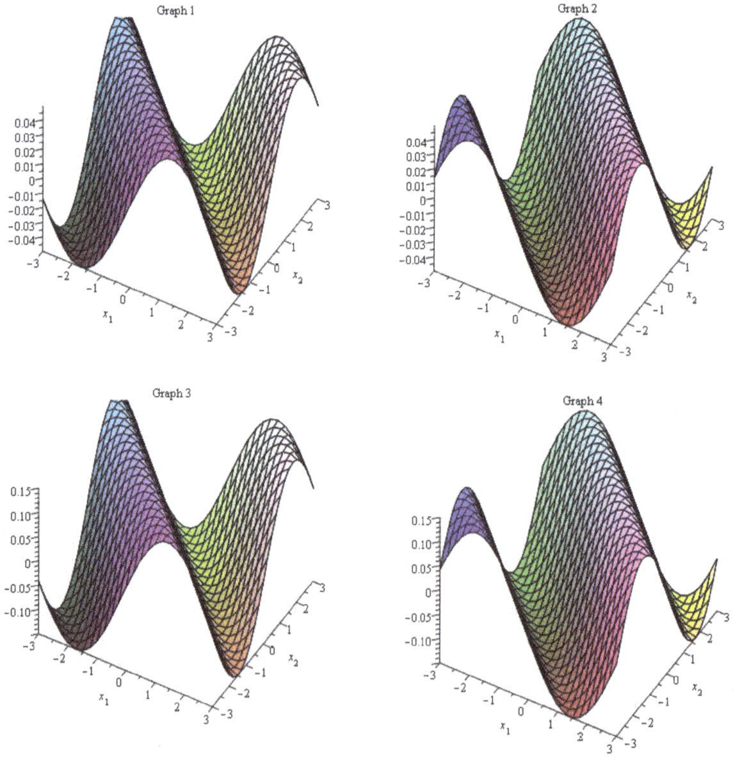


**Figure 1.** For example 1, the velocity profiles $f_1, f_2$ of NS equation at $\beta = 0.8$, $q = 0$, $\rho = 0.5, t = 3$.

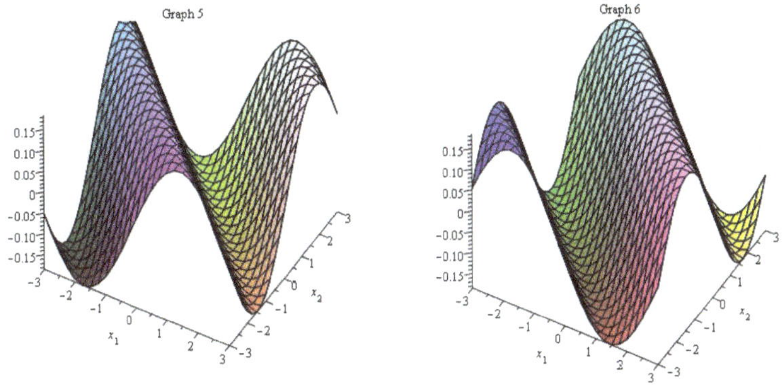


**Figure 2.** For example 1, the velocity profiles $f_1, f_2$ of NS equation at $\beta = 0.5$, $q = 0$, $\rho = 0.5$, $t = 3$.

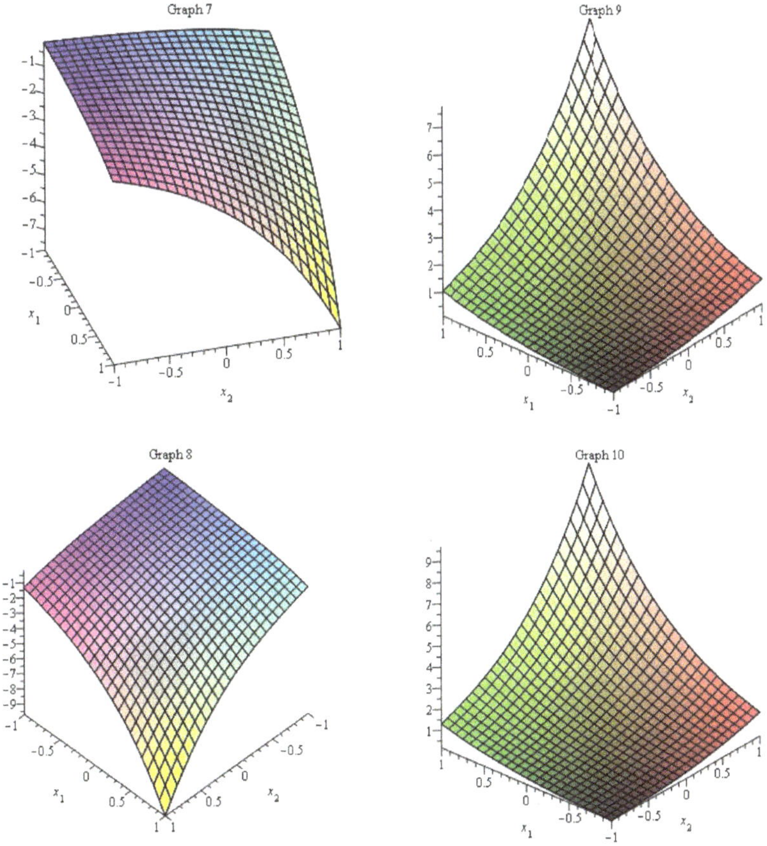


**Figure 3.** For example 2, the velocity profiles $f_1, f_2$ of NS equation at $\beta = 0.5$, $q = 0$, $\rho = 0.5$, $t = 0.05$.

**Example 2.** *The study of time fractional of order two dimensional Naiver-Stokes Equation (12) with initial conditions*

$$\begin{cases} u(x,y,0) = -e^{x_1+x_2}, \\ v(x,y,0) = e^{x_1+x_2}. \end{cases} \tag{20}$$

*Taking Laplace transform of (12)*

$$\begin{cases} \mathcal{L}(u_o) = \frac{-e^{x_1+x_2}}{s}, \\ \mathcal{L}(v_o) = \frac{e^{x_1+x_2}}{s}, \end{cases} \tag{21}$$

$$\begin{aligned} \mathcal{L}(u_1) &= -\frac{1}{s^\beta}\mathcal{L}\left[-e^{x_1+x_2}\frac{\partial(-e^{x_1+x_2})}{\partial x_1} + -e^{x_1+x_2}\frac{\partial(e^{x_1+x_2})}{\partial x_2}\right] \\ &+ \frac{\rho}{s^\beta}\mathcal{L}\left[\frac{\partial^2(-e^{x_1+x_2})}{\partial x_1^2} + \frac{\partial^2(-e^{x_1+x_2})}{\partial x_2^2}\right] + \frac{1}{s^\beta}\mathcal{L}(q), \\ \mathcal{L}(v_1) &= -\frac{1}{s^\beta}\mathcal{L}\left[-e^{x_1+x_2}\frac{\partial(e^{x_1+x_2})}{\partial x_1} + e^{x_1+x_2}\frac{\partial(e^{x_1+x_2})}{\partial x_2}\right] \\ &+ \frac{\rho}{s^\beta}\mathcal{L}\left[\frac{\partial^2(e^{x_1+x_2})}{\partial x_1^2} + \frac{\partial^2(e^{x_1+x_2})}{\partial x_2^2}\right] - \frac{1}{s^\beta}\mathcal{L}(q), \end{aligned} \tag{22}$$

$$\mathcal{L}(u_1) = \left[\frac{-2\rho e^{x_1+x_2}}{s^{\beta+1}}\right] + \frac{q}{s^{\beta+1}}, \qquad \mathcal{L}(v_1) = \left[\frac{2\rho e^{x_1+x_2}}{s^{\beta+1}}\right] - \frac{q}{s^{\beta+1}}, \tag{23}$$

$$\mathcal{L}(u_2) = \left[\frac{-4\rho^2 e^{x_1+x_2}}{s^{2\beta+2}}\right], \qquad \mathcal{L}(v_2) = \left[\frac{4\rho^2 e^{x_1+x_2}}{s^{2\beta+2}}\right]. \tag{24}$$

*Applying the inverse Laplace transform,*

$$\begin{aligned} u_o &= \mathcal{L}^{-1}\left[\frac{-e^{x_1+x_2}}{s}\right] = -e^{x_1+x_2}, \\ v_o &= \mathcal{L}^{-1}\left[\frac{e^{x_1+x_2}}{s}\right] = e^{x_1+x_2}, \\ u_1 &= \mathcal{L}^{-1}\left[\frac{-2\rho e^{x_1+x_2}}{s^{\beta+1}}\right] + \mathcal{L}^{-1}\left[\frac{q}{s^{\beta+1}}\right] = -2\rho e^{x_1+x_2}\frac{t^\beta}{\Gamma(\beta+1)} + \frac{q}{\Gamma(\beta+1)} \\ v_1 &= \mathcal{L}^{-1}\left[\frac{2\rho e^{x_1+x_2}}{s^{\beta+1}}\right] - \mathcal{L}^{-1}\left[\frac{q}{s^{\beta+1}}\right] = 2\rho e^{x_1+x_2}\frac{t^\beta}{\Gamma(\beta+1)} + \frac{q}{\Gamma(\beta+1)} \\ u_2 &= \mathcal{L}^{-1}\left[\frac{-4\rho^2 e^{x_1+x_2}}{s^{2\beta+2}}\right] = -(2\rho)^2 e^{x_1+x_2}\frac{t^{2\beta}}{\Gamma(2\beta+1)}, \\ v_2 &= \mathcal{L}^{-1}\left[\frac{4\rho^2 e^{x_1+x_2}}{s^{2\beta+2}}\right] = (2\rho)^2 e^{x_1+x_2}\frac{t^{2\beta}}{\Gamma(2\beta+1)}, \end{aligned}$$

*The LADM solution for example (2) is*

$$\begin{aligned} u(x_1,x_2,t) &= u_0(x_1,x_2,t) + u_1(x_1,x_2,t) + u_2(x_1,x_2,t) + u_3(x_1,x_2,t) + \ldots u_n(x_1,x_2,t), \\ v(x_1,x_2,t) &= v_0(x_1,x_2,t) + v_1(x_1,x_2,t) + v_2(x_1,x_2,t) + v_3(x_1,x_2,t) + \ldots v_n(x_1,x_2,t), \end{aligned}$$

$$f_1(x_1,x_2,t) = -e^{x_1+x_2} - 2\rho e^{x_1+x_2}\frac{t^\beta}{\Gamma(\beta+1)} + \frac{q}{\Gamma(\beta+1)}$$
$$-(2\rho)^2 e^{x_1+x_2}\frac{t^{2\beta}}{\Gamma(2\beta+1)} + \ldots$$
$$f_2(x_1,x_2,t) = e^{x_1+x_2} + 2\rho e^{x_1+x_2}\frac{t^\beta}{\Gamma(\beta+1)} - \frac{q}{\Gamma(\beta+1)} \quad (25)$$
$$+(2\rho)^2 e^{x_1+x_2}\frac{t^{2\beta}}{\Gamma(2\beta+1)} + \ldots$$

*when $\beta = 1$, then LADM solution is*

$$f_1(x_1,x_2,t) = -e^{x_1+x_2+2\rho t},$$
$$f_2(x_1,x_2,t) = e^{x_1+x_2+2\rho t}.$$

*The exact result of usual Navier-Stokes problem for the velocity profile. The activities of velocity profile of the Navier-Stokes problem is shown for $\beta = 1$ and 0.5 in Figure 4 correspondingly.*

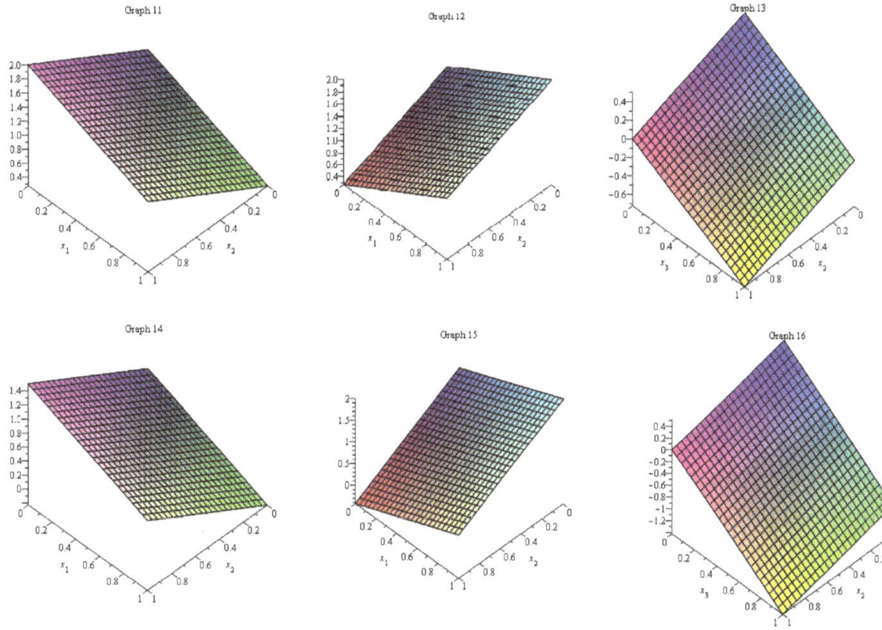


**Figure 4.** For example 3, the velocity profiles $f_1, f_2, f_3$ of NS equation at $\beta$ = 0.5, $x_3$ = 0.5, $t$ = 0.1.

**Example 3.** *The study time fractional order three dimensional Navier-Stokes Equation (3.1) by $q_1 = q_2 = q_3$ = 0, with initial conditions*

$$\begin{cases} u(x_1,x_2,x_3,0) = -0.5x_1 + x_2 + x_3, \\ v(x_1,x_2,x_3,0) = x_1 - 0.5x_2 + x_3, \\ w(x_1,x_2,x_3,0) = x_1 + x_2 - 0.5x_3. \end{cases} \quad (26)$$

*Taking Laplace transform of (1),*

$$\begin{aligned}\mathcal{L}(u_o) &= \frac{-0.5x_1 + x_2 + x_3}{s},\\ \mathcal{L}(v_o) &= \frac{x_1 - 0.5x_2 + x_3}{s},\\ \mathcal{L}(w_o) &= \frac{x_1 + x_2 - 0.5x_3}{s},\end{aligned} \tag{27}$$

$$\begin{aligned}\mathcal{L}(u_1) &= \frac{-2.25x_1}{s^{\beta+1}},\\ \mathcal{L}(v_1) &= \frac{-2.25x_2}{s^{\beta+1}},\\ \mathcal{L}(w_1) &= \frac{-2.25x_3}{s^{\beta+1}},\end{aligned} \tag{28}$$

$$\begin{aligned}\mathcal{L}(u_2) &= \frac{-(2.25)^2x_1}{s^{3\beta+3}},\\ \mathcal{L}(v_2) &= \frac{-(2.25)^2x_2}{s^{3\beta+3}},\\ \mathcal{L}(w_2) &= \frac{-(2.25)^2x_3}{s^{3\beta+3}}.\end{aligned} \tag{29}$$

*Applying the inverse Laplace transform,*

$$\begin{aligned}u_o &= \mathcal{L}^{-1}\left[\frac{-0.5x_1 + x_2 + x_3}{s}\right] = -0.5x_1 + x_2 + x_3,\\ v_o &= \mathcal{L}^{-1}\left[\frac{x_1 - 0.5x_2 + x_3}{s}\right] = x_1 - 0.5x_2 + x_3,\\ w_o &= \mathcal{L}^{-1}\left[\frac{x_1 + x_2 - 0.5x_3}{s}\right] = x_1 + x_2 - 0.5x_3,\end{aligned}$$

$$\begin{aligned}u_1 &= \mathcal{L}^{-1}\left[\frac{-2.25x_1}{s^{\beta+1}}\right] = -2.25x_1\frac{t^{\beta}}{\Gamma(\beta+1)},\\ v_1 &= \mathcal{L}^{-1}\left[\frac{-2.25x_2}{s^{\beta+1}}\right] = -2.25x_2\frac{t^{\beta}}{\Gamma(\beta+1)},\\ w_1 &= \mathcal{L}^{-1}\left[\frac{-2.25x_3}{s^{\beta+1}}\right] = -2.25x_3\frac{t^{\beta}}{\Gamma(\beta+1)},\\ u_2 &= \mathcal{L}^{-1}\left[\frac{-(2.25)^2x_1}{s^{3\beta+3}}\right] = -(2.25)^2x_1\frac{t^{3\beta}}{\Gamma(3\beta+1)},\\ v_2 &= \mathcal{L}^{-1}\left[\frac{-(2.25)^2x_2}{s^{3\beta+3}}\right] = -(2.25)^2x_2\frac{t^{3\beta}}{\Gamma(3\beta+1)},\\ w_2 &= \mathcal{L}^{-1}\left[\frac{-(2.25)^2x_3}{s^{3\beta+3}}\right] = -(2.25)^2x_3\frac{t^{3\beta}}{\Gamma(3\beta+1)},\end{aligned}$$

*The LADM solution for example (3) is*

$$\begin{aligned}f_1(x_1,x_2,x_3,t) &= u_0(x_1,x_2,x_3,0) + u_1(x_1,x_2,x_3,0) + u_2(x_1,x_2,x_3,0) + u_3(x_1,x_2,x_3,0)\\ &+ ...u_j(x_1,x_2,x_3,0),\\ f_2(x_1,x_2,x_3,t) &= v_0(x_1,x_2,x_3,0) + v_1(x_1,x_2,x_3,0) + v_2(x_1,x_2,x_3,0) + v_3(x_1,x_2,x_3,0)\\ &+ ...v_j(x_1,x_2,x_3,0),\\ f_3(x_1,x_2,x_3,t) &= w_0(x_1,x_2,x_3,0) + w_1(x_1,x_2,x_3,0) + w_2(x_1,x_2,x_3,0) + w_3(x_1,x_2,x_3,0)\\ &+ ...w_j(x_1,x_2,x_3,0),\end{aligned}$$

$$f_1(x_1,x_2,x_3,t) = -0.5x_1 + x_2 + x_3 - 2.25x_1\frac{t^{\beta}}{\Gamma(\beta+1)} - (2.25)^2x_1\frac{t^{3\beta}}{\Gamma(3\beta+1)} - (2.25)^4x_1\frac{t^{7\beta}}{\Gamma(7\beta+1)}$$
$$- (2.25)^6x_1\frac{t^{15\beta}}{\Gamma(15\beta+1)} + ...$$
$$f_2(x_1,x_2,x_3,t) = x_1 - 0.5x_2 + x_3 - 2.25x_2\frac{t^{\beta}}{\Gamma(\beta+1)} - (2.25)^2x_2\frac{t^{3\beta}}{\Gamma(3\beta+1)} - (2.25)^4x_2\frac{t^{7\beta}}{\Gamma(7\beta+1)}$$
$$- (2.25)^6x_2\frac{t^{15\beta}}{\Gamma(15\beta+1)} + ...$$
$$f_3(x_1,x_2,x_3,t) = x_1 + x_2 - 0.5x_3 - 2.25x_3\frac{t^{\beta}}{\Gamma(\beta+1)} - (2.25)^2x_3\frac{t^{3\beta}}{\Gamma(3\beta+1)} - (2.25)^4x_3\frac{t^{7\beta}}{\Gamma(7\beta+1)}$$
$$- (2.25)^6x_3\frac{t^{15\beta}}{\Gamma(15\beta+1)} + ...$$

*when $\beta = 1$, then LADM solution is*

$$f_1(x_1,x_2,x_3,t) = \frac{-0.5x_1 + x_2 + x_3 - 2.25x_1t}{1-2.25t^2},$$
$$f_2(x_1,x_2,x_3,t) = \frac{x_1 - 0.5x_2 + x_3 - 2.25x_2t}{1-2.25t^2},$$
$$f_3(x_1,x_2,x_3,t) = \frac{x_1 + x_2 - 0.5x_3 - 2.25x_3t}{1-2.25t^2}.$$

## 4. Description of Figures

Figure 1 is consists of two graphs namely Graph 1 and Graph 2. Graph 1 and Graph 2 represents the velocity profile $f_1$ and $f_2$ of the Navier-Stokes equation respectively in example 3.1 at $\beta = 1$.

Figure 2 is consists of two graphs namely Graph 3 and Graph 4. Graph 3 and Graph 4 represents the velocity profile $f_1$ and $f_2$ of the Navier-Stokes equation respectively in example 3.1 at $\beta = 0.8$.

Figure 3 is consists of two graphs namely Graph 5 and Graph 6. Graph 5 and Graph 6 represents the velocity profile $f_2$ and $f_2$ of the Navier-Stokes equation respectively in example 3.1 at $\beta = 0.5$.

Similarly in example 3.2, the plot of two velocity profiles $f_1$ and $f_2$ for the Navier-Stoke equation are represented by Graph 7 and Graph 9 at $\beta = 1$ and Graph 8 and Graph 10 at $\beta = 0.5$ respectively.

Also, in example 3.3, the plot of three velocity profiles $f_1$ , $f_2$ and $f_3$ for the Navier-Stoke equation are represented by Graph 11, Graph 12 and Graph 13 at $\beta = 1$ and Graph 14, Graph 15 and Graph 16 at $\beta = 0.5$ respectively.

## 5. Conclusions

In this paper, Laplace Adomian decomposition technique is assumed for the time-fractional classical Navier-Stokes solution of with given initial conditions. The analytical solution is given in for the power series for the given problem. The solution of the above three problems has shown, that the rate of convergence of the present method is overlapping or high than ADM and HAM. Moreover LADM have minimum calculations, simplifications as compared to ADM [12] and HPM [6].

**Author Contributions:** The authors have equally contributed to accomplish this research work.

**Funding:** Sarhad University of Science and Information Technology, Peshawar, Pakistan.

**Conflicts of Interest:** The authors have no conflict of interest.

## References


1. Khuri, S. A Laplace decomposition algorithm applied to a class of nonlinear differential equations. *J. Appl. Math.* **2001**, *1*, 141–155. [CrossRef]
2. Ahmed, H.; Bahgat, M.; Zaki, M. Numerical approaches to system of fractional partial differential equations. *J. Egypt. Math. Soc.* **2017**, *25*, 141–150. [CrossRef]
3. Jafari, H.; Khalique, C.; Nazari, M. Application of the Laplace decomposition method for solving linear and nonlinear fractional diffusion wave equations. *Appl. Math. Lett.* **2011**, *24*, 1799–1805. [CrossRef]
4. Ahmad Alhendi, F.; Alderremy, A. Numerical Solutions of Three-Dimensional Coupled Burgers' Equations by Using Some Numerical Methods. *J. Appl. Math. Phys.* **2016**, *4*, 2011–2030. [CrossRef]
5. Mishra, V.; Rani, D. Newton-Raphson based modified Laplace Adomian decomposition method for solving quadratic Riccati differential equations. In Proceedings of the 4th International Conference on Advancements in Engineering & Technology, Punjab, India, 18–19 March 2016; Volume 57.
6. Haq, F.; Shah, K.; Khan, A.; Shahzad, M.; Rahman, G. Numerical Solution of Fractional Order Epidemic Model of a Vector Born Disease by Laplace Adomian Decomposition Method. *Punjab Univ. J. Math.* **2017**, *49*, 13–22.
7. Qasim, A.F.; AL-Rawi, E.S. Adomian Decomposition Method with Modified Bernstein Polynomials for Solving Ordinary and Partial Differential Equations. *J. Appl. Math.* **2018**. [CrossRef]
8. Yousef, H.M.; Ismail, A.M. Application of the Laplace Adomian decomposition method for solution system of delay differential equations with initial value problem. In *AIP Conference Proceedings*; AIP Publishing: Melville, NY, USA, 2018; Volume 1974, p. 020038.
9. Rani, D.; Mishra, V. Modification of Laplace Adomian decomposition method for solving nonlinear Volterra integral and integro-differential equations based on Newton Raphson formula. *Eur. J. Pure Appl. Math.* **2018**, *11*, 202–214. [CrossRef]
10. Birajdar, G. Numerical Solution of Time Fractional Navier-Stokes Equation by Discrete Adomian decomposition method. *Nonlinear Eng.* **2014**, *3*,21–26. [CrossRef]
11. Carpinteri, A.; Mainardi, F. (Eds.) *Fractals and Fractional Calculus in Continuum Mechanics*; Springer: Berlin, Germany, 2014; Volume 378.
12. Momani, S.; Odibat, Z. Analytical solution of a time-fractional Navier–Stokes equation by Adomian decomposition method. *Appl. Math. Comput.* **2006**, *177*, 488–494. [CrossRef]
13. Zhou, Y.; Peng, L. Weak solutions of the time-fractional Navier-Stokes equations and optimal control. *Comput. Math. Appl.* **2017**, *73*, 1016–1027. [CrossRef]
14. Kumar, S.; Kumar, D.; Abbasbandy, S.; Rashidi, M. Analytical solution of fractional Navier-Stokes equation by using modified Laplace decomposition method. *Ain Shams Eng. J.* **2014**, *5*, 569–574 [CrossRef]
15. Edeki, S.O.; Akinlabi, G.O. Coupled Method for Solving Time-Fractional Navier-Stokes Equation. *Int. J. Circuits Syst. Signal Process.* **2018**, *12*, 27–34.
16. Singh, B.; Kumar, P. FRDTM for numerical simulation of multi-dimensional, time-fractional model of Navier-Stokes equation. *Ain Shams Eng. J.* **2016**, *9*, 827–834. [CrossRef]
17. Ganji, Z.; Ganji, D.; Ganji, A.; Rostamian, M. Analytical solution of time-fractional Navier-Stokes equation in polar coordinate by homotopy perturbation method. *Numer. Methods Part. Differ. Equ.* **2010**, *26*, 117–124. [CrossRef]
18. Hilfer, R. *Applications of Fractional Calculus in Physics*; World Science Publishing: River Edge, NJ, USA, 2000.
19. Miller, K.S.; Ross, B. An Introduction to the Fractional Calculus and Fractional Differential Equations. 1993. Available online: http://www.citeulike.org/group/14583/article/4204050 (accessed on 29 January 2019).
20. Podlubny, I. *Fractional Differential Equations: An Introduction to Fractional Derivatives, Fractional Differential Equations, to Methods of Their Solution and Some of Their Applications*; Elsevier: Amsterdam, The Netherlands, 1998; Volume 198.

symmetry

MDPI

Article

# Even Higher Order Fractional Initial Boundary Value Problem with Nonlocal Constraints of Purely Integral Type

**Said Mesloub** * and **Faten Aldosari**

Mathematics Department, College of Science, King Saud University, P.O. Box 2455, Riyadh 11451, Saudi Arabia; Faten@ksu.edu.sa

* Correspondence: mesloub@ksu.edu.sa

Received: 31 January 2019; Accepted: 24 February 2019; Published: 1 March 2019

**Abstract:** In this paper, the a priori estimate method, the so-called energy inequalities method based on some functional analysis tools is developed for a Caputo time fractional $2m$th order diffusion wave equation with purely nonlocal conditions of integral type. Existence and uniqueness of the solution are proved. The proofs of the results are based on some a priori estimates and on some density arguments.

**Keywords:** energy inequality; integral conditions; fractional wave equation; existence and uniqueness; initial boundary value problem

## 1. Introduction

Classical initial boundary value problems for partial differential equations with integer and noninteger order have been widely studied during the last three decades by using different methods. One of the most important methods used and applied to linear and nonlinear partial differential equations with integer order supplemented with classical conditions is the functional analysis method. However, for equations with Caputo time fractional order and nonlocal conditions, there are only a few results obtained by using the mentioned method. The Caputo fractional derivative is a nonlocal operator since it is an integral which is a nonlocal operator. Caputo time fractional derivative can be used to model systems with memory, since it requires all the past history. Time fractional order partial differential equations play a great role in reducing the errors coming from the neglected parameters while modeling real life phenomena.

One of the most important classes of the above equations are the fractional diffusion-wave equations that have been studied and used in different branches of Science. Problem (1) constitutes a large class of time fractional diffusion wave equations of even order such as second and fourth order time fractional wave equations that have numerous applications in physics and engineering as mentioned below. In our problem, local conditions at 0 and 1 are replaced by other conditions on the moments of order $1, 2, ..., 2m-1$ which are non-local integral conditions. Although mathematical models in two and three-dimensions are of big significance for applications, most of the recent research articles are devoted to the fractional order diffusion wave equations in one-dimensional settings. These equations model, for example, propagation of mechanical waves in viscoelastic media [1–4], a non-Markovian diffusion process with memory [5], and a model governing the propagation of mechanical diffusive waves in viscoelastic media that exhibit a power-law creep [2–4].

For various applications of fractional calculus, the reader could refer to [4,6–13].

In the literature, many researchers used the functional analysis method to investigate the well posedness of initial boundary value problems for partial differential equations with time and space integer order having nonlocal conditions—we cite, for example, the references [14–17]. For the fractional diffusion wave equations case with higher order derivatives and classical boundary conditions, there are only few papers dealing with the existence and uniqueness of solution such as [18–20]. In this paper, an initial boundary value problem with purely nonlocal constraints of integral type for a Caputo time fractional $2m$th order diffusion wave equation is studied by applying the functional analysis method, the so-called energy inequality method based mainly on some a priori estimates and on the density of the range of the operator generated by the studied problem. This work can be considered as a contribution to the development of the functional analysis method used to prove the well posedness of problems with fractional order. The obtained results show the efficiency of this method to study the existence and uniqueness of solution for the time fractional order differential equations with nonlocal conditions.

This paper is organized as follows: in Section 2, we set our fractional initial boundary value problem. In Section 3, we give some preliminaries concerning the used function spaces, some useful tools and write down the given problem in its operator form. In Section 4, we establish an a priori estimate for the solution and deduce some consequences about the uniqueness of the solution and its dependence on the free term and the given data. Section 5 provides proofs of the main result concerning the solvability of the posed problem. We end our problem with conclusions.

## 2. Problem Setting

In the domain $Q = (0,1) \times (0,T)$ where $0 \le T < \infty$, we consider the time fractional initial boundary problem of higher order with purely integral conditions

$$\begin{cases} \mathcal{L}v = \partial_t^{\alpha+1} v + (-1)^m \theta(t) \frac{\partial^{2m} v}{\partial x^{2m}} = f(x,t)\,, & x \in (0,1)\; t \in (0,T), \\ l_1 v = v(x,0) = g(x),\; l_2 v = v_t(x,0) = h(x), & x \in (0,1), \\ \int\limits_0^1 x^i v(x,t)dx = 0 \;,\; i = \overline{0, 2m-1}, & t \in (0,T), \end{cases} \tag{1}$$

where $\theta(t)$, $f(x,t)$, $g(x)$ and $h(x)$ are given functions that satisfy certain conditions which will be specified later on, and the operator $\partial_t^{\alpha+1}$ denotes the Caputo left fractional derivative of order $1+\alpha$ with $0 < \alpha < 1$ defined by (see [21])

$$\partial_t^{\alpha+1} v(x,t) = \frac{1}{\Gamma(1-\alpha)} \int\limits_0^t \frac{v_{\tau\tau}(x,\tau)}{(t-\tau)^{\alpha}} d\tau,\; t > 0, \tag{2}$$

where $\Gamma(1-\alpha)$ is the Gamma function.

The Riemann–Liouville integral of order $0 < \alpha < 1$ is defined by (see [21])

$$D_t^{-\alpha} v(t) = \frac{1}{\Gamma(\alpha)} \int\limits_0^t \frac{v(\tau)}{(t-\tau)^{1-\alpha}} d\tau. \tag{3}$$

Different properties of the Caputo fractional derivative and Riemann fractional-Liouville integral can be found in [21–23] and the references therein.

## 3. Preliminaries

In this section, we introduce some important lemmas and inequalities needed throughout the sequel, and write the posed problem in its operator form.

**Lemma 1** (Poincare type inequality). *For $m \in \mathbb{N}$, we have*

$$\|\mathcal{I}_x^{2m} v\|^2_{L^2(0,A)} \le (\frac{A}{2})^{2m} \|v\|^2_{L^2(0,A)}, \tag{4}$$

*where*

$$\mathcal{I}_x^{2m}v = \int_0^x \int_0^{\xi_1} ... \int_0^{\xi_{2m-1}} v(\eta,t)d\eta d\xi_{2m-1}...d\xi_1 = \int_0^x \frac{(x-\xi)^{2m-1}}{(2m-1)!}v(\xi,t)d\xi.$$

**Lemma 2** ([24]). *For any absolutely continuous function $J(s)$ on the interval $[0,T]$, the following inequality hold*

$$J(s)\,\partial_s^\alpha J(s) \geq \frac{1}{2}\,\partial_s^\alpha J^2(s), \quad 0<\alpha<1. \tag{5}$$

**Lemma 3** ([25]). *Let $\varphi(t)$ be nonnegative and absolutely continuous on $[0,T]$, and, for almost all $t\in[0,T]$, satisfies the inequality*

$$\frac{d\varphi}{dt} \leq C(t)\varphi(t) + B(t), \tag{6}$$

*where the functions $C(t)$ and $B(t)$ are summable and nonnegative on $[0,T]$. Then,*

$$\begin{aligned}\varphi(t) &\leq e^{\int_0^t C(\tau)d\tau}\left(\varphi(0) + \int_0^t B(\xi).e^{\int_0^\xi C(\tau)d\tau} d\xi\right)\\ &\leq e^{\int_0^t C(\tau)d\tau}\left(\varphi(0) + \int_0^t B(\tau)d\tau\right).\end{aligned} \tag{7}$$

Lemma 3 can be generalized as

**Lemma 4** ([24]). *Let a nonnegative absolutely continuous function $\mathcal{Z}(t)$ satisfy the inequality*

$$\partial_t^\alpha \mathcal{Z}(t) \leq c_1\mathcal{Z}(t) + c_2(t), \quad 0<\alpha<1, \tag{8}$$

*for almost all $t\in[0,T]$, where $c_1$ is a positive constant and $c_2(t)$ is an integrable nonnegative function on $[0,T]$. Then,*

$$\mathcal{Z}(t) \leq \mathcal{Z}(0)E_\alpha(c_1t^\alpha) + \Gamma(\alpha)E_{\alpha,\alpha}(c_1t^\alpha)D_t^{-\alpha}c_2(t), \tag{9}$$

*where*

$$E_\alpha(x) = \sum_{n=0}^\infty \frac{x^n}{\Gamma(\alpha n+1)} \text{ and } E_{\alpha,\alpha^*}(x) = \sum_{n=0}^\infty \frac{x^n}{\Gamma(\alpha n+\alpha^*)}, \tag{10}$$

*are Mittag–Leffler functions.*

**Lemma 5** ([14]). *Let $Z_i(\tau)$ $(i=1,2,3)$ be nonnegative functions on the interval $[0,T]$, $Z_1(\tau), Z_2(\tau)$ are integrable functions, and $Z_3(\tau)$ is nondecreasing. Then,*

$$\int_0^t Z_1(\tau)d(\tau) + Z_2(t) \leq Z_3(t) + C\int_0^t Z_2(\tau)d(\tau)$$

*implies*

$$\int_0^t Z_1(\tau)d(\tau) + Z_2(t) \leq e^{Ct}Z_3(t).$$

Young's inequality with $\varepsilon$: For any $\varepsilon>0$, we have the inequality

$$\lambda\beta \leq \frac{1}{p}|\varepsilon\lambda|^p + \frac{p-1}{p}\left|\frac{\beta}{\varepsilon}\right|^{\frac{p}{p-1}}, \quad \lambda,\beta\in\mathbb{R},\ p>1, \tag{11}$$

where $\lambda$ and $\beta$ are nonnegative numbers.

A special case of (11) is the Cauchy inequality with $\varepsilon$:

$$\lambda\beta \leq \frac{\varepsilon}{2}\lambda^2 + \frac{1}{2\varepsilon}\beta^2, \quad \varepsilon > 0, \tag{12}$$

The solution of the problem (1) can be regarded as the solution of operator equation

$$\mathcal{M}v = \mathcal{W} = (f, g, h), \tag{13}$$

where $\mathcal{M} = (\mathcal{L}, l_1, l_2)$, and $\mathcal{M} : \mathcal{B} \longrightarrow \mathcal{Y}$ is an unbounded operator with domain of definition

$$\mathcal{D}(\mathcal{M}) = \begin{cases} v \in L^2(Q),\ \partial_t^{\alpha+1} v,\ \frac{\partial^{2m} v}{\partial x^{2m}} \in L^2(Q), \\ \int\limits_0^1 x^i v dx = 0, \ \ i = \overline{0, 2m-1},\ t \in (0,T), \end{cases} \tag{14}$$

such that $v$ satisfies the initial conditions and where $\mathcal{B}$ is a Banach space of functions $v$ endowed with the finite norm

$$\|v\|_{\mathcal{B}}^2 = \sup_{0 \leq t \leq T} \left( D_t^{\alpha-1} \|\mathcal{I}_x^m v_t(x,t)\|_{L^2(0,1)}^2 + \int\limits_0^1 v^2(x,t)dx \right) \tag{15}$$

and $\mathcal{Y}$ is Hilbert space constituting of the elements $\mathcal{W} = (f, g, h)$ equipped with the norm

$$\|\mathcal{W}\|_Y^2 = \|g\|_{L^2(0,1)}^2 + \|h\|_{L^2(0,1)}^2 + \|f\|_{L^2(Q)}^2. \tag{16}$$

Here, $\mathcal{L}$ denotes the time fractional differential operator

$$\mathcal{L} = {}^C\partial_t^{\alpha+1} + (-1)^m \theta(t) \frac{\partial^{2m}}{\partial x^{2m}}.$$

## 4. A Priori Estimate for the Solution and Uniqueness

To prove the uniqueness of solution of problem (1), we establish an energy inequality for the solution from which we deduce the uniqueness of solution of the posed problem.

**Theorem 1.** *Assume that the function $\theta(t)$ satisfies the conditions*

$$i)\ c_2 \leq \theta(t) \leq c_1, \ \ ii)\ c_4 \leq \theta'(t) \leq c_3, \ \ \forall t \in [0,T], \tag{17}$$

*where $c_1$, $c_2$, $c_3$, and $c_4$ are positive constants. Then, for any $v \in \mathcal{D}(\mathcal{M})$, there exists a positive constant $\mathcal{K}$ such that the following a priori estimate is satisfied:*

$$\begin{aligned} &\sup_{0 \leq t \leq T} \left( D_t^{\alpha-1} \|\mathcal{I}_x^m v_t(x,t)\|_{L^2(0,1)}^2 + \int\limits_0^1 v^2(x,t)dx \right) \\ \leq\ & \mathcal{K} \left( \|g\|_{L^2(0,1)}^2 + \|h\|_{L^2(0,1)}^2 + \|f\|_{L^2(Q)}^2 \right), \end{aligned} \tag{18}$$

*where $\mathcal{K} = \mathcal{K}(\eta, \delta, \rho)$ is given by*

$$\mathcal{K} = \rho \left( 1 + \frac{T^\alpha}{\Gamma(1+\alpha)} \right), \tag{19}$$

*with $\eta, \delta$ and $\rho$ are respectively given by (30), (34) and (37).*

**Proof.** For $v \in D(L)$, we consider the scalar product in $L^2(0,1)$ of the differential equation in problem (1) and the integrodifferential operator $Nv = 2(-1)^m \mathcal{I}_x^{2m} v_t$, we have

$$\begin{aligned} & 2(-1)^m(\partial_t^{\alpha+1} v, \mathcal{I}_x^{2m} v_t)_{L^2(0,1)} + 2(\theta(t)\frac{\partial^{2m} v}{\partial x^{2m}}, \mathcal{I}_x^{2m} v_t)_{L^2(0,1)} \\ = \; & (\mathcal{L}v, Nv)_{L^2(0,1)}. \end{aligned} \tag{20}$$

We separately consider the inner products on the left-hand side of Equation (20) and we integrate by parts and taking into account boundary and initial conditions in Problem (1), we obtain

$$\begin{aligned} 2(-1)^m(\partial_t^{\alpha+1} v, \mathcal{I}_x^{2m} v_t)_{L^2(0,1)} &= 2(-1)^m(\partial_t^{\alpha} v_t, \mathcal{I}_x^{2m} v_t)_{L^2(0,1)} \\ &= 2(\partial_t^{\alpha}(\mathcal{I}_x^{m} v_t), \mathcal{I}_x^{m} v_t)_{L^2(0,1)}, \end{aligned} \tag{21}$$

$$\begin{aligned} 2\left(\theta(t)\frac{\partial^{2m} v}{\partial x^{2m}}, \mathcal{I}_x^{2m} v_t\right)_{L^2(0,1)} &= 2\int_0^1 \theta(t)\frac{\partial^{2m} v}{\partial x^{2m}} \mathcal{I}_x^{2m} v_t dx \\ &= 2(-1)^m(\theta(t)\frac{\partial^{m} v}{\partial x^{m}}, \mathcal{I}_x^{m} v_t)_{L^2(0,1)} \\ &= 2(\theta(t)v, v_t)_{L^2(0,1)}. \end{aligned} \tag{22}$$

Substitution of (21) and (22) into (20) yields

$$\begin{aligned} & 2(\partial_t^{\alpha}(\mathcal{I}_x^{m} v_t), \mathcal{I}_x^{m} v_t)_{L^2(0,1)} + 2(\theta(t)v, v_t)_{L^2(0,1} \\ = \; & (\mathcal{L}v, Nv)_{L^2(0,1)} = 2(-1)^m(f, \mathcal{I}_x^{2m} v_t)_{L^2(0,1)}. \end{aligned} \tag{23}$$

By Lemmas 1 and 2 and inequality (12), identity (23) reduces to

$$\partial_t^{\alpha}\|\mathcal{I}_x^{m} v_t\|_{L^2(0,1)}^2 + (\theta(t)v, v_t)_{L^2(0,1} \leq \|f\|_{L^2(0,1)}^2 + \frac{1}{2^m}\|\mathcal{I}_x^{m} v_t\|_{L^2(0,1)}^2. \tag{24}$$

Replacing $t$ by $\tau$ , integrating with respect to $\tau$ from zero to $t$ and using given conditions, we obtain

$$\begin{aligned} & \int_0^t \partial_\tau^{\alpha}\|\mathcal{I}_x^{m} v_\tau\|_{L^2(0,1)}^2 d\tau + \int_0^t\int_0^1 \theta(\tau) v v_\tau dx d\tau \\ \leq \; & \int_0^t \|f(x,\tau)\|_{L^2(0,1)}^2 d\tau + \frac{1}{2^m}\int_0^t \|\mathcal{I}_x^{m} v_\tau\|_{L^2(0,1)}^2 d\tau. \end{aligned} \tag{25}$$

The second term on the left-hand side can be evaluated as

$$\begin{aligned} 2\int_0^t\int_0^1 \theta(\tau) v v_\tau dx d\tau &= \int_0^1 \theta(t) v^2 dx - \theta(0)\int_0^1 g^2(x) dx \\ &- \int_0^t\int_0^1 \theta'^2 v^2 dx d\tau. \end{aligned} \tag{26}$$

Hence, inequality (25) becomes

$$
\begin{aligned}
&\int_0^t \partial_\tau^\alpha \|\mathcal{I}_x^m v_\tau\|^2_{L^2(0,1)} d\tau + \frac{1}{2}\int_0^1 \theta(t) v^2 dx \\
\leq\ & \theta(0)\int_0^1 g^2(x)dx + \int_0^t\int_0^1 \theta'^2 v^2 dx d\tau + \int_0^t \|f(x,\tau)\|^2_{L^2(0,1)} d\tau \\
&+\frac{1}{2^m}\int_0^t \|\mathcal{I}_x^m v_\tau\|^2_{L^2(0,1)} d\tau.
\end{aligned} \tag{27}
$$

Now, since

$$
\begin{aligned}
\int_0^t \partial_\tau^\alpha \|\mathcal{I}_x^m v_\tau\|^2_{L^2(0,1)} d\tau &= D^{\alpha-1}\|\mathcal{I}_x^m v_t\|^2_{L^2(0,1)} \\
&\quad -\frac{t^{1-\alpha}}{(1-\alpha)\Gamma(1-\alpha)}\|\mathcal{I}_x^m h\|^2_{L^2(0,1)},
\end{aligned} \tag{28}
$$

evoking conditions (17) and using (28), we infer from (26) that

$$
\begin{aligned}
&D^{\alpha-1}\|\mathcal{I}_x^m v_t\|^2_{L^2(0,1)} + \|v\|^2_{L^2(0,1)} \\
\leq\ & \eta\left(\|h\|^2_{L^2(0,1)} + \|g\|^2_{L^2(0,1)} + \int_0^t \|f\|^2_{L^2(0,1)} d\tau \right.\\
&\left. + \int_0^t \|\mathcal{I}_x^m v_\tau\|^2_{L^2(0,1)} d\tau + \int_0^t\int_0^1 v^2(x,\tau) dx d\tau\right),
\end{aligned} \tag{29}
$$

where

$$
\eta = \frac{\max\left(1, c_1, c_3^2, 2^{-m}, \frac{T^{1-\alpha}2^{-m}}{(1-\alpha)\Gamma(1-\alpha)}\right)}{\min(1, c_2)}. \tag{30}
$$

If, in Lemma 3, we set

$$
\varphi(t) = \int_0^t\int_0^1 v^2(x,\tau)dxd\tau,\ \varphi'(t) = \|v\|^2_{L^2(0,1)},\ \text{and}\ \varphi(0) = 0, \tag{31}
$$

then it yields

$$
\begin{aligned}
\int_0^t\int_0^1 v^2(x,\tau)dxd\tau &\leq Te^{\eta T}\eta\left(\int_0^t \|f(x,\tau)\|^2_{L^2(0,1)} d\tau + \int_0^t \|\mathcal{I}_x^m v_\tau\|^2_{L^2(0,1)} d\tau\right. \\
&\quad \left. +\|g(x)\|^2_{L^2(0,1)} + \|h(x)\|^2_{L^2(0,1)}\right).
\end{aligned} \tag{32}
$$

Consequently, (30) transforms to

$$
\begin{aligned}
& D^{\alpha-1}\|\mathcal{I}_x^m v_t\|^2_{L^2(0,1)} + \|v\|^2_{L^2(0,1)} \\
\leq\ & \delta \left( \int_0^t \|f(x,\tau)\|^2_{L^2(0,1)} d\tau + \int_0^t \|\mathcal{I}_x^m v_\tau\|^2_{L^2(0,1)} d\tau \right. \\
& \left. + \|g(x)\|^2_{L^2(0,1)} + \|h(x)\|^2_{L^2(0,1)} \right), \qquad (33)
\end{aligned}
$$

where

$$
\delta = \max\left\{\eta, \eta^2 T e^{\eta}\right\}. \qquad (34)
$$

Now, by dropping the second term on the left-hand side of (33) then setting $\mathcal{Z}(t) = \int_0^t \|\mathcal{I}_x^m v_\tau\|^2_{L^2(0,1)} d\tau$, $\partial_t^\alpha \mathcal{Z}(t) = D^{\alpha-1}\|\mathcal{I}_x^m v_t\|^2_{L^2(0,1)}$ in Lemma 4, we obtain

$$
\begin{aligned}
& \int_0^t \|\mathcal{I}_x^m v_\tau\|^2_{L^2(0,1)} d\tau \\
\leq\ & \delta\Gamma(\alpha) E_{\alpha,\alpha}(\delta t^\alpha) \left( D_t^{-\alpha-1}\|f\|^2_{L^2(0,1)} + \frac{T}{\alpha\Gamma(\alpha)}\|h\|^2_{L^2(0,1)} + \frac{T}{\alpha\Gamma(\alpha)}\|g\|^2_{L^2(0,1)} \right) \qquad (35) \\
\leq\ & \delta\Gamma(\alpha) E_{\alpha,\alpha}(\delta t^\alpha) \max\left\{1, \frac{T}{\alpha\Gamma(\alpha)}\right\} \\
& \times \left( D_t^{-\alpha-1}\|f\|^2_{L^2(0,1)} + \|h\|^2_{L^2(0,1)} + \|g\|^2_{L^2(0,1)} \right).
\end{aligned}
$$

Combination of (33) and (36) leads to

$$
\begin{aligned}
& D^{\alpha-1}\|\mathcal{I}_x^m v_t\|^2_{L^2(0,1)} + \|v\|^2_{L^2(0,1)} \\
\leq\ & \rho \left( \int_0^t \|f(x,\tau)\|^2_{L^2(0,1)} d\tau + D_t^{-\alpha-1}\|f\|^2_{L^2(0,1)} \right. \qquad (36) \\
& \left. + \|h\|^2_{L^2(0,1)} + \|g\|^2_{L^2(0,1)} \right),
\end{aligned}
$$

where

$$
\rho = \delta \max\left(1, \delta\Gamma(\alpha) E_{\alpha,\alpha}(\delta t^\alpha) \max\left\{1, \frac{T}{\alpha\Gamma(\alpha)}\right\}\right). \qquad (37)
$$

It is obvious that

$$
\begin{aligned}
D_t^{-\alpha-1}\|f\|^2_{L^2(0,1)} &\leq \frac{t^\alpha}{\Gamma(1+\alpha)} \int_0^t \|f\|^2_{L^2(0,1)} dt, \\
&\leq \frac{T^\alpha}{\Gamma(1+\alpha)} \int_0^T \|f\|^2_{L^2(0,1)} dt. \qquad (38)
\end{aligned}
$$

Then, it follows from (37) and (38) that

$$
\begin{aligned}
& D^{\alpha-1}\|\mathcal{I}_x^m v_t\|^2_{L^2(0,1)} + \|v\|^2_{L^2(0,1)} \\
\leq\ & \mathcal{K}\left( \|f\|^2_{L^2(Q)} + \|h\|^2_{L^2(0,1)} + \|g\|^2_{L^2(0,1)} \right), \qquad (39)
\end{aligned}
$$

where

$$
\mathcal{K} = \rho\left(1 + \frac{T^\alpha}{\Gamma(1+\alpha)}\right).
$$

Observe that the right-hand side of (39) is independent of the variable $t$, so we are allowed to take the least upper bound of the left-hand side with respect to $t$ over $[0, T]$, and the a priori estimate (18) then follows and from which we deduce the uniqueness and continuous dependence of the solution on the input data of problem (1). □

## 5. Existence of Solution

In this section, we prove the main result concerning the existence of solution of problem (1) The a priori estimate (18) shows that the unbounded operator $\mathcal{M}$ has an inverse $\mathcal{M}^{-1} : R(\mathcal{M}) \to \mathcal{B}$. Since $R(\mathcal{M})$ is a subset of $\mathcal{Y}$, we therefore can construct its closure $\overline{\mathcal{M}}$ in a manner that the estimate (18) holds for this extension and $R(\overline{\mathcal{M}})$ coincides with the whole space $\mathcal{B}$. Hence, the following:

**Corollary 1.** *The operator $\mathcal{M} : \mathcal{B} \to \mathcal{Y}$ admits a closure (the proof is similar to that in [14]).*
*Estimate (18) can be then extended to*

$$\begin{aligned} &\sup_{0\le t\le T}\left(D_t^{\alpha-1}\|\mathcal{I}_x^m v_t(x,t)\|^2_{L^2(0,1)} + \int_0^1 v^2(x,t)dx\right) \\ \le\ &\mathcal{K}\left(\|g\|^2_{L^2(0,1)} + \|h\|^2_{L^2(0,1)} + \|f\|^2_{L^2(Q)}\right) \end{aligned} \tag{40}$$

*for all $v \in D(\overline{\mathcal{M}})$.*

It follows from (40) that the strong solution of problem (1) is unique, that is, $\overline{\mathcal{M}}v = \mathcal{Y}$. We also deduce from estimate (40) the following:

**Corollary 2.** *$R(\overline{\mathcal{M}})$ is a closed subset in $\mathcal{Y}$ and $R(\mathcal{M}) = R(\overline{\mathcal{M}})$ and $\overline{\mathcal{M}}^{-1} = \overline{\mathcal{M}^{-1}}$.*

We are now ready to give the result of existence of the solution of problem (1).

**Theorem 2.** *Suppose that conditions of Theorem 4.1 are satisfied. Then, for all $\mathcal{W} = (f, g, h) \in \mathcal{Y}$, there exists a unique strong solution $v = \overline{\mathcal{M}}^{-1}\mathcal{W} = \overline{\mathcal{M}^{-1}}\mathcal{W}$ of problem (1).*

**Proof.** Estimate (40) asserts that, if a strong solution of (1) exists, it is unique and depends continuously on the data. Corollary 2 says that, in order to prove that problem (1) admits a strong solution for any $\mathcal{W} = (f, g, h) \in \mathcal{Y}$, it suffices to show that the closure of the range of the operator $\mathcal{M}$ is dense in $\mathcal{Y}$. To establish the existence of the strong solution of problem (1), we use a density argument. That is, we show that the range $R(\mathcal{M})$ of the operator $\mathcal{M}$ is dense in the space $\mathcal{Y}$ for every element $v$ in the Banach space $\mathcal{B}$. For this, we consider the following special case of density. □

**Theorem 3.** *Suppose that conditions of Theorem 1 are satisfied. If for all functions $v \in \mathcal{D}(\mathcal{M})$ such that $l_1 v = v(x,0) = 0$, $l_2 v = v_t(x,0) = 0$ and for some function $\psi \in L^2(Q)$, we have*

$$\int_0^T (\mathcal{L}v, \psi)_{L^2(0,1)}\,dt = 0, \tag{41}$$

*then $\psi$ vanishes a.e in $Q$.*

**Proof.** Identity (40) is equivalent to

$$\int_0^T \left(\partial_t^{\alpha+1} v + (-1)^m \theta(t)\frac{\partial^{2m} v}{\partial x^{2m}}, \psi\right)_{L^2(0,1)} dt = 0. \tag{42}$$

Assume that a function $\sigma(x,t)$ verifies conditions boundary and initial conditions in (1) and such that $\sigma, \sigma_x, \mathcal{I}_t\sigma, \mathcal{I}_t\mathcal{I}_x^{2m}\sigma, \mathcal{I}_t^2\sigma$ and $\partial_t^{\beta+1}\sigma \in L^2(Q)$, we then set

$$v(x,t) = \mathcal{I}_t^2\sigma = \int_0^t \int_0^s \sigma(x,z)dzds. \tag{43}$$

Equation (42) then becomes

$$\int_0^T \left(\partial_t^{\alpha+1}\mathcal{I}_t^2\sigma + (-1)^m\theta(t).\frac{\partial^{2m}}{\partial x^{2m}}(\mathcal{I}_t^2\sigma), \psi\right)_{L^2(0,1)} dt = 0. \tag{44}$$

We now introduce the function

$$\psi(x,t) = \mathcal{I}_t\sigma + (-1)^m\mathcal{I}_x^{2m}\mathcal{I}_t\sigma. \tag{45}$$

Equation (44) then reduces to

$$\begin{aligned} &\int_0^T \left(\partial_t^{\alpha+1}\mathcal{I}_t^2\sigma, \mathcal{I}_t\sigma\right)_{L^2(0,1)} dt + \int_0^T \left(\partial_t^{\alpha+1}\mathcal{I}_t^2\sigma, (-1)^m\mathcal{I}_x^{2m}\mathcal{I}_t\sigma\right)_{L^2(0,1)} dt \\ &+ \int_0^T \left((-1)^m\theta(t).\frac{\partial^{2m}}{\partial x^{2m}}(\mathcal{I}_t^2\sigma), \mathcal{I}_t\sigma\right)_{L^2(0,1)} dt \\ &+ \int_0^T \left(\theta(t).\frac{\partial^{2m}}{\partial x^{2m}}(\mathcal{I}_t^2\sigma), \mathcal{I}_x^{2m}\mathcal{I}_t\sigma\right)_{L^2(0,1)} dt \\ = \;& 0. \end{aligned} \tag{46}$$

Recall that the function $\sigma$ satisfies boundary conditions in (1) and then, computing the inner products in (45), one has

$$\begin{aligned} \left(\partial_t^{\alpha+1}\mathcal{I}_t^2\sigma, \mathcal{I}_t\sigma\right)_{L^2(0,1)} &= (\partial_t^{\alpha}\mathcal{I}_t\sigma, \mathcal{I}_t\sigma)_{L^2(0,1)}, \\ &\geq \frac{1}{2}\partial_t^{\alpha}\|\mathcal{I}_t\sigma\|_{L^2(0,1)}^2, \end{aligned} \tag{47}$$

$$\begin{aligned} \left(\partial_t^{\alpha+1}\mathcal{I}_t^2\sigma, (-1)^m\mathcal{I}_x^{2m}\mathcal{I}_t\sigma\right)_{L^2(0,1)} &= (\partial_t^{\alpha}\mathcal{I}_x^m\mathcal{I}_t\sigma, \mathcal{I}_x^m\mathcal{I}_t\sigma)_{L^2(0,1)}, \\ \geq \;& \frac{1}{2}\partial_t^{\alpha}\|\mathcal{I}_x^m\mathcal{I}_t\sigma\|_{L^2(0,1)}^2, \end{aligned} \tag{48}$$

$$\begin{aligned} &\left((-1)^m\theta(t).\frac{\partial^{2m}}{\partial x^{2m}}(\mathcal{I}_t^2\sigma), \mathcal{I}_t\sigma\right)_{L^2(0,1)} \\ = \;& \left(\theta(t)\frac{\partial^m}{\partial x^m}(\mathcal{I}_t^2\sigma), \frac{\partial^m}{\partial x^m}(\mathcal{I}_tP)\right)_{L^2(0,1)} \\ = \;& \frac{1}{2}\frac{d}{dt}\int_0^1 \theta(t)\left(\frac{\partial^m}{\partial x^m}(\mathcal{I}_t^2\sigma)\right)^2 dx - \frac{1}{2}\int_0^1 \theta'(t)\left(\frac{\partial^m}{\partial x^m}(\mathcal{I}_t^2\sigma)\right)^2 dx, \end{aligned} \tag{49}$$

$$\left(\theta(t).\frac{\partial^{2m}}{\partial x^{2m}}(\mathcal{I}_t^2\sigma),\mathcal{I}_x^{2m}\mathcal{I}_t\sigma\right)_{L^2(0,1)}$$
$$= \left(\theta(t)(\mathcal{I}_t^2\sigma),\mathcal{I}_t\sigma\right)_{L^2(0,1)} = \frac{1}{2}\frac{d}{dt}\int_0^1 \theta(t)(\mathcal{I}_t^2\sigma)^2dx \tag{50}$$
$$-\frac{1}{2}\int_0^1 \theta'(t)(\mathcal{I}_t^2\sigma)^2dx.$$

Evoking (47)–(51), replace $t$ by $\tau$, integrating with respect to $\tau$ from zero to $t$ and using conditions (17), we obtain

$$D_t^{\alpha-1}\|\mathcal{I}_t\sigma\|^2_{L^2(0,1)} + D_t^{\alpha-1}\|\mathcal{I}_x^m\mathcal{I}_t\sigma\|^2_{L^2(0,1)} + \int_0^1\left(\frac{\partial^m}{\partial x^m}(\mathcal{I}_t^2\sigma)\right)^2dx$$
$$+\int_0^1(\mathcal{I}_t^2\sigma)^2dx \tag{51}$$
$$\leq \frac{c_3}{\min(1,c_1)}\left(\int_0^t\int_0^1\left(\frac{\partial^m}{\partial x^m}(\mathcal{I}_\tau^2\sigma)\right)^2dxd\tau + \int_0^t\int_0^1(\mathcal{I}_\tau^2\sigma)^2dxd\tau\right).$$

By dropping the first two terms on the left-hand side of (50) and applying Gronwall's Lemma 5, by setting $Z_1(t) = 0$, $Z_3(t) = 0$ and

$$Z_2(t) = \int_0^1\left(\frac{\partial^m}{\partial x^m}(\mathcal{I}_t^2\sigma)\right)^2dx + \int_0^1(\mathcal{I}_t^2\sigma)^2dx,$$

we have

$$\int_0^1\left(\frac{\partial^m}{\partial x^m}(\mathcal{I}_t^2\sigma)\right)^2dx + \int_0^1(\mathcal{I}_t^2\sigma)^2dx \leq 0 \tag{52}$$

for all $t \in [0,T]$. Hence, $\psi = 0$ a.e in $Q$. □

To complete the proof of Theorem 2, and prove the density ($\overline{R(\mathcal{M})} = \mathcal{Y}$) in a general case, suppose that, for some element $(F_1,\theta_1,\theta_2) \in R(\mathcal{M})^\perp$, we have

$$\int_0^T(\mathcal{L}v,F_1)_{L^2(0,1)}\,ds + (l_1v,\theta_1)_{L^2(0,1)} + (l_2v,\theta_2)_{L^2(0,1)} = 0, \tag{53}$$

and then we prove that $F_1 = 0$, $\theta_1 = 0$, $\theta_2 = 0$. If we put $v \in D(\mathcal{M})$ satisfying conditions $l_1v = v(x,0) = 0$ and $l_2v = v_t(x,0) = 0$ into (53), we obtain

$$\int_0^T(\mathcal{L}v,F_1)_{L^2(0,1)}\,ds = 0,\ \forall v \in D(\mathcal{M}). \tag{54}$$

By Theorem 3, Equation (54) implies that $F_1 = 0$ a.e in $Q$. Then, (53) becomes

$$(l_1v,\theta_1)_{L^2(0,1)} + (l_2v,\theta_2)_{L^2(0,1)} = 0\ \ \forall\theta \in D(\mathcal{M}). \tag{55}$$

Since the range of the trace operator $l_1$ and $l_2$ is dense in $L^2(0,1)$, it follows then from (55) that $\theta_1 = 0$, $\theta_2 = 0$. This ends the proof of Theorem 2.

## 6. Conclusions

The existence and uniqueness of a generalized solution for a higher order fractional diffusion wave equation in Caputo sense subject to initial and weighted integral boundary conditions are established. It is found that the method of energy inequalities is successfully applied to obtaining a priori bounds for the solution of fractional initial boundary value problems of higher order with nonlocal constraints as in the classical case. The obtained results will contribute to the development of the functional analysis method and enrich the existing non-extensive literature on the non local fractional mixed problems in the Caputo sense.

**Author Contributions:** Both authors have contributed equally to this paper.

**Funding:** The authors would like to extend their sincere appreciation to the Deanship of Scientific Research at King Saud University for its funding this Research group No. (RG-117).

**Conflicts of Interest:** The authors declare no conflict of interest.

## References

1. Daftardar-Gejji, V.; Jafari, H. Boundary value problems for fractional diffusion-wave equations. *Aust. J. Math. Anal. Appl.* **2006**, *3*, 1–18.
2. Mainardi, F. Fractional relaxation-oscillation and fractional diffusion wave phenomena. *Chaos Solitons Fractals* **1996**, *7*, 1461–1477. [CrossRef]
3. Mainardi, F.; Luchko, Y.; Pagnini, G. The fundamental solution of the space-time fractional diffusion equation. *Fract. Calc. Appl. Anal.* **2001**, *4*, 153–192.
4. Mainardi, F.; Paradisi, P. Fractional diffusive waves. *J. Comput. Acoust.* **2001**, *9*, 1417–1436. [CrossRef]
5. Metzler, R.; Klafter, J. Boundary value problems for fractional diffusion equations. *Physica A* **2000**, *298*, 107–125. [CrossRef]
6. Carlson, G.E.; Halijak, C.A. Simulation of the fractional derivative operator vG and the fractional integral operator 1/v/S. *Kansas State Univ. Bull.* **1961**, *45*, 21–22.
7. Engheta, N. On fractional calculus and fractional multipoles in electromagnetism. *IEEE Trans. Antennas Propag.* **1996**, *44*, 554–566. [CrossRef]
8. Gorenflo, R. Abel integral equations with special emphasis on applications. In *Lectures in Mathematical Sciences*; University of Tokyo: Tokyo, Japan, 1996; Volume 13.
9. Hilfer, R. *Applications of Fractional Calculus in Physics*; World Scientific: Singapore, 2000.
10. Kaplan, T.; Gray, L.J.; Liu, S.H. Self-affine fractal model for a metal-electrolyte interface. *Phys. Rev. B* **1987**, *35*, 5379–5381. [CrossRef]
11. Nigmatullin, R. Realization of the generalized transfer equation in a medium with fractional geometry. *Phys. Status B Basic Res.* **1986**, *133*, 425–430. [CrossRef]
12. Marin, M.; Ochsner, A. The effect of a dipolar structure on the Holder stability in Green Naghdi thermoelasticity. *Contin. Mech. Thermodyn.* **2017**, *29*, 1365–1374. [CrossRef]
13. Anastassiou, G.; Argyros, I. *Intelligent Numerical Methods: Applications to Fractional Calculus, Studies in Computational Intelligence*; Springer: Heidelberg, Germany, 2016.
14. Mesloub, S. A nonlinear nonlocal mixed problem for a second order parabolic equation. *J. Math. Anal. Appl* **2006**, *316*, 189–209. [CrossRef]
15. Mesloub, S. On a singular two dimensional nonlinear evolution equation with non local conditions. *Nonlinear Anal.* **2008**, *68*, 2594–2607. [CrossRef]
16. Mesloub, S.; Bouziani, A. On a class of singular hyperbolic equations with a weighted integral condition. *Int. J. Math. Math. Sci.* **1999**, *22*, 511–519. [CrossRef]
17. Yurchuk, N.I. Mixed problem with an integral condition for certain parabolic equations. *Differential'nye Uravneniya* **1986**, *22*, 2117–2126.
18. Amanov, D. Solvability of boundary value problems for higher order differential equation with fractional derivative. *Probl. Comput. Appl. Math. Tashkent Uzbikistan* **2009**, *121*, 55–62.
19. Amanov, D.; Ashyralev, A. Initial-Boundary Value Problem for Fractional Partial Differential Equations of Higher Order. *Abstr. Appl. Anal.* **2012**, *2012*, 973102. [CrossRef]

20. Mesloub, S.; Obaid, A. On a singular nonlocal time fractional order mixed problem with a memory term. *Math. Methods Appl. Sci.* **2018**, *41*, 4676–4690. [CrossRef]
21. Podlubny, I. *Fractional Differential Equations*; Academic Press: San Diego, CA, USA, 1999.
22. Kilbas, A.A.; Srivastava, H.M.; Trujillo, J.J. *Theory and Applications of Fractional Differential Equations*; Elsevier: Amsterdam, The Netherlands, 2006.
23. Samko, S.G.; Kilbas, A.A.; Marichev, O.I. Fractional Integrals and Derivatives, Theory and Applications. *Minsk Nauka I Tekhnika* **1993**, 3, 397–414.
24. Alikhanov, A.A. A Priori Estimates for Solutions of Boundary Value Problems for Fractional-Order Equations. *arXiv* **2010**, arXiv:1105.4592v1.
25. Ladyzhenskaya, O.A. *The Boundary Value Problems of Mathematical Physics*; Springer: New York, NY, USA, 1985.

*symmetry*

MDPI

*Article*

# On Some Fractional Integral Inequalities of Hermite-Hadamard's Type through Convexity

**Shahid Qaisar [1], Jamshed Nasir [2], Saad Ihsan Butt [2] and Sabir Hussain [3,*]**

1 Department of Mathematics, COMSATS University Islamabad, Sahiwal Campus, Sahiwal 57000, Pakistan; shahidqaisar90@cuisahiwal.edu.pk
2 Department of Mathematics, COMSATS University Islamabad, Lahore Campus, Lahore 54000, Pakistan; jnasir143@gmail.com (J.N.); saadihsanbutt@gmail.com (S.I.B.)
3 Department of Mathematics, College of Science, Qassim University, P.O. Box 6644, Buraydah 51482, Saudi Arabia
* Correspondence: sabiriub@yahoo.com or sh.hussain@qu.edu.sa

Received: 7 January 2019; Accepted: 22 January 2019; Published: 26 January 2019

**Abstract:** In this paper, we incorporate the notion of convex function and establish new integral inequalities of type Hermite–Hadamard via Riemann—Liouville fractional integrals. It is worth mentioning that the obtained inequalities generalize Hermite–Hadamard type inequalities presented by Özdemir, M.E. et. al. (2013) and Sarikaya, M.Z. et. al. (2011).

**Keywords:** Hermite–Hadamard's Inequality; Convex Functions; Power-mean Inequality; Jenson Integral Inequality; Riemann—Liouville Fractional Integration

**MSC:** 26A15; 26A51; 26D10

---

## 1. Introduction and Preliminaries

One of the generalizations of classical differentiation and integration is fractional calculus. The contribution of fractional calculus presents in diverse fields, such as pure mathematics, economics, and physical and engineering sciences. The role of inequalities found to be very significant in all fields of mathematics and an attractive and active field of research. Recently, convexity has become the major part in different fields of science. A function $g : I \subset \mathbb{R} \to \mathbb{R}$ is named as convex, if the inequality

$$g(\omega x + (1-\omega) y) \leq \omega g(x) + (1-\omega) g(y)$$

holds for all $x, y \in I$ and $\omega \in [0,1]$. In fact, large number of articles has been written on inequalities using classical convexity, but one of the most important and well known is Hermite– Hadamard's inequality. In [1], this double inequality is stated as: Let $g : I \subset \mathbb{R} \to \mathbb{R}$ be a convex function on the interval $I$ of real numbers and $x, y \in I$ with $x < y$. Then,

$$g\left(\frac{x+y}{2}\right) \leq \frac{1}{y-x}\int_x^y g(t)\,dt \leq \frac{g(x)+g(y)}{2}.$$

Both inequalities hold in the reversed direction for $g$ to be concave. In the field of mathematical inequalities, Hermite–Hadamard's inequality has been given more attention by many mathematician due to its applicability and usefulness. Many researchers have extended the Hermite–Hadamard's inequality, to different forms using the classical convex function. For further details involving Hermite–Hadamard's type inequality on different concept of convex function and generalizations, the interested reader is referred to [2–12] and references therein.

First, we recall some important definitions and results that will be used in the sequel.

**Definition 1.** *For $g \in L^1[x,y]$. The left-sided and right-sided Riemann–Liouville fractional integrals of order $\alpha > 0$ with $a \geq 0$ are defined as $J_{a^+}^{\alpha} g(x) = \frac{1}{\Gamma(\alpha)} \int_a^x (x-t)^{\alpha-1} g(t)dt, for a < x$, and $J_{b^-}^{\alpha} g(x) = \frac{1}{\Gamma(\alpha)} \int_x^b (t-x)^{\alpha-1} g(t)dt$, for $x < b$, respectively, where $\Gamma(.)$ is Gamma function and is defined as $\Gamma(\alpha) = \int_0^{\infty} e^{-u} u^{\alpha-1} du$. It is to be noted that $J_{a^+}^{0} g(x) = J_{b^-}^{0} g(x) = g(x)$.*

In the case of $\alpha = 1$, the fractional integral reduces to the classical integral.

Properties relating to these operators can be found in [7]. For useful details on Hermite–Hadamard type inequalities connected with fractional integral inequalities, the readers are directed to [8–14].

In [15], Özdemir et. al proved some inequalities related to Hermite–Hadamard's inequalities for functions whose second derivatives in absolute value at certain powers are $s$-convex functions as follows:

**Theorem 1.** *Let $f : I \subset [0,\infty) \to \mathbb{R}$ be a twice differentiable mapping on $I^0$ (where $I^0$ is the interior of I) such that $f'' \in L[a,b]$, where $a,b \in I$ with $a < b$. If $|f''|$ is s-convex on $[a,b]$, for some fixed $s \in (0,1]$, then the following inequality holds:*

$$\begin{aligned}\left| f\left(\frac{a+b}{2}\right) - \frac{1}{b-a}\int_a^b f(x)dx \right| &\leq \frac{(b-a)^2}{8(s+1)(s+2)(s+3)} \times \\ &\left\{ |f''(a)| + (s+1)(s+2)|f''(\frac{a+b}{2})| + |f''(b)| \right\} \\ &\leq \frac{\left[1+(s+2)2^{1-s}\right](b-a)^2}{8(s+1)(s+2)(s+3)} \left\{ |f''(a)| + |f''(b)| \right\}.\end{aligned}$$

**Corollary 1.** *Under the assumptions of Theorem 1, if $s = 1$, then we get*

$$\left| f\left(\frac{a+b}{2}\right) - \frac{1}{b-a}\int_a^b f(x)dx \right| \leq \frac{(b-a)^2 \left[|f''(a)| + |f''(b)|\right]}{48}. \tag{1}$$

**Theorem 2.** *Let $f : I \subset [0,\infty) \to \mathbb{R}$ be a twice differentiable mapping on $I^0$ (where $I^0$ is the interior of I) such that $f'' \in L[a,b]$, where $a,b \in I$ with $a < b$. If $|f''|$ is s-concave on $[a,b]$, for some fixed $s \in (0,1]$ and for $q > 1$ with $\frac{1}{p} + \frac{1}{q}$, then the following inequality holds:*

$$\left| \frac{1}{b-a}\int_a^b f(x)dx - f\left(\frac{a+b}{2}\right) \right| \leq \frac{(b-a)^2}{16} \frac{2^{\frac{s}{q}}}{(2p+1)^{1/p}} \left[ \left| f''\left(\frac{3a+b}{4}\right) \right| + \left| f''\left(\frac{a+3b}{4}\right) \right| \right]. \tag{2}$$

**Corollary 2.** *Under the assumptions of Theorem 2, if we choose $s = 1$ and $\frac{1}{3} < \frac{1}{(2p+1)^{1/p}} < 1$, for $p > 1$, we have*

$$\left| \frac{1}{b-a}\int_a^b f(x)dx - f\left(\frac{a+b}{2}\right) \right| \leq \frac{(b-a)^2}{16} \left[ \left| f''\left(\frac{3a+b}{4}\right) \right| + \left| f''\left(\frac{a+3b}{4}\right) \right| \right]. \tag{3}$$

In [6], Sarikaya et al. proved some inequalities related to Hermite–Hadamard's inequalities for functions whose derivatives in absolute value at certain powers are convex as follows:

**Theorem 3.** *Let $I \subset \mathbb{R}$ be an open interval, $a, b \in I$ with $a < b$ and $f : [a,b] \to \mathbb{R}$ be a twice differentiable function such that $f''$ is integrable and $0 < \lambda \leq 1$ on $(a,b)$ with $a < b$. If $|f''|^q$ is convex on $[a,b]$, for $q \geq 1$, then the following inequality holds:*

$$\left|(\lambda-1) f\left(\frac{a+b}{2}\right) - \lambda\frac{f(a)+f(b)}{2} + \frac{1}{b-a}\int_a^b f(x)dx\right|$$
$$\leq \begin{cases} \frac{(b-a)^2}{2}\left(\frac{\lambda^3}{3}+\frac{1-3\lambda}{24}\right)^{1-1/q} \\ \times\left\{\left(\left[\frac{\lambda^4}{6}+\frac{3-8\lambda}{3\times 2^6}\right]|f''(a)|^q + \left[\frac{(2-\lambda)\lambda^3}{6}+\frac{5-16\lambda}{3\times 2^6}\right]|f''(b)|^q\right)^{1/q}\right. \\ \left.+\left(\left[\frac{(1+\lambda)(1-\lambda)^3}{6}+\frac{48\lambda-27}{3\times 2^6}\right]|f''(a)|^q + \left[\frac{\lambda^4}{6}+\frac{3-8\lambda}{3\times 2^6}\right]|f''(b)|^q\right)^{1/q}\right\}, & \text{for } 0 \leq \lambda \leq 1/2; \\ \frac{(b-a)^2}{2}\left(\frac{3\lambda-1}{24}\right)^{1-1/q}\left\{\left(\frac{8\lambda-3}{3\times 2^6}|f''(a)|^q + \frac{16\lambda-5}{3\times 2^6}|f''(b)|^q\right)^{1/q}\right. \\ \left.+\left(\frac{16\lambda-5}{3\times 2^6}|f''(a)|^q + \frac{8\lambda-3}{3\times 2^6}|f''(b)|^q\right)^{1/q}\right\}, & \text{for } 1/2 \leq \lambda \leq 1. \end{cases}$$

**Corollary 3.** *Under the assumptions of Theorem 3, if $\lambda = 0$, then we get the following inequality,*

$$\begin{aligned} \left|\frac{1}{b-a}\int_a^b f(x)dx - f\left(\frac{a+b}{2}\right)\right| &\leq \frac{(b-a)^2}{48}\left(\frac{5|f''(a)|^q+3|f''(b)|^q}{8}\right)^{1/q} \\ &+ \frac{(b-a)^2}{48}\left(\frac{3|f''(a)|^q+5|f''(a)|^q}{8}\right)^{1/q}. \end{aligned} \tag{4}$$

The aim of this article is to establish Hermite–Hadamard type inequalities for Riemann-Liouville fractional integral using the convexity as well as concavity, for functions whose absolute values of second derivative are convex. We derive a general integral inequality for Riemann–Liouville fractional integral.

## 2. Main Results

To prove our main results, we need to prove the following lemma, which plays the key role in the next developments:

**Lemma 1.** *Let $f : [a,b] \to \mathbb{R}$ be a twice differentiable function on $(a,b)$ with $a < b$. If $f'' \in L[a,b]$ and $n \in N$, then the following equality for fractional integrals holds with $0 < \alpha \leq 1$:*

$$\begin{aligned} &\frac{\Gamma(\alpha+2)}{(n+1)^2(b-a)^\alpha}\left[J^\alpha_{\left(\frac{n}{n+1}a+\frac{1}{n+1}b\right)^-} f(a) + J^\alpha_{\left(\frac{1}{n+1}a+\frac{n}{n+1}b\right)^+} f(b)\right] - f\left(\frac{n}{n+1}a+\frac{1}{n+1}b\right) \\ &= \frac{(b-a)^2}{(n+1)^{\alpha+3}(\alpha+1)}\left[\int_0^1 (1-\omega)^{\alpha+1}\left(f''(\frac{n+\omega}{n+1}a+\frac{1-\omega}{n+1}b+f''(\frac{1-\omega}{n+1}a+\frac{n+\omega}{n+1}b)\right)d\omega\right. \\ &\left.\int_0^1 ((1+\omega)^{\alpha+1} - 2^\alpha(1+\omega)+\alpha 2^\alpha(1-\omega))f''(\frac{1-\omega}{n+1}a+\frac{n+\omega}{n+1}b)+f''(\frac{n+\omega}{n+1}a+\frac{1-\omega}{n+1}b)d\omega\right]. \end{aligned}$$

**Proof.** To compute each integral, we use integration by parts successively and get

$$
\begin{aligned}
P_1 &= \int_0^1 (1-\omega)^{\alpha+1} f''(\frac{n+\omega}{n+1}a + \frac{1-\omega}{n+1}b)d\omega \\
&= \frac{(n+1)(1-\omega)^{\alpha+1} f'(\frac{n+\omega}{n+1}a + \frac{1-\omega}{n+1}b)d\omega}{a-b}\Big|_0^1 \\
&\quad + \frac{(n+1)(a+1)}{a-b}\int_0^1 (1-\omega)^{\alpha} f'(\frac{n+\omega}{n+1}a + \frac{1-\omega}{n+1}b)d\omega \\
&= \frac{(n+1)}{b-a} f'\left(\frac{n}{n+1}a + \frac{1}{n+1}b\right) + \frac{(n+1)(\alpha+1)}{a-b}\left[\frac{2(1-\omega)^{\alpha} f'(\frac{1+w}{n+1}a + \frac{1-\omega}{n+1}b)d\omega}{a-b}\Big|_0^1\right. \\
&\quad \left. + \frac{(n+1)\alpha}{a-b}\int_0^1 (1-\omega)^{\alpha} f(\frac{n+\omega}{n+1}a + \frac{1-\omega}{n+1}b)d\omega\right] \\
&= \frac{(n+1)}{b-a} f'\left(\frac{n}{n+1}a + \frac{1}{n+1}b\right) + \frac{(n+1)(\alpha+1)}{a-b}\left[\frac{-2}{a-b} f\left(\frac{n}{n+1}a + \frac{1}{n+1}b\right)\right. \\
&\quad \left. - \frac{(n+1)\alpha}{b-a}\int_0^1 (1-\omega)^{\alpha-1} f(\frac{n+\omega}{n+1}a + \frac{1-\omega}{n+1}b)d\omega\right]
\end{aligned}
$$

$$
\begin{aligned}
&= \frac{(n+1)}{b-a} f'\left(\frac{n}{n+1}a + \frac{1}{n+1}b\right) - \frac{(n+1)^2(\alpha+1)}{(b-a)^2} f\left(\frac{n}{n+1}a + \frac{1}{n+1}b\right) \\
&+ \frac{(n+1)^{\alpha+2}}{(b-a)^{\alpha+2}}\Gamma(\alpha+2) J^{\alpha}_{\frac{n}{n+1}a + \frac{1}{n+1}b^-} f(a). \\
P_2 &= \int_0^1 ((1+\omega)^{\alpha+1} - 2^{\alpha}(1+\omega) + \alpha 2^{\alpha}(1-\omega)) f''(\frac{1-\omega}{n+1}a + \frac{n+\omega}{n+1}b)d\omega \\
&= \frac{2\left[(1+\omega)^{\alpha+1} - 2^{\alpha}(1+\omega) + \alpha 2^{\alpha}(1-\omega)\right] f'(\frac{1-\omega}{n+1}a + \frac{n+\omega}{n+1}b)d\omega)}{b-a}\Big|_0^1 \\
&\quad + \frac{2(\alpha+1)}{b-a}\int_0^1 \left[(\alpha+1)(1+\omega)^{\alpha} - 2^{\alpha} - \alpha 2^{\alpha}\right] f'(\frac{1-\omega}{n+1}a + \frac{n+\omega}{n+1}b)d\omega) \\
&= -\frac{n+1}{b-a}\left[1 + 2^{\alpha}(\alpha-1)\right] f'\left(\frac{1}{n+1}a + \frac{n}{n+1}b\right) \\
&\quad + \frac{(n+1)(\alpha+1)}{b-a}\int_0^1 \left[(\alpha+1)(1+\omega)^{\alpha} - 2^{\alpha} - \alpha 2^{\alpha}\right] f'(\frac{1-\omega}{n+1}a + \frac{n+\omega}{n+1}b)d\omega) \\
&= -\frac{n+1}{b-a}\left[1 + 2^{\alpha}(\alpha-1)\right] f'\left(\frac{1}{n+1}a + \frac{n}{n+1}b\right) \\
&\quad + \frac{(n+1)(\alpha+1)}{b-a}\left[\frac{(n+1)\left[(\alpha+1)(1+\omega)^{\alpha} - 2^{\alpha} - \alpha 2^{\alpha}\right] f'(\frac{1-\omega}{n+1}a + \frac{n+\omega}{n+1}b)d\omega)}{b-a}\Big|_0^1\right. \\
&\qquad \left. - \frac{\alpha(n+1)(\alpha+1)}{b-a}\int_0^1 (1+\omega)^{\alpha-1} f\left(\frac{1-\omega}{n+1}a + \frac{n+\omega}{n+1}b\right) d\omega\right] \\
&= -\frac{n+1}{b-a}\left[1 + 2^{\alpha}(\alpha-1)\right] f'\left(\frac{1}{n+1}a + \frac{n}{n+1}b\right) + \frac{(n+1)^2(\alpha+1)}{(b-a)^2}(2^{\alpha}-1) \\
&\times f\left(\frac{1}{n+1}a + \frac{n}{n+1}b\right) - \frac{(n+1)^{\alpha+2}}{(b-a)^{\alpha+2}}\Gamma(\alpha+2) J^{\alpha}_{\frac{1}{n+1}a + \frac{n}{n+1}b^-} f(b).
\end{aligned}
$$

Analogously: $P_3 = -\frac{n+1}{b-a} f'\left(\frac{n}{n+1}a + \frac{1}{n+1}b\right) - \frac{(n+1)^2(\alpha+1)}{(b-a)^2} f\left(\frac{n}{n+1}a + \frac{1}{n+1}b\right) + \frac{(n+1)^{\alpha+2}}{(b-a)^{\alpha+2}}\Gamma(\alpha + 2) J^{\alpha}_{\left(\frac{n}{n+1}a + \frac{1}{n+1}b\right)^-} f(a)$.

$P_4 = -\frac{n+1}{b-a}\left[2^{\alpha}(1-\alpha) - 1\right] f'\left(\frac{1}{n+1}a + \frac{n}{n+1}b\right) - \frac{(n+1)^2(\alpha+1)}{(b-a)^2}$

$\times \left(2^{\alpha}-1\right) f\left(\frac{1}{n+1} a+\frac{n}{n+1} b\right)+\frac{(n+1)^{\alpha+2}}{(b-a)^{\alpha+2}} \Gamma(\alpha+2) J_{\left(\frac{1}{n+1} a+\frac{n}{n+1} b\right)^{+}}^{\alpha} f(b)$.

Adding above equalities, we get

$$\begin{aligned} & \frac{n+1}{b-a} f\left(\frac{n}{n+1} a+\frac{1}{n+1} b\right)-\frac{\Gamma(\alpha+2)}{(n+1)^{2}(b-a)^{\alpha}}\left[J_{\left(\frac{n}{n+1} a+\frac{1}{n+1} b\right)^{-}}^{\alpha} f(a)+J_{\left(\frac{1}{n+1} a+\frac{n}{n+1} b\right)^{+}}^{\alpha} f(b)\right] \\ = & P_{1}+P_{2}+P_{3}+P_{4} . \end{aligned}$$

This completes the proof. □

**Theorem 4.** *Let $f:[a, b] \rightarrow \mathbb{R}$ be a twice differentiable function on $(a, b)$ with $a<b$ and $n \in N^{*}$. If $f^{\prime \prime} \in L[a, b]$ and $\left|f^{\prime \prime}\right|$ is convex on $[a, b]$, then the following inequality for Riemann–Liouville fractional integrals holds:*

$$\begin{aligned} & \left|\frac{\Gamma(\alpha+2)}{(n+1)(b-a)^{\alpha}}\left[J_{\left(\frac{n}{n+1} a+\frac{1}{n+1} b\right)^{-}}^{\alpha} f(a)+J_{\left(\frac{1}{n+1} a+\frac{n}{n+1} b\right)^{+}}^{\alpha} f(b)\right]-f\left(\frac{n}{n+1} a+\frac{1}{n+1} b\right)\right| \\ & \leq(b-a)^{2} \frac{2^{\alpha-1}(2+(\alpha-1) \alpha)}{(n+1)^{\alpha+3}(\alpha+1)(\alpha+2)}\left(\left|f^{\prime \prime}(a)\right|+\left|f^{\prime \prime}(b)\right|\right) . \end{aligned} \tag{5}$$

**Proof.** Using Lemma 1 and properties of modulus, we have

$$\begin{aligned} & \left|\frac{\Gamma(\alpha+2)}{(n+1)(b-a)^{\alpha}}\left[J_{\left(\frac{n}{n+1} a+\frac{1}{n+1} b\right)^{-}}^{\alpha} f(a)+J_{\left(\frac{1}{n+1} a+\frac{n}{n+1} b\right)^{+}}^{\alpha} f(b)\right]-f\left(\frac{n}{n+1} a+\frac{1}{n+1} b\right)\right| \\ \leq & \frac{(b-a)^{2}}{(n+1)^{\alpha+3}(\alpha+1)} \sum_{i=1}^{4}\left|P_{i}\right| . \end{aligned}$$

Now, using convexity of $\left|f^{\prime \prime}\right|$, we have

$$\begin{aligned} & \left|\frac{\Gamma(\alpha+2)}{(n+1)(b-a)^{\alpha}}\left[J_{\left(\frac{n}{n+1} a+\frac{1}{n+1} b\right)^{-}}^{\alpha} f(a)+J_{\left(\frac{1}{n+1} a+\frac{n}{n+1} b\right)^{+}}^{\alpha} f(b)\right]-f\left(\frac{n}{n+1} a+\frac{1}{n+1} b\right)\right| \\ \leq & \left|\frac{(b-a)^{2}}{(n+1)^{\alpha+3}(\alpha+1)} \int_{0}^{1}(1-\omega)^{\alpha+1} f^{\prime \prime}\left(\frac{n+\omega}{n+1} a+\frac{1-\omega}{n+1} b\right) d \omega\right| \\ + & \left|\frac{(b-a)^{2}}{(n+1)^{\alpha+3}(\alpha+1)} \int_{0}^{1}(1-\omega)^{\alpha+1} f^{\prime \prime}\left(\frac{1-\omega}{n+1} a+\frac{n+\omega}{n+1} b\right) d \omega\right| \\ + & \left|\frac{(b-a)^{2}}{(n+1)^{\alpha+3}(\alpha+1)} \int_{0}^{1}\left((1+\omega)^{\alpha+1}-2^{\alpha}(1+\omega)+\alpha 2^{\alpha}(1-\omega)\right) f^{\prime \prime}\left(\frac{n+\omega}{n+1} a+\frac{1-\omega}{n+1} b\right) d \omega\right| \\ + & \left|\frac{(b-a)^{2}}{(n+1)^{\alpha+3}(\alpha+1)} \int_{0}^{1}\left((1+\omega)^{\alpha+1}-2^{\alpha}(1+\omega)+\alpha 2^{\alpha}(1-\omega)\right) f^{\prime \prime}\left(\frac{1-\omega}{n+1} a+\frac{n+\omega}{n+1} b\right) d \omega\right| \\ = & \frac{(bv-a)^{2}}{(n+1)^{\alpha+3}(\alpha+1)} \int_{0}^{1}(1-\omega)^{\alpha+1}\left|f^{\prime \prime}\left(\frac{n+\omega}{n+1} a+\frac{1-\omega}{n+1} b\right)\right| d \omega \\ + & \frac{(b-a)^{2}}{(n+1)^{\alpha+3}(\alpha+1)} \int_{0}^{1}(1-\omega)^{\alpha+1}\left|f^{\prime \prime}\left(\frac{1-\omega}{n+1} a+\frac{n+\omega}{n+1} b\right)\right| d \omega \\ + & \frac{(b-a)^{2}}{(n+1)^{\alpha+3}(\alpha+1)} \int_{0}^{1}\left((1+\omega)^{\alpha+1}-2^{\alpha}(1+\omega)+\alpha 2^{\alpha}(1-\omega)\right)\left|f^{\prime \prime}\left(\frac{n+\omega}{n+1} a+\frac{1-\omega}{n+1} b\right)\right| d \omega \end{aligned}$$

$$+\frac{(b-a)^2}{(n+1)^{\alpha+3}(\alpha+1)}\int_0^1((1+\omega)^{\alpha+1}-2^\alpha(1+\omega)+\alpha 2^\alpha(1-\omega))|f''(\frac{1-\omega}{n+1}a+\frac{n+\omega}{n+1}b)|d\omega$$

$$\leq\frac{(b-a)^2}{(n+1)^{\alpha+3}(\alpha+1)}\int_0^1(1-\omega)^{\alpha+1}\left[\left(\frac{n+\omega}{n+1}\right)|f''(a)|+\left(\frac{1-\omega}{n+1}\right)|f''(b)|\right]d\omega$$

$$+\frac{(b-a)^2}{(n+1)^{\alpha+3}(\alpha+1)}\int_0^1(1-\omega)^{\alpha+1}\left[\left(\frac{1-\omega}{n+1}\right)|f''(a)|+\left(\frac{n+\omega}{n+1}\right)|f''(b)|\right]d\omega$$

$$+\frac{(b-a)^2}{(n+1)^{\alpha+3}(\alpha+1)}\int_0^1((1+\omega)^{\alpha+1}-2^\alpha(1+\omega)+\alpha 2^\alpha(1-\omega))$$

$$\times\left[\left(\frac{n+\omega}{n+1}\right)|f''(a)|+\left(\frac{1-\omega}{n+1}\right)|f''(b)|\right]d\omega$$

$$+\frac{(b-a)^2}{(n+1)^{\alpha+3}(\alpha+1)}\int_0^1((1+\omega)^{\alpha+1}-2^\alpha(1+\omega)+\alpha 2^\alpha(1-\omega))$$

$$\times\left[\left(\frac{1-\omega}{n+1}\right)|f''(a)|+\left(\frac{n+\omega}{n+1}\right)|f''(b)|\right]d\omega.$$

This completes the proof. □

**Remark 1.** *If we take $\alpha = n = 1$ in Theorem 4, then the inequality (5) reduces to the inequality (1). The inequality (1) was obtained by Ozdemir [15].*

The corresponding version for powers of the absolute value of the derivative is incorporated in the following theorem.

**Theorem 5.** *Let $f:[a,b]\to\mathbb{R}$ be a twice differentiable function on $(a,b)$ with $a<b$ and $n\in N$. If $f''\in L[a,b]$ and $|f''|^q$ is convex on $[a,b]$, then the following inequality for Riemann–Liouville fractional integrals holds with $0<\alpha\leq 1$:*

$$\left|\frac{\Gamma(\alpha+2)}{(n+1)^2(b-a)^\alpha}\left[J^\alpha_{\left(\frac{n}{n+1}a+\frac{1}{n+1}b\right)^-}f(a)+J^\alpha_{\left(\frac{1}{n+1}a+\frac{n}{n+1}b\right)^+}f(b)\right]-f\left(\frac{n}{n+1}a+\frac{1}{n+1}b\right)\right|$$
$$\leq\frac{(b-a)^2}{(n+1)^{\alpha+3}(\alpha+1)}\left[(U_5)^{1-1/q}\left\{\left(\frac{U_1|f''(a)|^q+U_2|f''(b)|^q}{(n+1)}\right)^{1/q}+\left(\frac{U_2|f''(a)|^q+U_1|f''(b)|^q}{(n+1)}\right)^{1/q}\right\}+\right. \quad (6)$$
$$\left.(U_6)^{1-1/q}\left\{\left(\frac{U_3|f''(a)|^q+U_4|f''(b)|^q}{(n+1)}\right)^{1/q}+\left(\frac{U_4|f''(a)|^q+U_3|f''(b)|^q}{(n+1)}\right)^{1/q}\right\}\right].$$

*where*

$$U_1 = \frac{n\alpha+3n+1}{(\alpha+2)(\alpha+3)},\quad U_2=\frac{1}{\alpha+3},$$
$$U_3 = \frac{6+2^\alpha(\alpha^3+5\alpha-6)+3n(\alpha+3)(-2+2^\alpha(2+(\alpha-1)\alpha))}{6(\alpha+2)(\alpha+3)}$$
$$U_4 = \frac{-3(\alpha+4)+2^\alpha(12+(\alpha-1)\alpha(\alpha+4))}{3(\alpha+2)(\alpha+3)},\qquad U_5=\frac{1}{\alpha+2},$$
$$\text{and}\quad U_6 = \frac{2^{\alpha+2}-1}{\alpha+2}-2^{\alpha+1}+\alpha 2^{\alpha-1}+2^{\alpha-1}-3\alpha 2^{\alpha-1}.$$

**Proof.** Using Lemma 1, well-known power-mean integral inequality and the fact that $|f''|^q$ is convex , we have

$$\left|\frac{\Gamma(\alpha+2)}{(n+1)^2(b-a)^\alpha}\left[J^\alpha_{\left(\frac{n}{n+1}a+\frac{1}{n+1}b\right)^-}f(a)+J^\alpha_{\left(\frac{1}{n+1}a+\frac{n}{n+1}b\right)^+}f(b)\right]-f\left(\frac{n}{n+1}a+\frac{1}{n+1}b\right)\right|$$

$$\begin{aligned}
\leq\ & \frac{(b-a)^2}{(n+1)^{\alpha+3}(\alpha+1)}\left(\int_0^1(1-\omega)^{\alpha+1}d\omega\right)^{1-1/q} \\
&\times\left(\int_0^1(1-\omega)^{\alpha+1}\left|f''\left(\frac{n+\omega}{n+1}a+\frac{1-\omega}{n+1}b\right)\right|^q d\omega\right)^{1/q} \\
&+\frac{(b-a)^2}{(n+1)^{\alpha+3}(\alpha+1)}\left(\int_0^1(1-\omega)^{\alpha+1}d\omega\right)^{1-1/q} \\
&\times\left(\int_0^1(1-\omega)^{\alpha+1}\left|f''\left(\frac{1-\omega}{n+1}a+\frac{n+\omega}{n+1}b\right)\right|^q d\omega\right)^{1/q} \\
&+\frac{(b-a)^2}{(n+1)^{\alpha+3}(\alpha+1)}\left(\int_0^1((1+\omega)^{\alpha+1}-2^\alpha(1+\omega)+\alpha 2^\alpha(1-\omega))d\omega\right)^{1-1/q} \\
&\times\left(\int_0^1((1+\omega)^{\alpha+1}-2^\alpha(1+\omega)+\alpha 2^\alpha(1-\omega))\left|f''\left(\frac{n+\omega}{n+1}a+\frac{1-\omega}{n+1}b\right)\right|^q d\omega\right)^{1/q} \\
&+\frac{(b-a)^2}{(n+1)^{\alpha+3}(\alpha+1)}\left(\int_0^1((1+\omega)^{\alpha+1}-2^\alpha(1+\omega)+\alpha 2^\alpha(1-\omega))d\omega\right)^{1-1/q} \\
&\times\left(\int_0^1((1+\omega)^{\alpha+1}-2^\alpha(1+\omega)+\alpha 2^\alpha(1-\omega))\left|f''\left(\frac{1-\omega}{n+1}a+\frac{n+\omega}{n+1}b\right)\right|^q d\omega\right)^{1/q}
\end{aligned}$$

$$\begin{aligned}
=\ & \frac{(b-a)^2}{(n+1)^{\alpha+3}(\alpha+1)}\left(\int_0^1(1-\omega)^{\alpha+1}d\omega\right)^{1-1/q}\left(\int_0^1(1-\omega)^{\alpha+1}\left|f''\left(\frac{n+\omega}{n+1}a+\frac{1-\omega}{n+1}b\right)\right|^q d\omega\right)^{1/q} \\
&+\frac{(b-a)^2}{(n+1)^{\alpha+3}(\alpha+1)}\left(\int_0^1(1-\omega)^{\alpha+1}d\omega\right)^{1-1/q}\left(\int_0^1(1-\omega)^{\alpha+1}\left|f''\left(\frac{1-\omega}{n+1}a+\frac{n+\omega}{n+1}b\right)\right|^q d\omega\right)^{1/q} \\
&+\frac{(b-a)^2}{(n+1)^{\alpha+3}(\alpha+1)}\left(\int_0^1((1+\omega)^{\alpha+1}-2^\alpha(1+\omega)+\alpha 2^\alpha(1-\omega))d\omega\right)^{1-1/q} \\
&\times\left(\int_0^1((1+\omega)^{\alpha+1}-2^\alpha(1+\omega)+\alpha 2^\alpha(1-\omega))\left|f''\left(\frac{n+\omega}{n+1}a+\frac{1-\omega}{n+1}b\right)\right|^q d\omega\right)^{1/q} \\
&+\frac{(b-a)^2}{(n+1)^{\alpha+3}(\alpha+1)}\left(\int_0^1((1+\omega)^{\alpha+1}-2^\alpha(1+\omega)+\alpha 2^\alpha(1-\omega))d\omega\right)^{1-1/q} \\
&\times\left(\int_0^1((1+\omega)^{\alpha+1}-2^\alpha(1+\omega)+\alpha 2^\alpha(1-\omega))\left|f''\left(\frac{1-\omega}{n+1}a+\frac{n+\omega}{n+1}b\right)\right|^q d\omega\right)^{1/q}
\end{aligned}$$

$$\leq \frac{(b-a)^2}{(n+1)^{\frac{\alpha+3}{q}}(\alpha+1)}\left(\frac{1}{\alpha+2}\right)^{1-1/q}\left(\int_0^1 (1-\omega)^{\alpha+1}\left((n+\omega)|f''(a)|^q+(1-\omega)|f''(b)|^q\right)d\omega\right)^{1/q}$$
$$+\frac{(b-a)^2}{(n+1)^{\frac{\alpha+3}{q}}(\alpha+1)}\left(\frac{1}{\alpha+2}\right)^{1-1/q}\left(\int_0^1 (1-\omega)^{\alpha+1}\left((1-\omega)|f''(a)|^q+(n+\omega)|f''(b)|^q\right)d\omega\right)^{1/q}$$
$$+\frac{(b-a)^2}{(n+1)^{\frac{\alpha+3}{q}}(\alpha+1)}\left(\frac{2^{\alpha+2}-1}{\alpha+2}-2^{\alpha+1}+\alpha 2^{\alpha-1}+2^{\alpha-1}-3\alpha 2^{\alpha-1}\right)^{1-1/q}$$
$$\times\left(\int_0^1 ((1+\omega)^{\alpha+1}-2^{\alpha}(1+\omega)+\alpha 2^{\alpha}(1-\omega))\left((n+\omega)|f''(a)|^q+(1-\omega)|f''(b)|^q\right)d\omega\right)^{1/q}$$
$$+\frac{(b-a)^2}{(n+1)^{\frac{\alpha+3}{q}}(\alpha+1)}\left(\frac{2^{\alpha+2}-1}{\alpha+2}-2^{\alpha+1}+\alpha 2^{\alpha-1}+2^{\alpha-1}-3\alpha 2^{\alpha-1}\right)^{1-1/q}$$
$$\times\left(\int_0^1 ((1+\omega)^{\alpha+1}-2^{\alpha}(1+\omega)+\alpha 2^{\alpha}(1-\omega))\left((n+\omega)|f''(a)|^q+(1-\omega)|f''(b)|^q\right)d\omega\right)^{1/q}.$$

Simple computations give

$$\int_0^1 (1-\omega)^{\alpha+1}(n+\omega)\,d\omega = \frac{n\alpha+3n+1}{(\alpha+2)(\alpha+3)} = U_1,$$
$$\int_0^1 (1-\omega)^{\alpha+1}(1-\omega)\,d\omega = \frac{1}{\alpha+3} = U_2,$$

$$\int_0^1 ((1+\omega)^{\alpha+1}-2^{\alpha}(1+\omega)+\alpha 2^{\alpha}(1-\omega))(n+\omega)\,d\omega$$
$$= \frac{6+2^{\alpha}(\alpha^3+5\alpha-6)+3n(\alpha+3)(-2+2^{\alpha}(2+(\alpha-1)\alpha))}{6(\alpha+2)(\alpha+3)} = U_3,$$

$$\int_0^1 ((1+\omega)^{\alpha+1}-2^{\alpha}(1+\omega)+\alpha 2^{\alpha}(1-\omega))(1-\omega)\,d\omega$$
$$= \frac{-3(\alpha+4)+2^{\alpha}(12+(\alpha-1)\alpha(\alpha+4))}{3(\alpha+2)(\alpha+3)} = U_4,$$
$$\int_0^1 (1-\omega)^{\alpha+1}\,d\omega = \frac{1}{\alpha+2} = U_5,$$
$$\int_0^1 ((1+\omega)^{\alpha+1}-2^{\alpha}(1+\omega)+\alpha 2^{\alpha}(1-\omega))d\omega$$
$$= \frac{2^{\alpha+2}-1}{\alpha+2}-2^{\alpha+1}+\alpha 2^{\alpha-1}+2^{\alpha-1}-3\alpha 2^{\alpha-1} = U_6.$$

This completes the proof. □

**Remark 2.** *If we take $\alpha = n = 1$ in Theorem 5, then the inequality (6) reduces to the inequality (4). The inequality in (1) was obtained by Sarikaya [6].*

In the following theorem, we obtain estimate of Hermite–Hadamard inequality for concave function.

**Theorem 6.** *Let $f : [a,b] \to \mathbb{R}$ be a twice differentiable function on $(a,b)$ with $a < b$ and $n \in N$. If $f'' \in L[a,b]$ and $|f''|^q$ is concave on $[a,b]$, then the following inequality for Riemann–Liouville fractional integrals holds with $0 < \alpha \leq 1$:*

$$\left|\frac{\Gamma(\alpha+2)}{(n+1)^2(b-a)^\alpha}\left[J^\alpha_{\left(\frac{n}{n+1}a+\frac{1}{n+1}b\right)^-}f(a)+J^\alpha_{\left(\frac{1}{n+1}a+\frac{n}{n+1}b\right)^+}f(b)\right]-f\left(\frac{n}{n+1}a+\frac{1}{n+1}b\right)\right|$$
$$\leq \frac{(b-a)^2}{(n+1)^{\alpha+3}(\alpha+1)}\times\left[U_5\left\{\left|f''\left(\left\{\frac{U_1(a)+U_2(b)}{U_5(n+1)}\right\}\right)\right|+\left|f''\left(\left\{\frac{U_1(b)+U_2(a)}{U_5(n+1)}\right\}\right)\right|\right\}\right. \tag{7}$$
$$\left.+U_6\left|f''\left(\left\{\frac{(U_3(a)+U_4(b)}{U_6(n+1)}\right\}\right)\right|+\left|f''\left(\left\{\frac{(U_3(b)+U_4(a)}{U_6(n+1)}\right\}\right)\right|\right].$$

**Proof.** Using the concavity of $|f''|^q$ and the power-mean inequality, we obtain

$$\begin{aligned}|f''(\lambda a+(1-\lambda)b)|^q &> \lambda|f''(a)|^q+(1-\lambda)|f''(b)|^q\\ &\geq (\lambda|f''(a)|+(1-\lambda)|f''(b)|)^q.\end{aligned}$$

Hence,

$$|f''(\lambda a+(1-\lambda)b)| \geq \lambda|f''(a)|+(1-\lambda)|f''(b)|,$$

thus $|f''|$ is also concave. Using Jensen integral inequality, we have

$$\left|\frac{\Gamma(\alpha+2)}{(n+1)^2(b-a)^\alpha}\left[J^\alpha_{\left(\frac{n}{n+1}a+\frac{1}{n+1}b\right)^-}f(a)+J^\alpha_{\left(\frac{1}{n+1}a+\frac{n}{n+1}b\right)^+}f(b)\right]-f\left(\frac{n}{n+1}a+\frac{1}{n+1}b\right)\right|$$
$$\leq \frac{(b-a)^2}{(n+1)^{\alpha+3}(\alpha+1)}\left(\int_0^1(1-\omega)^{\alpha+1}d\omega\right)\left|f''\left(\frac{\int_0^1(1-\omega)^{\alpha+1}\left|\left(\frac{n+\omega}{n+1}a+\frac{1-\omega}{n+1}b\right)\right|^q d\omega}{\int_0^1(1-\omega)^{\alpha+1}d\omega}\right)\right|^q$$
$$+\frac{(b-a)^2}{(n+1)^{\alpha+3}(\alpha+1)}\left(\int_0^1(1-\omega)^{\alpha+1}d\omega\right)\left|f''\left(\frac{\int_0^1(1-\omega)^{\alpha+1}\left|\left(\frac{1-\omega}{n+1}a+\frac{n+\omega}{n+1}b\right)\right|^q d\omega}{\int_0^1(1-\omega)^{\alpha+1}d\omega}\right)\right|^q$$
$$+\frac{(b-a)^2}{(n+1)^{\alpha+3}(\alpha+1)}\left(\int_0^1((1+\omega)^{\alpha+1}-2^\alpha(1+\omega)+\alpha 2^\alpha(1-\omega))d\omega\right)$$
$$\times\left|f''\left(\frac{\int_0^1((1+\omega)^{\alpha+1}-2^\alpha(1+\omega)+\alpha 2^\alpha(1-\omega))(\frac{n+\omega}{n+1}a+\frac{1-\omega}{n+1}b)d\omega}{\int_0^1((1+\omega)^{\alpha+1}-2^\alpha 1+\omega+\alpha 2^\alpha(1-\omega))d\omega}\right)\right|^q$$
$$+\frac{(b-a)^2}{(n+1)^{\alpha+3}(\alpha+1)}\left(\int_0^1((1+\omega)^{\alpha+1}-2^\alpha(1+\omega)+\alpha 2^\alpha(1-\omega))d\omega\right)$$
$$\times\left|f''\left(\frac{\int_0^1((1+\omega)^{\alpha+1}-2^\alpha(1+\omega)+\alpha 2^\alpha(1-\omega))(\frac{1-\omega}{n+1}a+\frac{n+\omega}{n+1}b)d\omega}{\int_0^1((1+\omega)^{\alpha+1}-2^\alpha(1-\omega)+\alpha 2^\alpha(1-\omega))d\omega}\right)\right|^q$$
$$\leq \frac{(b-a)^2}{(n+1)^{\alpha+3}(\alpha+1)}(U_5)\left|f''\left(\frac{U_1(a)+U_2(b)}{U_5(n+1)}\right)\right|^q+\frac{(b-a)^2}{(n+1)^{\alpha+3}(\alpha+1)}(U_5)\left|f''\left(\frac{U_1(b)+U_2(a)}{U_5(n+1)}\right)\right|^q$$
$$+\frac{(b-a)^2}{(n+1)^{\alpha+3}(\alpha+1)}(U_6)\left|f''\left(\frac{(U_3(a)+U_4(b)}{U_6(n+1)}\right)\right|^q+\frac{(b-a)^2}{(n+1)^{\alpha+3}(\alpha+1)}(U_6)\left|f''\left(\frac{(U_3(b)+U_4(a)}{U_6(n+1)}\right)\right|^q$$

$$\left|\frac{\Gamma(\alpha+2)}{(n+1)^2(b-a)^\alpha}\left[J^\alpha_{\left(\frac{n}{n+1}a+\frac{1}{n+1}b\right)^-}f(a)+J^\alpha_{\left(\frac{1}{n+1}a+\frac{n}{n+1}b\right)^+}f(b)\right]-f\left(\frac{n}{n+1}a+\frac{1}{n+1}b\right)\right|$$
$$\leq \frac{(b-a)^2}{(n+1)^{\alpha+3}(\alpha+1)}\times\left[U_5\left\{\left|f''\left(\left\{\frac{U_1(a)+U_2(b)}{U_5(n+1)}\right\}\right)\right|+\left|f''\left(\left\{\frac{U_1(b)+U_2(a)}{U_5(n+1)}\right\}\right)\right|\right\}\right.$$
$$\left.+U_6\left|f''\left(\left\{\frac{(U_3(a)+U_4(b)}{U_6(n+1)}\right\}\right)\right|+\left|f''\left(\left\{\frac{(U_3(b)+U_4(a)}{U_6(n+1)}\right\}\right)\right|\right].$$

The proof is completed. □

**Corollary 4.** *On letting $\alpha = n = 1$ in Theorem 6, the inequality in Equation (8) becomes:*

$$\left|\frac{1}{b-a}\int_a^b f(x)dx - f\left(\frac{a+b}{2}\right)\right| \le \frac{(b-a)^2}{48}\left|f''\left(\frac{5a+3b}{8}\right)\right| + \left|f''\left(\frac{3a+5b}{8}\right)\right|. \tag{8}$$

**Remark 3.** *The inequality in (8) is an improvement of the obtained inequality in Corollary 4 of [15]. This gives us a comparatively better estimate.*

## 3. Conclusions

We have derived some inequalities of Hermite–Hadamard type by establishing more general inequalities for functions that possesses second derivative on interior of an interval of real numbers, by using the Holder inequality and the assumptions that the mappings $|(f'')|^q$, for $q \ge 1$ are convex, as well as concave. The results presented here, certainly, provided refinements of those results proved in [6,15], since, by putting $\alpha = n = 1$ in our obtained inequalities, we achieve the already-presented inequalities in [6,15].

**Author Contributions:** All authors contributed equally for writing and analyzing this paper.

**Funding:** This research was partially supported by the Higher Education Commission of Pakistan [Grant number No. 5325/Federal/NRPU/$R\&D$/HEC/2016].

**Acknowledgments:** The first author is grateful to S. M. Junaid Zaidi, Executive Director and Raheel Qamar Rector, COMSATS University Islamabad, Pakistan for providing excellent research facilities.

**Conflicts of Interest:** Authors declare no conflict of interest.

## References

1. Dragomir, S.S.; Pearce, C.E.M. Selected Topics on Hermite−Hadamard Inequalities and Applications, RGMIA Monographs, Victoria University, 2000. Available online: http://www.sta.vu.edu.au/RGMIA/monographs/hermite_hadamard.html (accessed on 12 December 2018).
2. Dahmani, Z. New inequalities in fractional integrals. *Int. J. Nonlinear Sci.* **2000**, *9*, 493–497.
3. Dahmani, Z.; Tabharit, L.; Taf, S. Some fractional integral inequalities. *Nonlinear Sci. Lett. A* **2010**, *1*, 155–160.
4. Dragomir, S.S.; Bhatti, M.I.; Iqbal, M.; Muddassar, M. Some new Hermite–Hadamard type fractional Integral inequalities. *J. Comput. Anal. Appl.* **2015**, *18*, 655–661.
5. Iqbal, M.; Qaisar, S.; Muddassar, M. A short note on integral inequality of type Hermite-Hadamard through convexity. *J. Comput. Anal. Appl.* **2016**, *21*, 946–953.
6. Sarikaya, M.Z.; Aktan, N. On the generalization of some integral inequalities and their applications. *Math. Comput. Model.* **2011**, *54*, 2175–2182. [CrossRef]
7. Gorenflo, R.; Mainardi, F. *Fractional Calculus: Integral and Differential Equations of Fractional Order*; Springer: Wien, Austria, 1997; pp. 223–276.
8. Anastassiou, G.; Hooshmandasl, M.R.; Ghasemi, A.; Moftakharzadeh, F. Montogomery identities for fractional integrals and related fractional inequalities. *J. Ineq. Pure Appl. Math.* **2009**, *10*, 97.
9. Belarbi, S.; Dahmani, Z. On some new fractional integral inequalities. *J. Ineq. Pure Appl. Math.* **2009**, *10*, 86.
10. Dahmani, Z. On Minkowski and Hermite-Hadamard integral inequalities via fractional integration. *Ann. Funct. Anal.* **2010**, *1*, 51–58. [CrossRef]
11. Dahmani, Z.; Tabharit, L.; Taf, S. New generalizations of Gruss inequality using Riemann–Liouville fractional integrals. *Bull. Math. Anal. Appl.* **2010**, *2*, 93–99.
12. Bahtti, M.I.; Iqbal, M.; Dragomir, S.S. Some new integral ineuqalities of the type of Hermite-Hadamard's for the mappings whose absolute values of their derivativeare convex. *J. Comput. Anal. Appl.* **2014**, *16*, 643–653.
13. Sarikaya, M.Z.; Set, E.; Yaldiz, H.; Basak, N. Hermite-Hadamard's inequalities for fractional integrals and related fractional inequalities. *Math. Comput. Model.* **2013**, *57*, 2403–2407. [CrossRef]

14. Qaisar, S.; Iqbal, M.; Muddassar, M. New Hermite-Hadamard's inequalities for pre-invex function via fractional integrals. *J. Comput. Anal. Appl.* **2016**, *20*, 1318–1328.
15. Özdemir, M.E.; Yıldız, C.; Akdemir, A.O.; Set, E. On some inequalities for s-convex functions and applications. *J. Ineq. Appl.* **2013**, *2013*, 333. [CrossRef]

symmetry

MDPI

*Article*

# Balanced Truncation Model Order Reduction in Limited Frequency and Time Intervals for Discrete-Time Commensurate Fractional-Order Systems

**Marek Rydel *, Rafał Stanisławski and and Krzysztof J. Latawiec**

Department of Electrical, Control and Computer Engineering, Opole University of Technology, ul. Prószkowska 76, 45-758 Opole, Poland; r.stanislawski@po.opole.pl (R.S.); k.latawiec@po.opole.pl (K.J.L.)

* Correspondence: m.rydel@po.opole.pl

Received: 5 January 2019; Accepted: 18 February 2019; Published: 19 February 2019

**Abstract:** In this paper we investigate an implementation of new model order reduction techniques to linear time-invariant discrete-time commensurate fractional-order state space systems to obtain lower dimensional fractional-order models. Since the models of physical systems correctly approximate the physical phenomena of the modeled systems for restricted time and frequency ranges only, a special attention is given to time- and frequency-limited balanced truncation and frequency-weighted methods. Mathematical formulas for calculation of the time- and frequency-limited, as well as frequency-weighted controllability and observability Gramians, are extended to fractional-order systems. An instructive simulation experiment corroborates the potential of the introduced methodology.

**Keywords:** fractional-order system; model order reduction; controllability and observability Gramians

## 1. Introduction

In the field of modeling and simulation of fractional-order systems there are two different approaches to application of model order reduction (MOR) techniques: (1) Approximation of fractional-order systems by high integer-order models and their reduction to the low integer-order ones, and (2) reduction of the fractional-order systems without changing the class of the model, i.e., the reduced model is also the fractional-order one.

The first approach can be implemented by either determination of the fractional-order derivative/difference approximators involved in a fractional-order system [1,2] or by selection of integer-order approximators to the whole fractional-order systems [3–8]. In both approaches, a very high integer-order model is usually obtained, which is not effective from the computational point of view due to large memory requirements and long simulation times. For these reasons, classical model reduction techniques can be used to reduce integer-order model dimensions [8–10]. Therefore, the term MOR for fractional-order systems is usually related to order reduction of integer-order approximators to fractional-order systems.

This paper tackles the issue of MOR for commensurate fractional-order systems where a final result of using the MOR technique is the fractional-order model of lower dimensions. This issue has not been systematically studied and only a few approaches for reduction of continuous-time fractional-order systems have been published [11–14].

The classical Balanced Truncation (BT) method introduced for classical integer-order systems has been extended to discrete-time fractional-order systems [15]. The reduction paradigms used by the BT method enforce an accurate approximation for the whole range of frequencies. However, models of physical systems, e.g., models for mechanical, electrical and biological systems, characterize a certain adequacy scope, determined by the frequency range, for which the models correctly approximate the physical phenomena of the modeled systems. Likewise, when the reduced model is used to carry out a simulation in the determined time interval, an appropriate approximation accuracy of the output signal $y(t)$ is required only for $t$ lower than a specified final time of simulation. For these reasons, the reduction aims to determine such a reduced model which is particularly accurate in the given frequency range $[\omega_{min}, \omega_{max}]$ and/or time interval $[t_{min}, t_{max}]$. Such an approach allows for larger errors outside these specified intervals, without negative impact on the usefulness of the obtained reduced model. The time and frequency boundaries can be applied either by using frequency- and time-limited controllability and observability Gramians [16–19] or by frequency-weighted functions connected to the model which are the subject to the reduction process [20–26]. In this paper, we focus on the generalization of such approaches to reduction and an accurate approximation in given frequency and time intervals for the discrete-time commensurate fractional-order systems.

The remainder of this paper is structured as follows. A description of fractional-order state space systems considered in the paper is introduced in Section 2. Section 3 includes fundamentals of the MOR concept, in particular for the BT method. Section 4 contains the main result of the paper concerning definitions of controllability and observability Gramians in the time- and frequency-domains for fractional-order systems. Numerical examples of Section 5 illustrate the use of the introduced Gramians in the model reduction process. The paper is completed with the conclusion section.

## 2. System Representation

Consider a linear discrete-time commensurate fractional-order (DTCFO) state space system $G = \{A_f, B, C, D\}$ described by

$$\begin{aligned} \Delta^{\alpha} x(k+1) &= A_f x(k) + Bu(k), \quad x(0) = x_0 \\ y(k) &= Cx(k) + Du(k) \end{aligned} \tag{1}$$

where $k = 0, 1, \ldots,$ is the discrete time, $x(k) \in \Re^n$ is the state vector, $u(k)$ and $y(k)$ are the input and output signals, respectively. Matrices $A_f \in \Re^{n\times n}$, $B \in \Re^{n\times n_u}$, $C \in \Re^{n_y\times n}$, $D \in \Re^{n_y\times n_u}$ describe the system properties, with $n_u$ and $n_y$ being the numbers of inputs and outputs, respectively. $\Delta^{\alpha} x(k+1)$ defines the fractional-order difference of order $\alpha \in (0, 2)$ which can be represented by the Grünwald-Letnikov fractional difference ([27], ch. 3.5)

$$\Delta^{\alpha} x(k+1) = \sum_{j=0}^{k+1} (-1)^j \binom{\alpha}{j} x(k-j+1) \quad k = 0, 1, \ldots \tag{2}$$

with

$$\binom{\alpha}{j} = \begin{cases} 1 & j = 0 \\ \frac{\alpha(\alpha-1)\ldots(\alpha-j+1)}{j!} & j > 0 \end{cases}$$

Assuming the zero initial condition, that is $\Delta^{\alpha} x(0) = 0$, the $\mathcal{Z}$-transform of the system (1) is given by

$$\begin{aligned} w(z)X(z) &= A_f X(z) + BU(z) \\ Y(z) &= CX(z) + DU(z) \end{aligned} \tag{3}$$

where $w(z)$ is the $\mathcal{Z}$-transform of the "forward-shifted" fractional-order difference

$$w(z) = z(1 - z^{-1})^{\alpha} = \sum_{j=0}^{k+1} (-1)^j \binom{\alpha}{j} z^{-j+1} \tag{4}$$

**Remark 1.** *The above formulation can be extended to discretized models of continuous time systems. In the case of use of the forward-shifted Euler discretization operator ([27], ch. 3.5) we have*

$$s^{\alpha} \approx \frac{1}{h^{\alpha}} w(z) \tag{5}$$

*where $s^{\alpha}$ is the Laplace transform of the fractional-order derivative and $h$ is the sampling period. Discretized models of continuous-time fractional-order systems*

$$\begin{aligned} \Delta^{\alpha} x(t+h) &= h^{\alpha} \bar{A}_f x(t) + h^{\alpha} \bar{B} u(t), \quad x(0) = x_0 \\ y(t) &= C x(t) + D u(t) \end{aligned} \tag{6}$$

*where $t = kh$, can simply be transferred to the system (1) by using the following substitutions*

$$A_f \rightarrow h^{\alpha} \bar{A}_f \quad B \rightarrow h^{\alpha} \bar{B} \tag{7}$$

Implementation of the Grünwald–Letnikov difference (2) results in increasing computational burden at each time step, which finally becomes computationally infeasible. Therefore, in practical implementations, finite-length expansions are used, for instance finite fractional difference [6,8] ([28], ch. 7)

$$\Delta^{\alpha} x(k+1) \approx \sum_{j=0}^{L} (-1)^j \binom{\alpha}{j} x(k-j+1) \quad k = 0, 1, \ldots \tag{8}$$

with $x(l) = 0 \,\forall\, l < 0$ and $L$ being the implementation length.

It is worth emphasizing that precise approximation of the Grünwald–Letnikov difference with the finite fractional difference needs a very high length $L$ [6].

## 3. Model Order Reduction

Let us shift now to the MOR problem for the DTCFO system (1). The fractional-order MOR issue aims towards finding a DTCFO model $\tilde{G}$ with reduced dimension $r < n$

$$\begin{aligned} \Delta^{\alpha} \tilde{x}(k+1) &= \tilde{A}_f \tilde{x}(k) + \tilde{B} u(k), \quad \tilde{x}(0) = \tilde{x}_0 \\ \tilde{y}(k) &= \tilde{C} \tilde{x}(k) + \tilde{D} u(k) \end{aligned} \tag{9}$$

where $\tilde{A}_f \in \Re^{r \times r}$, $\tilde{B} \in \Re^{r \times n_u}$, $\tilde{C} \in \Re^{n_y \times r}$, $\tilde{D} \in \Re^{n_y \times n_u}$, $\tilde{x}(k) \in \Re^{r}$ are such that the approximation errors both in the time domain $\|y(k) - \tilde{y}(k)\|$ and in the frequency domain $\|G(z) - \tilde{G}(z)\|$ are small for the chosen norm $\|\cdot\|$.

In this paper, we concentrate on the BT technique, which is based on the concept of the balanced model realization [29,30], ([31], ch. 7.1). In order to arrive at the balanced system, the linear state transformation $x \rightarrow Tx$ is applied to diagonalize the controllability $P$ and observability $Q$ Gramians of the system

$$TPT^T = (T^T)^{-1} Q T^{-1} = \operatorname{diag}(\sigma_1, \ldots, \sigma_n) \tag{10}$$

where $\sigma_i$, $i = 1, ..., n$, are the square roots of the eigenvalues for the product of $P$ and $Q$, which are called the Hankel singular values (HSV). The magnitude of HSV classifies a degree of reachability and observability of states in the system. On this basis, reduction eliminates states corresponding to small values $\sigma_i$, which means that they have a weak impact on the system properties. The balanced model is obtained by applying transformation matrices in the following way:

$$TA_fT^{-1} = \begin{bmatrix} \tilde{A}_f & \tilde{A}_{12} \\ \tilde{A}_{21} & \tilde{A}_{22} \end{bmatrix}, \quad TB = \begin{bmatrix} \tilde{B} \\ \tilde{B}_2 \end{bmatrix}, \quad CT^{-1} = \begin{bmatrix} \tilde{C} & \tilde{C}_2 \end{bmatrix}, \quad \tilde{D} = D \tag{11}$$

The order $r$ can be selected on the basis of an approximation error of the reduced system. For the BT method, the $\mathcal{H}_\infty$ norm of the approximation error is upper bounded as follows [30], ([31], ch. 7.2):

$$\left\|G(z) - \tilde{G}(z)\right\|_{\mathcal{H}_\infty} \leq 2 \sum_{j=r+1}^{n} \sigma_j \tag{12}$$

Calculation of the transformation matrix $T$ is not a unique operation. The exemplary algorithms can be found in References [23,30] ([31], ch. 7.3), [32].

The BT reduction method enforces an accurate approximation for all times $t \in (0, \infty)$ and frequencies $\omega \in \Re$. If it is necessary to determine a model (9) which is particularly accurate in a given frequency range $[\omega_{min}, \omega_{max}]$ and/or time interval $[k_{min}, k_{max}]$, then frequency- or time-limited controllability and observability Gramians can be used to calculate the transformation matrix $T$ instead of infinite ones. Higher model accuracy can be obtained now, especially when the optimal values of the parameters of the weighting functions and frequency-/time-intervals are used [33].

## 4. Controllability and Observability Gramians for Discrete-Time Fractional-Order Systems

In this section, the definitions of controllability and observability Gramians both in the time- and frequency-domains are recalled. Based on the definitions for integer-order systems, the Gramian generalizations to fractional-order systems are derived here.

### *4.1. Gramians in the Time Domain*

The definition of the controllability Gramian is connected to the minimal energy required for the transfer of the system from the zero initial state $x(0) = 0$ to the final state $x(k) = x_p$, whereas the observability Gramian is defined with relation to the energy generated by the nonzero initial state $x(0) = x_0$ with the zero input signal $u(k) = 0$. For asymptotically stable systems, the infinite controllability and observability Gramians are respectively defined as

$$P = \sum_{k=0}^{\infty} \xi(k)\xi^T(k), \qquad Q = \sum_{k=0}^{\infty} \eta^T(k)\eta(k) \tag{13}$$

where $\xi(k)$ is the state of the discrete-time system resulting from the input in the form of the Kronecker delta, and $\eta(k)$ is the output of the system produced by the nonzero initial conditions and the zero input signal.

For asymptotically stable integer-order systems, we obtain the well-known formulas for the controllability and observability Gramians, respectively:

$$P = \sum_{i=0}^{\infty} A^i BB^T \left(A^T\right)^i, \quad Q = \sum_{i=0}^{\infty} \left(A^T\right)^i C^T C A^i \tag{14}$$

where $A = A_f + I$. Finally, the Gramians $P$ and $Q$ are the solutions to the discrete-time Lyapunov equations, respectively, are

$$APA^T - P = -BB^T, \quad A^T QA - Q = -C^T C. \tag{15}$$

Based on definitions (13), it is easy to formulate the generalized form of the controllability and observability Gramians for the fractional-order systems.

**Lemma 1.** *Consider an asymptotically stable discrete-time commensurate fractional-order state space system (1), with the Grünwald–Letnikov fractional-order difference (2). Then the controllability and observability Gramians at finite time $k_L < \infty$ are as follows*

$$P(k_L) = \sum_{k=1}^{k_L} \phi(k-1)BB^T\phi^T(k-1), \qquad Q(k_L) = \sum_{k=0}^{k_L} \phi^T(k)\, C^T C\, \phi(k) \tag{16}$$

*where $\phi(k)$, $k = 0, 1, ...,$ are calculated as*

$$\phi(k) = \begin{cases} I & k = 0 \\ (A_f + \alpha I)\phi(k-1) - \sum\limits_{j=2}^{k} (-1)^j \binom{\alpha}{j} \phi(k-j) & k > 0 \end{cases}$$

**Proof.** The state space equations of the system as in Equation (1) can be rewritten as ([27], ch. 3.5)

$$\begin{aligned} x(k+1) &= (A_f + \alpha I)x(k) - \sum_{i=2}^{k+1} (-1)^i \binom{\alpha}{i} x(k+1-i) + Bu(k) \\ y(k) &= Cx(k) + Du(k) \end{aligned} \tag{17}$$

The state response for the DTCFO system (17), with the zero initial condition and the Kronecker delta input signal, is as follows:

$$\xi(k) = \begin{cases} 0 & k = 0 \\ B & k = 1 \\ (A_f + \alpha I)\xi(k-1) - \sum\limits_{j=2}^{k} (-1)^j \binom{\alpha}{j} \xi(k-j) & k > 1 \end{cases} \tag{18}$$

The output response for the DTCFO system (17), with the nonzero initial condition $x(0) = x_0$ and the zero input signal, is now

$$y(k) = \eta(k)x(0) \tag{19}$$

where

$$\eta(k) = \begin{cases} C & k = 0 \\ C\left( (A_f + \alpha I)x(k-1) - \sum\limits_{j=2}^{k} (-1)^j \binom{\alpha}{j} x(k-j) \right) & k > 0 \end{cases} \tag{20}$$

The responses (18) and (20) immediately result in Equation (16), which completes the proof. □

**Remark 2.** *As mentioned before, both the minimal energy required for the transfer of the system from the zero initial state to $x(k) = x_p$ and the energy generated by the nonzero initial state are obtained for $k_L \to \infty$. Implementation of Equation (16) in order to calculate controllability and observability Gramians implies the infinite number of elements $\phi(k)$. Furthermore each of $\phi(k)$ requires the determination of the Grünwald–Letnikov difference calculated from 0 to $k+1$, which is computationally infeasible. Therefore, in contrast to integer-order systems for which the solution of Equation (13) can be determined by solving Lyapunov Equation (15), the Gramians for the fractional-order systems can be calculated for the finite length only.*

If the response of the reduced model (9) is expected to match the original fractional-order model (1) in some interval $[k_1, k_2]$, then the balancing of the model can be executed on the basis of the Gramians calculated within this restricted time interval. As the controllability and observability Gramians for the asymptotically stable systems are positive definite, then it can be shown that

$$P(k_2) \geq P(k_1) \quad Q(k_2) \geq Q(k_1) \quad \text{for} \quad k_2 \geq k_1 \tag{21}$$

Therefore, the time-limited Gramians are also positive definite, and within the restricted time interval $[k_1, k_2]$ they can be calculated as follows:

$$P(k_1, k_2) = P(k_2) - P(k_1), \qquad Q(k_1, k_2) = Q(k_2) - Q(k_1) \tag{22}$$

where $P(k)$ and $Q(k)$ are given from Equation (16).

## 4.2. *Gramians in the Frequency Domain*

In addition to the definition (13) given in the time domain, the Gramians can also be expressed in the frequency domain. The connection can be made on the basis of the Plancherel's theorem, which states that the integral of the inner product of two functions in the time domain is equal to the integral of their frequency spectrum. In particular, for the asymptotically stable discrete-time integer-order systems, applying the Plancherel's theorem to Equation (14) yields ([31], ch. 4.3)

$$\begin{aligned} P &= \frac{1}{2\pi}\int_0^{2\pi}\left(e^{j\theta}I - A\right)^{-1} BB^T \left((e^{-j\theta}I - A^T\right)^{-1} d\theta \\ Q &= \frac{1}{2\pi}\int_0^{2\pi}\left(e^{-j\theta}I - A^T\right)^{-1} C^T C \left(e^{j\theta}I - A\right)^{-1} d\theta \end{aligned} \tag{23}$$

Based on definitions (23), it is easy to formulate the generalized form of the controllability and observability Gramians for the fractional-order systems.

**Lemma 2.** *Consider an asymptotically stable discrete-time commensurate fractional-order state space system* (1), *with the Grünwald–Letnikov fractional-order difference* (2). *Then the infinite controllability and observability Gramians of the fractional-order system are respectively given as*

$$\begin{aligned} P &= \frac{1}{2\pi}\int_{-\pi}^{+\pi}\left(w(z)\, I - A_f\right)^{-1} BB^T \left(w^*(z)\, I - A_f^T\right)^{-1} d\theta \\ Q &= \frac{1}{2\pi}\int_{-\pi}^{+\pi}\left(w^*(z)\, I - A_f^T\right)^{-1} C^T C \left(w(z)\, I - A_f\right)^{-1} d\theta \end{aligned} \tag{24}$$

*where $w(z)$ is as in Equation* (4), *with $z = e^{i\theta}$, $\theta \in [-\pi, \pi]$, and $^*$ denotes the complex conjugate transpose.*

**Proof.** For continuous-time fractional-order systems referred to in Remark 1, the input-to-state map becomes $(s^\alpha I - \bar{A}_f)^{-1}\bar{B}$, while the state-to-output map is $C(s^\alpha I - \bar{A}_f)^{-1}$. Then the controllability and observability Gramians are as follows:

$$\begin{aligned} P &= \frac{1}{2\pi}\int_{-\infty}^{\infty}\left(s^\alpha I - \bar{A}_f\right)^{-1} \bar{B}\bar{B}^T \left((s^\alpha)^* I - \bar{A}_f^T\right)^{-1} d\omega \\ Q &= \frac{1}{2\pi}\int_{-\infty}^{\infty}\left((s^\alpha)^* I - \bar{A}_f^T\right)^{-1} C^T C \left(s^\alpha I - \bar{A}_f\right)^{-1} d\omega \end{aligned} \tag{25}$$

where $s = i\omega$, $\omega \in (-\infty, \infty)$. Using the forward-shifted Euler discretization operator as referred to in Remark 1 results immediately in (24), which completes the proof. □

If the response of the reduced model is expected to match the full fractional-order model output within a restricted frequency range, then the balancing of the model can be executed on the basis of the Gramians calculated for that specific interval. Similarly, as for time-limited Gramians, the frequency-limited Gramians in restricted frequency interval $[\Theta_1, \Theta_2]$ can be calculated as follows:

$$P(\Theta_1, \Theta_2) = P(\Theta_2) - P(\Theta_1), \qquad Q(\Theta_1, \Theta_2) = Q(\Theta_2) - Q(\Theta_1) \tag{26}$$

where $P(\Theta)$ and $Q(\Theta)$ are given as

$$\begin{aligned} P(\Theta) &= \frac{1}{2\pi}\int_{-\Theta}^{+\Theta}\left(w(z)\,I - A_f\right)^{-1} BB^T\left(w^*(z)I - A_f^T\right)^{-1} d\theta \\ Q(\Theta) &= \frac{1}{2\pi}\int_{-\Theta}^{+\Theta}\left(w^*(z)\,I - A_f^T\right)^{-1} C^TC\left(w(z)I - A_f\right)^{-1} d\theta \end{aligned} \tag{27}$$

### 4.3. Frequency Weighted Gramians

The required approximation accuracy in the given frequency interval $[\Theta_1, \Theta_2]$ can also be achieved by implementation of the frequency weighting functions in a form of the external systems connected to the inputs and/or outputs of the full fractional-order model ([31], ch. 7.6). Such an approach, called Frequency Weighted (FW) method, is a generalization to the BT method designed for asymptotically stable models with asymptotically stable input and output weighting functions with minimal realizations.

Consider an asymptotically stable discrete-time integer- or fractional-order LTI MIMO state space system as an input weighting function $H_i = \{A_i, B_i, C_i, D_i\}$

$$\begin{aligned} \Delta^{\alpha_i} x_i(k+1) &= A_i x_i(k) + B_i u_i(k), \\ y_i(k) &= C_i x_i(k) + D_i u_i(k) \end{aligned} \tag{28}$$

and as an output weighting function $H_o = \{A_o, B_o, C_o, D_o\}$

$$\begin{aligned} \Delta^{\alpha_o} x_o(k+1) &= A_o x_o(k) + B_o u_o(k), \\ y_o(k) &= C_o x_o(k) + D_o u_o(k) \end{aligned} \tag{29}$$

of orders $n_i$ and $n_o$, respectively. Assuming that no pole-zero cancellations occur during the design of the augmented systems $GH_i$ and $H_oG$, we arrive at [22,25], ([31], ch. 7.6)

$$\begin{aligned} GH_i &= \left[\begin{array}{c|c} \tilde{A}_i & \tilde{B}_i \\ \hline \tilde{C}_i & \tilde{D}_i \end{array}\right] = \left[\begin{array}{cc|c} A_f & BC_i & BD_i \\ 0 & A_i & B_i \\ \hline C & DC_i & DD_i \end{array}\right] \\ H_oG &= \left[\begin{array}{c|c} \tilde{A}_o & \tilde{B}_o \\ \hline \tilde{C}_o & \tilde{D}_o \end{array}\right] = \left[\begin{array}{cc|c} A_f & 0 & B \\ B_oC & A_o & B_oD \\ \hline D_oC & C_o & D_oD \end{array}\right] \end{aligned} \tag{30}$$

It is well known that the frequency weighted controllability and observability Gramians are computed on the basis of the system connected to the input weight $GH_i$ and to the output weight $H_oG$, respectively.

**Lemma 3.** *Consider asymptotically stable augmented systems $GH_i$ and $H_oG$ as in Equation (30) consisting of asymptotically stable discrete-time commensurate fractional-order state space system (1), with the Grünwald–Letnikov fractional-order difference (2) and the weighting functions as in Equations (28) and (29). The controllability and observability Gramians of such systems at finite time $k_L < \infty$ are as follows:*

$$P_i(k_L) = \sum_{k=1}^{k_L} \phi_i(k-1)\,\tilde{B}_i\tilde{B}_i^T\,\phi_i^T(k-1), \qquad Q_o(k_L) = \sum_{k=0}^{k_L} \phi_o^T(k)\,\tilde{C}_o^T\tilde{C}_o\,\phi_o(k) \tag{31}$$

*where* $\phi_i(k)$, $\phi_o(k)$, $k = 0, 1, ...$, *are calculated in a recurrent way:*

$$\phi_i(k) = \begin{cases} I & k = 0 \\ (\tilde{A}_i + \tilde{\alpha}_i)\,\phi_i(k-1) - \sum_{j=2}^{k} \psi_i(j)\phi_i(k-j) & k > 0 \end{cases}$$

$$\phi_o(k) = \begin{cases} I & k = 0 \\ (\tilde{A}_o + \tilde{\alpha}_o)\,\phi_o(k-1) - \sum_{j=2}^{k} \psi_o(j)\phi_o(k-j) & k > 0 \end{cases}$$

*with* $\tilde{\alpha}_i = diag(\underbrace{\alpha, \ldots, \alpha}_{n}, \underbrace{\alpha_i, \ldots, \alpha_i}_{n_i})$, $\tilde{\alpha}_o = diag(\underbrace{\alpha, \ldots, \alpha}_{n}, \underbrace{\alpha_o, \ldots, \alpha_o}_{n_o})$ *and*

$$\psi_i(j) = (-1)^j diag\left(\underbrace{\binom{\alpha}{j}, \ldots, \binom{\alpha}{j}}_{n}, \underbrace{\binom{\alpha_i}{j}, \ldots, \binom{\alpha_i}{j}}_{n_i}\right)$$

$$\psi_o(j) = (-1)^j diag\left(\underbrace{\binom{\alpha}{j}, \ldots, \binom{\alpha}{j}}_{n}, \underbrace{\binom{\alpha_o}{j}, \ldots, \binom{\alpha_o}{j}}_{n_o}\right)$$

**Proof.** The proof directly stems from proof for Lemma 1 with substitution of the system matrices by the matrices of the augmented systems $GH_i$ and $H_oG$. □

Note that the application of the weighting functions influences the order of the controllability $P_i$ and observability $Q_o$ Gramians for the augmented systems. Therefore, they are partitioned into two-by-two blocks, so that the dimension of the $P_{11} \in \Re^{n \times n}$ and $Q_{11} \in \Re^{n \times n}$ are the same as the state matrix $A_f$

$$P_i = \begin{bmatrix} P_{11} & P_{12} \\ P_{12}^T & P_{22} \end{bmatrix}, \qquad Q_o = \begin{bmatrix} Q_{11} & Q_{12} \\ Q_{12}^T & Q_{22} \end{bmatrix} \tag{32}$$

Finally, the frequency-weighted controllability $\tilde{P}$ and observability $\tilde{Q}$ Gramians of dimensions $n \times n$ can be assumed as [20]

$$\tilde{P} = P_{11}, \qquad \tilde{Q} = Q_{11} \tag{33}$$

The proposed solution, despite its simplicity, may lead to the instability of the reduced model in case of two-sided weighting. Therefore, several modifications to this approach have been proposed to cope with this problem [21–25].

The frequency-weighted Gramians as in Equation (33) can also be defined in the frequency domain. Given that the input-to-state map and the state-to-output map of the fractional-order system are modified by the connected weighting functions, it is possible to generalize the definitions of the infinite controllability and observability Gramians for the fractional-order system (Lemma 2) to the frequency-weighted Gramians.

**Lemma 4.** *Consider asymptotically stable augmented systems* $GH_i$ *and* $H_oG$ *as in Equation (30), consisting of asymptotically stable discrete-time commensurate fractional-order state space system (1), with the Grünwald–Letnikov fractional-order difference (2) and the weighting functions as in Equations (28) and (29). Then the frequency-weighted controllability* $\tilde{P}$ *and observability* $\tilde{Q}$ *Gramians are defined as*

$$\tilde{P} = \frac{1}{2\pi}\int_{-\pi}^{\pi}\left(w(z)I - A_f\right)^{-1} B\, H_i(w_i(z))\, H_i^T\,(w_i^*(z))\, B^T\left(w^*(z)I - A_f^T\right)^{-1} d\theta$$
$$\tilde{Q} = \frac{1}{2\pi}\int_{-\pi}^{\pi}\left(w^*(z)I - A_f^T\right)^{-1} C^T\, H_o^T(w_o^*(z))\; H_o(w_o(z))\; C\left(w(z)I - A_f\right)^{-1} d\theta \tag{34}$$

*where* $H_i(w_i(z)) = D_i + C_i(w_i(z)I - A_i)^{-1}B_i$, $H_o(w_o(z)) = D_o + C_o(w_o(z)I - A_o)^{-1}B_o$ *and* $w(z) = z(1-z^{-1})^{\alpha}$, $w_i(z) = z(1-z^{-1})^{\alpha_i}$, $w_o(z) = z(1-z^{-1})^{\alpha_o}$, *with* $z = e^{i\theta}$, $\theta \in [-\pi, \pi]$.

**Proof.** Given that the input-to-state map for the augmented system $GH_i$ and the state-to-output map for $H_oG$ become $\left(\tilde{w}_i(z) - \tilde{A}_i\right)^{-1}\tilde{B}_i$ and $\tilde{C}_o\left(\tilde{w}_o(z) - \tilde{A}_o\right)^{-1}$, where $\tilde{w}_i(z) = \text{diag}\left(\underbrace{w(z),\ldots,w(z)}_{n},\underbrace{w_i(z),\ldots,w_i(z)}_{n_i}\right)$ and $\tilde{w}_o(z) = \text{diag}\left(\underbrace{w(z),\ldots,w(z)}_{n},\underbrace{w_o(z),\ldots,w_o(z)}_{n_o}\right)$, respectively. The $\tilde{P}$ and $\tilde{Q}$ are the submatrices of the Gramians for the augmented systems (32) partitioned into two-by-two blocks. Therefore, the proof for the frequency-weighted controllability Gramian follows by noticing that

$$\begin{aligned}
&\begin{pmatrix} I & 0 \end{pmatrix}\left(\begin{bmatrix} w(z)I & 0 \\ 0 & w_i(z)I \end{bmatrix} - \begin{bmatrix} A_f & BC_i \\ 0 & A_i \end{bmatrix}\right)^{-1}\begin{bmatrix} BD_i \\ B_i \end{bmatrix} = \\
&\begin{pmatrix} I & 0 \end{pmatrix}\begin{bmatrix} w(z)I - A_f & -BC_i \\ 0 & w_i(z)I - A_i \end{bmatrix}^{-1}\begin{bmatrix} BD_i \\ B_i \end{bmatrix} = \\
&\begin{pmatrix} I & 0 \end{pmatrix}\begin{bmatrix} (w(z)I - A_f)^{-1} & (w(z)I - A_f)^{-1}BC_i(w_i(z)I - A_i)^{-1} \\ 0 & (w_i(z)I - A_i)^{-1} \end{bmatrix}\begin{bmatrix} BD_i \\ B_i \end{bmatrix} = \\
&(w(z)I - A_f)^{-1}B(D_i + C_i(w_i(z)I - A_i)^{-1}B_i) = (w(z)I - A_f)^{-1}BH_i(w_i(z))
\end{aligned}$$

while for the frequency-weighted observability Gramian,

$$\begin{aligned}
&\begin{bmatrix} D_oC & C_o \end{bmatrix}\left(\begin{bmatrix} w(z)I & 0 \\ 0 & w_o(z)I \end{bmatrix} - \begin{bmatrix} A_f & 0 \\ B_oC & A_o \end{bmatrix}\right)^{-1}\begin{pmatrix} I \\ 0 \end{pmatrix} = \\
&\begin{bmatrix} D_oC & C_o \end{bmatrix}\begin{bmatrix} w(z)I - A_f & 0 \\ -B_oC & w_o(z)I - A_o \end{bmatrix}^{-1}\begin{pmatrix} I \\ 0 \end{pmatrix} = \\
&\begin{bmatrix} D_oC & C_o \end{bmatrix}\begin{bmatrix} (w(z)I - A_f)^{-1} & 0 \\ (w_o(z)I - A_o)^{-1}B_oC(w(z)I - A_f)^{-1} & (w_o(z)I - A_o)^{-1} \end{bmatrix}\begin{pmatrix} I \\ 0 \end{pmatrix} = \\
&(D_o + C_o(w_o(z)I - A_o)^{-1}B_o)(w(z)I - A_f)^{-1}B = H_o(w_o(z))(w(z)I - A_f)^{-1}B
\end{aligned}$$

Therefore, the frequency-weighted Gramians are the blocks $P_{11}$ and $Q_{11}$ of the controllability $P_i$ and observability $Q_o$ Gramians for the augmented systems, respectively, which completes the proof. □

**Remark 3.** *It is important to note that Lemmas 1 to 4 introduce various definitions of controllability and observability Gramians. However, the calculations of the Gramians directly from the above definitions for large-scale DTCFO systems are computationally demanding. In particular, time-domain Gramian definitions as in Lemma 1 and 3 are infeasible for large-scale systems due to the requirement of calculation of the Grünwald–Letnikov difference from 0 to* $k+1$. *The common practice for integer order systems when* $n_u, n_y << n$ *is to compute low-rank approximations of the Gramians such that* $P \approx SYS^T$ *with* $S \in \Re^{n\times l}$, $Y \in \Re^{l\times l}$, $l << n$. *It is motivated by the typical rapid decay of HSV [34,35], which can also be assumed for fractional-order systems. There exists various algorithms for calculating the low-rank Gramian factorizations for integer-order* *systems [17,19,35,36]. Therefore, future works will be carried out towards the extension of the low-rank approach to increase the efficiency of Gramians calculations for DTCFO systems.*

## 5. Simulation Examples

In this section, examples of model order reduction for a fractional-order system are presented. All reduced models were obtained by using the BT method with the Gramians calculated within various time- and frequency-intervals, as well as with different frequency-weighted functions. In particular, to calculate transformation matrix $T$ the following Gramians are selected: (1) indefinite Gramians defined in the frequency domain as in Equation (24) - denoted as the BT, (2) frequency-limited Gramians as in Equation (26), denoted as the FLBT, (3) time-limited Gramians as in Equation (16), denoted as the TLBT, (4) frequency-weighted Gramians as in Equation (34), denoted as the FW. The examined discrete-time fractional-order model is an extension of the continuous-time model for a simple mechanical system presented in [16] and (moderate) large-scale dynamical system of order 1006 as in [36].

**Example 1.** *Consider the DTCFO state space system (6) with the sampling period $h = 0.01$, fractional order $\alpha = 0.85$ and*

$$\left[\begin{array}{c|c} \bar{A}_f & \bar{B} \\ \hline C & D \end{array}\right] = \left[\begin{array}{cccccc|c} 0 & 0 & 0 & 1 & 0 & 0 & 0 \\ 0 & 0 & 0 & 0 & 1 & 0 & 0 \\ 0 & 0 & 0 & 0 & 0 & 1 & 0 \\ -5.4545 & 4.5455 & 0 & -0.0545 & 0.0455 & 0 & 0.0909 \\ 10 & -21 & 11 & 0.1 & -0.21 & 0.11 & 0.4 \\ 0 & 5.5 & -6.5 & 0 & 0.055 & -0.065 & -0.5 \\ \hline 2 & -2 & 3 & 0 & 0 & 0 & 0 \end{array}\right]$$

*Model order reduction is performed from the original six states variables to the reduced four ones. Frequency responses for the full fractional-order system and the reduced models in addition to approximation errors are depicted in Figure 1. Figure 2 shows step responses and their approximation errors for the same models. It is clearly visible that the reduction based on infinite and time-limited Gramians for the time interval $t \in [0, 10]$ (s) cannot properly approximate the low-frequency properties of the system. For this reason, in order to improve the approximation for low frequencies, the frequency-interval for frequency-limited Gramians is chosen as $\Theta \in [0, 0.01]$ (rad/s). For the same purpose, the low-pass Butterworth filter of order $n_f = 5$ and cut-off frequency $\omega_f = 0.01$ (rad/s) is selected as a frequency weighted function. Table 1 presents approximation errors for the considered models, where DCE is the steady state approximation error, $MSE_\omega$ is the mean square approximation error for the frequency responses in the frequency range $\omega \in [10^{-3}, 1]$ (rad/s), $\mathcal{H}_\infty$ is the norm approximation error and $MSE_t$ is the mean square error for system step response in the discrete-time range $t \in [0, 100]$ (s).*

*Figure 3 presents frequency responses and approximation errors for the reduced models obtained in order to improve the quality of approximation for high frequencies. In particular, the third resonance frequency $\omega = 6.5$ (rad/s) is considered. Figure 4 shows the impulse responses and their approximation errors for the same models. For this purpose, the time interval $t \in [0, 0.1]$ (s) and frequency interval $\Theta \in [0.1, 100\pi]$ (rad/s) are selected for time- and frequency-limited Gramians. Similarly, the frequency weighted function is chosen in a form of a high-pass Butterworth filter of order $n_f = 5$ and cut-off frequency $\omega_f = 0.1$ (rad/s). Table 2 presents approximation errors for the analyzed models, in terms of $MSE_\omega$ for $\omega > 3$ (rad/s), $\mathcal{H}_\infty$-norm and $MSE_t$ for system impulse responses within $t \in [0, 1]$ (s).*

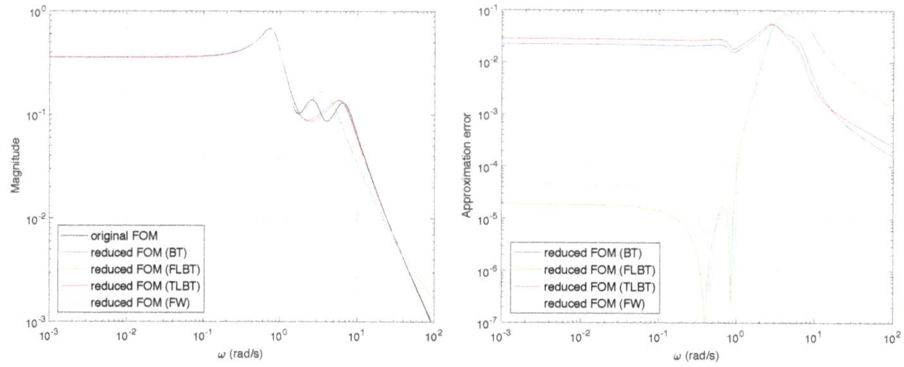


**Figure 1.** (**left**) Frequency responses for full- and reduced-order models and (**right**) approximation errors.

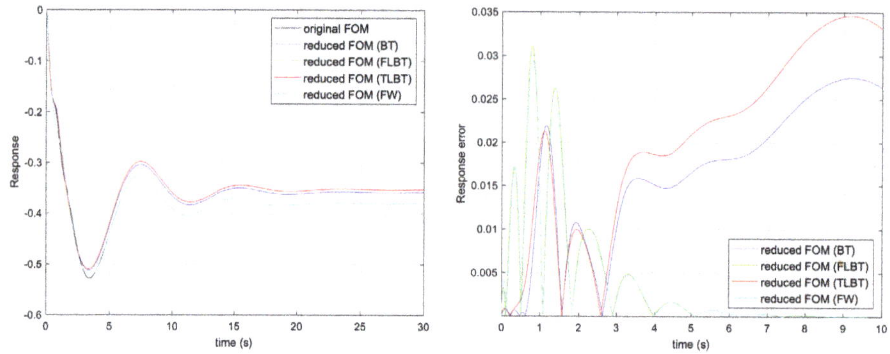


**Figure 2.** (**left**) Step responses for full- and reduced-order models and (**right**) approximation errors.

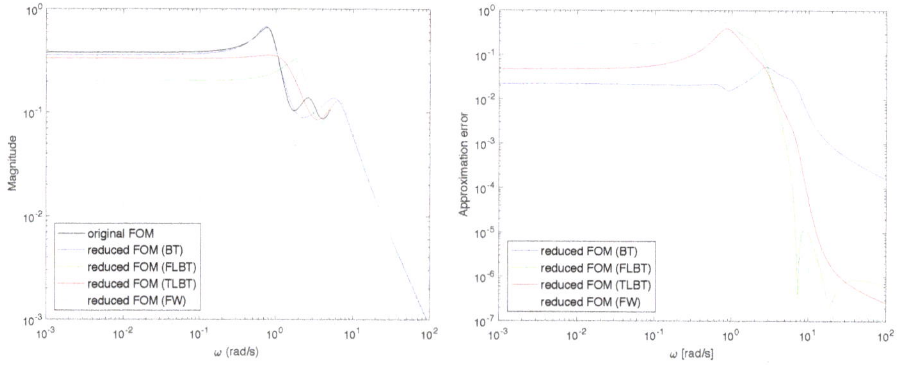


**Figure 3.** (**left**) Frequency responses for full- and reduced-order models and (**right**) approximation errors.

**Table 1.** Frequency and step approximation errors for the reduced models.

| | **DCE** | $MSE_\omega$ | $\mathcal{H}_\infty$ | $MSE_t$ |
|---|---|---|---|---|
| BT | $22.2 \times 10^{-3}$ | $4.58 \times 10^{-4}$ | $53.0 \times 10^{-3}$ | $4.62 \times 10^{-4}$ |
| FLBT | $1.88 \times 10^{-5}$ | $2.68 \times 10^{-10}$ | $88.8 \times 10^{-3}$ | $6.37 \times 10^{-6}$ |
| TLBT | $28.1 \times 10^{-3}$ | $7.33 \times 10^{-4}$ | $54.9 \times 10^{-3}$ | $7.40 \times 10^{-4}$ |
| FW | $4.52 \times 10^{-5}$ | $1.42 \times 10^{-9}$ | $83.7 \times 10^{-3}$ | $5.60 \times 10^{-6}$ |

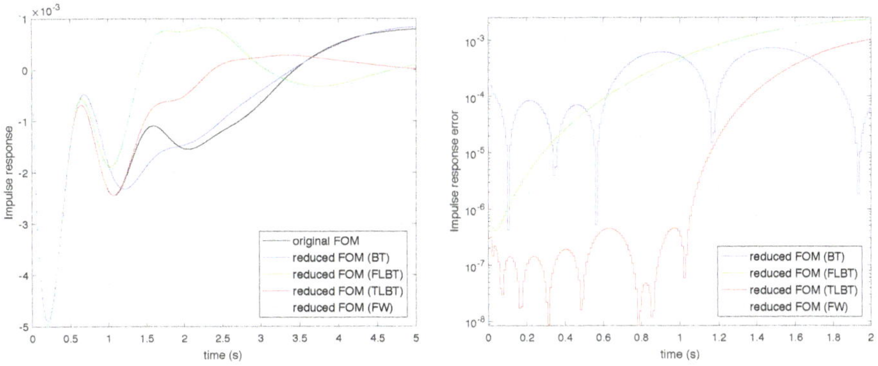

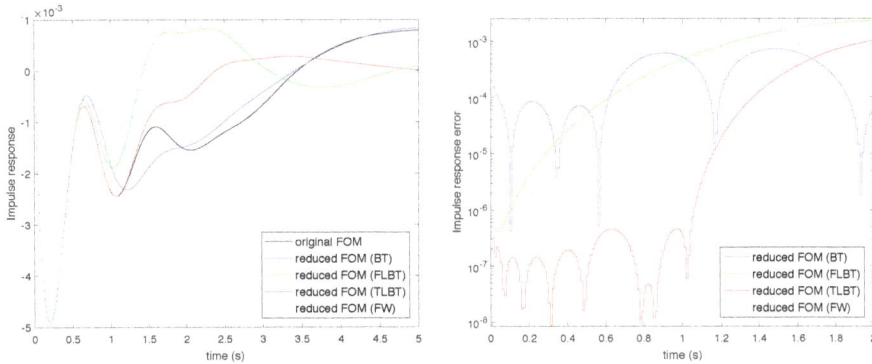

**Figure 4.** (**left**) Impulse responses for full- and reduced-order models and (**right**) approximation errors.

**Table 2.** Frequency and impulse approximation errors for the reduced models.

| | $MSE_\omega$ | $\mathcal{H}_\infty$ | $MSE_t$ |
|---|---|---|---|
| BT | $2.54 \times 10^{-4}$ | 0.053 | $9.21 \times 10^{-8}$ |
| FLBT | $3.46 \times 10^{-5}$ | 0.538 | $3.23 \times 10^{-8}$ |
| TLBT | $3.68 \times 10^{-5}$ | 0.385 | $5.05 \times 10^{-14}$ |
| FW | $4.31 \times 10^{-5}$ | 0.539 | $2.24 \times 10^{-8}$ |

**Example 2.** *Consider the system as in reference [36] with fractional-order $\alpha = 0.95$ which is discretized using the sampling period $h = 0.002$. The calculation of controllability and observability Gramians in the time domain is very computationally demanding for systems of order $n = 1006$. Therefore, the reduced models obtained by using only the BT, FLBT and FW methods are compared. All reduced models are of order $r = 6$. Frequency responses for the full fractional-order system and the reduced models as well as approximation errors are presented in Figure 5. Like in Example 1, it is clearly visible that the reduction based on the infinite Gramians cannot properly approximate the low-frequency properties of the system. For this reason, the frequency-interval for frequency-limited Gramians and frequency weighting functions are chosen the same as in Example 1. In Table 3, approximation errors are listed for the analyzed models, in terms of DCE, $MSE_\omega$ and $\mathcal{H}_\infty$-norm.*

**Table 3.** Frequency and step approximation errors for the reduced models.

| | *DCE* | $MSE_\omega$ | $\mathcal{H}_\infty$ |
|---|---|---|---|
| BT | 5.538 | 29.93 | 5.538 |
| FLBT | $4.94 \times 10^{-4}$ | $1.90 \times 10^{-7}$ | 11.12 |
| FW | $6.30 \times 10^{-4}$ | $3.14 \times 10^{-7}$ | 11.35 |

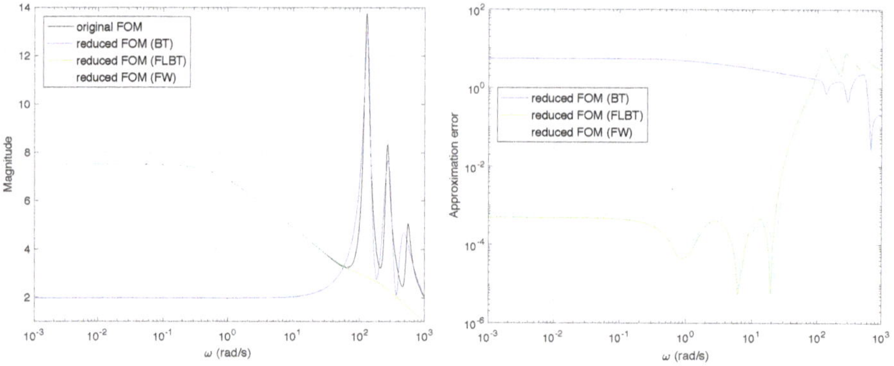


**Figure 5.** (**left**) Frequency responses for full- and reduced-order models and (**right**) approximation errors.

The Matlab scripts used to compute the presented results can be obtained from Supplementary Materials.

## 6. Conclusions

This paper presents new results in BT model order reduction in limited time- and frequency-intervals for DTCFO systems. The main contribution of the paper is an introduction of new definitions for controllability and observability Gramians for the fractional-order systems both in the time and frequency domains. These results enable new implementations of the Gramians in the balanced truncation model order reduction method in limited time and frequency intervals as well as in the frequency weighted reduction method. As a result of the reduction process, accurate low-dimension fractional-order approximators in given frequency and/or time intervals can be calculated. Simulation examples confirm the effectiveness of the introduced methodology for order reduction of DTCFO systems.

**Supplementary Materials:** The Matlab scripts used to compute the presented results can be obtained from: http://doi.org/10.5281/zenodo.2322833.

**Author Contributions:** The individual contribution and responsibilities of the authors were as follows: Conceptualization: M.R. and R.S.; Methodology and Investigation: M.R.; Software Implementation: M.R. and R.S.; Validation: R.S. and K.J.L.; Visualization: M.R.; Writing—original draft: M.R.; Writing—review & editing: R.S. and K.J.L.; All authors have read and approved the manuscript.

**Funding:** This research received no external funding.

**Conflicts of Interest:** The authors declare no conflict of interest.

## References

1. Ferdi, Y. Computation of Fractional Order Derivative and Integral via Power Series Expansion and Signal Modelling. *Nonlinear Dyn.* **2006**, *46*, 1–15. [CrossRef]
2. Dhabale, A.S.; Dive, R.; Aware, M.V.; Das, S. A New Method for Getting Rational Approximation for Fractional Order Differintegrals. *Asian J. Control* **2015**, *17*, 2143–2152. [CrossRef]
3. Vinagre, B.M.; Podlubny, I.; Hernandez, A.; Feliu, V. Some approximations of fractional order operators used in control theory and applications. *Fractional Calculus Appl. Anal.* **2000**, *3*, 231–248.
4. Xue, D.; Zhao, C.; Chen, Y. A Modified Approximation Method of Fractional Order System. In Proceedings of the 2006 International Conference on Mechatronics and Automation, Luoyang, China, 25–28 June 2006; pp. 1043–1048.
5. Dzieliński, A.; Sierociuk, D. Stability of Discrete Fractional Order State-space Systems. *J. Vib. Control* **2008**, *14*, 1543–1556. [CrossRef]

6. Stanisławski, R.; Latawiec, K.J. Fractional-order discrete-time Laguerre filters—A new tool for modeling and stability analysis of fractional-order LTI SISO systems. *Discrete Dyn. Nat. Soc.* **2016**, *2016*, 1–9. [CrossRef]
7. Du, B.; Wei, Y.; Liang, S.; Wang, Y. Rational approximation of fractional order systems by vector fitting method. *Int. J. Control Autom. Syst.* **2017**, *15*, 186–195. [CrossRef]
8. Stanisławski, R.; Rydel, M.; Latawiec, K.J. Modeling of discrete-time fractional-order state space systems using the balanced truncation method. *J. Franklin Inst.* **2017**, *354*, 3008–3020. [CrossRef]
9. Krajewski, W.; Viaro, U. A method for the integer-order approximation of fractional-order systems. *J. Franklin Inst.* **2014**, *351*, 555–564. [CrossRef]
10. Rydel, M. New integer-order approximations of discrete-time non-commensurate fractional-order systems using the cross Gramian. *Adv. Comput. Math.* **2018**. [CrossRef]
11. Tavakoli-Kakhki, M.; Haeri, M. Model reduction in commensurate fractional-order linear systems. *Proc. Inst. Mech. Eng. Part I J. Syst. Control Eng.* **2009**, *223*, 493–505. [CrossRef]
12. Garrappa, R.; Maione, G. Model order reduction on Krylov subspaces for fractional linear systems. *IFAC Proc. Volumes* **2013**, *46*, 143–148. [CrossRef]
13. Shen, J.; Lam, J. $H_\infty$ Model Reduction for Positive Fractional Order Systems. *Asian J. Control* **2014**, *16*, 441–450. [CrossRef]
14. Jiang, Y.L.; Xiao, Z.H. Arnoldi-based model reduction for fractional order linear systems. *Int. J. Syst. Sci.* **2015**, *46*, 1411–1420. [CrossRef]
15. Rydel, M.; Stanisławski, R.; Latawiec, K.J.; Gałek, M. Model order reduction of commensurate linear discrete-time fractional-order systems. *IFAC PapersOnLine* **2018**, *51*, 536–541. [CrossRef]
16. Gawronski, W.; Juang, J. Model reduction in limited time and frequency intervals. *Int. J. Syst. Sci.* **1990**, *21*, 349–376. [CrossRef]
17. Benner, P.; Kürschner, P.; Saak, J. Frequency-Limited Balanced Truncation with Low-Rank Approximations. *SIAM J. Sci. Comput.* **2016**, *38*, A471–A499. [CrossRef]
18. Zulfiqar, U.; Imran, M.; Ghafoor, A.; Liaquat, M. A New Frequency-Limited Interval Gramians-Based Model Order Reduction Technique. *IEEE Trans. Circuits Syst. Express Briefs* **2017**, *64*, 680–684. [CrossRef]
19. Kürschner, P. Balanced truncation model order reduction in limited time intervals for large systems. *Adv. Comput. Math.* **2018**, *44*, 1821–1844. [CrossRef]
20. Enns, D. Model reduction with balanced realizations: An error bound and frequency weighted generalization. In Proceedings of the 23rd IEEE Conference on Decision and Control, Las Vegas, NV, USA, 12–14 December 1984; pp. 127–132.
21. Lin, C.; Chiu, T. Model reduction via frequency weighted balanced realization. *Theory Adv. Technol.* **1992**, *8*, 341–451.
22. Wang, G.; Sreeram, V.; Liu, W.Q. A new frequency-weighted balanced truncation method and an error bound. *IEEE Trans. Autom. Control* **1999**, *44*, 1734–1737. [CrossRef]
23. Varga, A.; Anderson, B. Accuracy-enhancing methods for balancing-related frequency-weighted model and controller reduction. *Automatica* **2003**, *39*, 919–927. [CrossRef]
24. Sreeram, V.; Sahlan, S. Improved results on frequency-weighted balanced truncation and error bounds. *Int. J. Robust Nonlin.* **2012**, *22*, 1195–1211. [CrossRef]
25. Imran, M.; Ghafoor, A.; Sreeram, V. A frequency weighted model order reduction technique and error bounds. *Automatica* **2014**, *50*, 3304–3309. [CrossRef]
26. Rydel, M.; Stanisławski, R. A new frequency weighted Fourier-based method for model order reduction. *Automatica* **2018**, *88*, 107–112. [CrossRef]
27. Monje, C.; Chen, Y.; Vinagre, B.; Xue, D.; Feliu-Batlle, V. *Fractional-order Systems and Controls: Fundamentals and Applications*; Series on Advances in Industrial Control; Springer: London, UK, 2010.
28. Podlubny, I. *Fractional Differential Equations*; Academic Press: Orlando, FL, USA, 1999.
29. Moore, B. Principal component analysis in linear systems: controllability, observability and model reduction. *IEEE Trans. Autom. Control* **1981**, *AC-26*, 17–32. [CrossRef]
30. Safonov, M.G.; Chiang, R.Y. A Schur Method for Balanced-Truncation Model Reduction. *IEEE Trans. Autom. Control* **1989**, *34*, 729–733. [CrossRef]
31. Antoulas, A. *Approximation of Large-Scale Dynamical System*; SIAM: Philadelphia, PA, USA, 2005.

32. Laub, A.; Heath, M.; Paige, C.; Ward, R. Computation of system balancing transformations and other applications of simultaneous diagonalization algorithms. *IEEE Trans. Autom. Control* **1987**, *32*, 115–122. [CrossRef]
33. Rydel, M.; Stanisławski, W. Selection of reduction parameters for complex plant MIMO LTI models using the evolutionary algorithm. *Math. Comput. Simul* **2017**, *140*, 94–106. [CrossRef]
34. Penzl, T. Eigenvalue decay bounds for solutions of Lyapunov equations: The symmetric case. *Syst. Control Lett.* **2000**, *40*, 139–144. [CrossRef]
35. Sadkane, M. A low-rank Krylov squared Smith method for large-scale discrete-time Lyapunov equations. *Linear Algebra Appl.* **2012**, *436*, 2807–2827. [CrossRef]
36. Penzl, T. Algorithms for Model Reduction of Large Dynamical Systems. *Linear Algebra Appl.* **2006**, *415*, 322–343. [CrossRef]

symmetry

MDPI

Article

# Positive Solutions for a Three-Point Boundary Value Problem of Fractional Q-Difference Equations

**Chen Yang**

Basic Course Department, Business College of Shanxi University, Taiyuan 030031, China; yangchen0809@126.com

Received: 25 July 2018; Accepted: 20 August 2018; Published: 23 August 2018

**Abstract:** In this work, a three-point boundary value problem of fractional $q$-difference equations is discussed. By using fixed point theorems on mixed monotone operators, some sufficient conditions that guarantee the existence and uniqueness of positive solutions are given. In addition, an iterative scheme can be made to approximate the unique solution. Finally, some interesting examples are provided to illustrate the main results.

**Keywords:** fractional $q$-difference equation; existence and uniqueness; positive solutions; fixed point theorem on mixed monotone operators

**2010 MR Subject Classification:** 26A33; 34B15; 33D05; 39A13

## 1. Introduction

We will deal with a fractional $q$-difference equation subject to three-point boundary conditions

$$\begin{cases} D_q^\alpha x(t) + f(t, x(t), x(t)) + g(t, x(t)) = 0, & 0 < t < 1, \ 2 < \alpha < 3, \\ x(0) = D_q x(0) = 0, \ D_q x(1) = \beta D_q x(\eta), \end{cases} \tag{1}$$

where $0 < \beta\eta^{\alpha-2} < 1$, $0 < q < 1$, $D_q^\alpha$ is the Riemann–Liouville fractional $q$-derivative of order $\alpha$.

Due to fast development in fractional calculus, many researchers studied $q$-difference calculus or quantum calculus. For this topic, the earlier results can be seen in Al-Salam [1] and Agarwal [2], and some recent results related to $q$-difference calculus in [3–15] and some references therein. Nowadays, fractional $q$-difference calculus has been given in wide applications of different science areas, which include basic hyper-geometric functions, mechanics, the theory of relativity, combinatorics and discrete mathematics. So many mathematical models have been abstracted out(see [16–18]) and problem (1) is one of the models. Therefore, fractional $q$-difference calculus has been of great interest and many good results can be found in [5–8] and references therein. Recently, the fruits about fractional $q$-difference equation boundary value problems emerge continuously. For different problems of fractional $q$-difference equations, the existence and the uniqueness of solutions have been always considered in literature. To solve these boundary value problems, some techniques have been applied, such as the monotone iterative technique, the lower-upper solution method, the Schauder fixed point theorem and the Krasnoselskii fixed point theorem. For details, one can see [13–15,19–25].

In [15], Liang and Zhang considered the existence and uniqueness of positive nondecreasing solutions for a fractional $q$-difference equation involving three-point boundary conditions

$$\begin{cases} D_q^\alpha x(t) + f(t, x(t)) = 0, & 2 < \alpha < 3, \ 0 < t < 1, \\ x(0) = D_q x(0) = 0, \ D_q x(1) = \beta D_q x(\eta), \end{cases} \tag{2}$$

where $0 < \beta\eta^{\alpha-2} < 1$. They gave some sufficient conditions for Label (2), and their tool is a fixed point theorem in partially ordered sets.

In [19], Sriphanomwan et al. investigated the problem of fractional $q$-difference equations

$$\begin{cases} D_q^\alpha(D_q^\beta(1+p(t)))x(t) = f(t, x(t), D_\theta^\mu x(t), \Psi_\omega^v x(t)), \\ x(0) = x(\eta), \quad I_r^\gamma x(T) = \int_0^T \frac{(T-rs)^{(\gamma-1)}}{\Gamma_r(\gamma)} x(s) d_r s = g(x), \end{cases} \tag{3}$$

where $t \in I_\chi^T := \{\chi^k T : k \in \mathbf{N} \cup \{0, T\}\}, 0 < \alpha, \beta, \mu \le 1, 1 < \alpha + \beta \le 2, v, \gamma > 0, \eta \in I_\chi^T - \{0, T\}$, and $p, q, r, \theta, \omega$ are simple fractions. The existence and uniqueness of solutions for Label (3) was obtained. The used methods are the Banach contraction mapping principle and Krasnosel'skii fixed point theorem.

By using Schauder fixed point theorem and the Banach fixed point theorem, Yang [25] discussed a fractional $q$-difference equation with three-point boundary conditions:

$$\begin{cases} D_q^\alpha x(t) + f(t, x(t)) = 0, \quad 0 \le t \le 1, \ 1 < \alpha \le 2, \\ x(0) = 0, \ x(1) = \beta x(\xi), \end{cases} \tag{4}$$

where $0 < \beta\xi^{\alpha-1} < 1$, $0 < \xi < 1$. The author gave the existence and uniqueness of positive solutions for Label (4).

In a very recent paper [24], the authors considered a special fractional $q$-difference equation with a three-point problem

$$\begin{cases} D_q^\alpha u(t) + f(t, u(t)) = b, \quad 0 < t < 1, \ 2 < \alpha < 3, \\ u(0) = D_q u(0) = 0, \ D_q u(1) = \beta D_q u(\eta), \end{cases} \tag{5}$$

where $0 < \beta\eta^{\alpha-2} < 1$, $0 < q < 1$, $b \ge 0$ is a constant. The existence and uniqueness of solutions for Label (5) by using fixed point theorems for $\psi$-$(h, r)$-concave operators.

Motivated by [15,26], we consider the existence and uniqueness of positive solutions for Label (1). Different from the methods mentioned above, our tools are two fixed point theorems for mixed monotone operators. To the authors' knowledge, Label (1) is a new form of fractional $q$-difference equations. We can give the existence and uniqueness of solutions for Label (1). Furthermore, we can make an iteration to approximate the unique solution.

## 2. Preliminaries

Here, we list some concepts and lemmas of fractional $q$-calculus. One can see [1–8], for example. For $0 < q < 1$ and $f$ defined on $[a, b]$, let

$$(I_q f)(t) = \int_0^t f(s) d_q s = (1-q) \sum_{n=0}^{\infty} f(tq^n) t q^n, \ \ t \in [0, b].$$

Then,

$$\int_a^b f(t) d_q t = \int_a^c f(t) d_q t + \int_c^b f(t) d_q t, \ \ \forall c \in [a, b].$$

**Definition 1.** *(See [3]). $\alpha \ge 0$ and $f$ is defined on $[0, 1]$. The Riemann–Liouville fractional $q$-integral is $(I_q^0 f)(t) = f(t)$ and*

$$(I_q^\alpha f)(t) = \frac{1}{\Gamma_q(\alpha)} \int_0^t (t - qs)^{(\alpha-1)} f(s) d_q s, \ \alpha > 0.$$

*Clearly, $(I_q^\alpha f)(t) = (I_q f)(t)$ when $\alpha = 1$.*

**Lemma 1.** *(See [22]). If $f, g$ are continuous on $[0, s]$ and $f(t) \leq g(t)$ for $t \in [0, s]$, then*

*(i)* $\int_0^s f(t)d_qt \leq \int_0^s g(t)d_qt$. *In addition, if* $\alpha > 1$, *then* $I_q^\alpha f(s) \leq I_q^\alpha g(s)$, $t \in [0, s]$,
*(ii)* $|\int_0^s f(t)d_qt| \leq \int_0^s |f(t)|d_qt$, $t \in [0, s]$.

**Definition 2.** *(See [3]). The Riemann–Liouville fractional q-derivative of order* $\alpha \geq 0$ *is*

$$(D_q^\alpha f)(t) = (D_q^n I_q^{n-\alpha} f)(t), \ \alpha > 0, \ t \in [0, 1],$$

*where n denotes the smallest integer greater than or equal to* $\alpha$.

*When* $\alpha = 1$, $(D_q^\alpha f)(t) = D_q f(t)$. *Furthermore,*

$$(I_q^\alpha D_q^p f)(t) = (D_q^p I_q^\alpha f)(t) - \sum_{n=0}^{p-1} \frac{t^{\alpha-p+n}}{\Gamma_q(\alpha - p + n + 1)}(D_q^n f)(0), \ p \in \mathbf{N}.$$

**Lemma 2.** *If $f(t)$ is continuous with $f(t) \geq 0$ for $t \in [0, 1]$, and there is $t_0 \in (0, 1)$ such that $f(t_0) \neq 0$. Then,*

$$\int_0^1 f(t)d_qt > 0, \ t \in [0, 1],$$

*where*

$$\int_0^1 f(t)d_qt = (1 - q)\sum_{n=0}^{\infty} q^n f(q^n), \ q \in (0, 1).$$

**Proof.** Because $f(t) \geq 0$ and $f(t_0) \neq 0$, there is $n_0 \in \mathbf{N}$ such that $t_0 = q^{n_0}$, then

$$f(q^{n_0})q^{n_0} > 0, 0 < q < 1,$$

and thus

$$(1 - q)\sum_{n=0}^{\infty} q^n f(q^n) \geq (1 - q)f(q^{n_0})q^{n_0} = (1 - q)f(t_0)t_0 > 0.$$

Hence, we have $\int_0^1 f(t)d_qt > 0$. □

Here, we list other facts that are important in the sequel. See [26–30] for instance.

$(X, \|\cdot\|)$ is a real Banach space, its partial order induced by a cone $K$ of $X$, i.e., $x \leq y$ if and only if $y - x \in K$. If there is $N > 0$ such that $\|x\| \leq N\|y\|$ for $\theta \leq x \leq y$, $x, y \in X$, then $K$ is called normal, where $\theta$ denotes the zero element of $X$. The notation $x$–$y$ denotes that there exist $\mu, \nu > 0$ such that $\mu x \leq y \leq \nu x$, $\forall\, x, y \in X$. For fixed $h > \theta$, define a set $K_h = \{x \in E \mid x \sim h\}$. Then, $K_h \subset K$.

**Definition 3.** *(See [27]). Suppose $T : K \to K$ is a given operator. If*

$$T(tx) \geq tTx, \ \forall t \in (0, 1), \ x \in K, \tag{6}$$

*then T is said to be sub-homogeneous.*

**Definition 4.** *(See [27]). Let $0 \leq \gamma < 1$. An operator $T : K \to K$ satisfies*

$$T(tx) \geq t^\gamma Tx, \ \forall t \in (0, 1), \ x \in K. \tag{7}$$

*Then, T is said to be $\gamma$-concave.*

**Lemma 3.** *(See [27]). Let $h > \theta$, $0 < \gamma < 1$, $T_1 : K \times K \to K$ be a mixed monotone operator and*

$$T_1(tx, t^{-1}y) \geq t^{\gamma} T_1(x, y), \ \forall t \in (0,1), \ x, y \in K. \tag{8}$$

*$T_2 : K \to K$ is an increasing sub-homogeneous operator. Moreover,*

(i) *there exists $h_0 \in K_h$ such that $T_1(h_0, h_0), T_2 h_0 \in K_h$;*
(ii) *there exists $\sigma > 0$ such that $T_1(x, y) \geq \sigma T_2 x, \ x, y \in K$.*

*Then:*

(a) *$T_1 : K_h \times K_h \to K_h$ and $T_2 : K_h \to K_h$;*
(b) *there are $u_0, v_0 \in K_h$ and $\tau \in (0,1)$ satisfying*

$$\tau v_0 \leq u_0 < v_0, \ u_0 \leq T_1(u_0, v_0) + T_2 u_0 \leq T_1(v_0, u_0) + T_2 v_0 \leq v_0;$$

(c) *$T_1(x, x) + T_2 x = x$ exists a unique solution $x^*$ in $K_h$;*
(d) *for $x_0, y_0 \in K_h$, set*

$$x_n = T_1(x_{n-1}, y_{n-1}) + T_2 x_{n-1}, \ y_n = T_1(y_{n-1}, x_{n-1}) + T_2 y_{n-1}, \ n = 1, 2, \ldots,$$

*then $x_n \to x^*$, $y_n \to x^*$ as $n \to \infty$.*

**Lemma 4.** *(See [27]). Let $h > \theta$, $0 < \gamma < 1$, $T_1 : K \times K \to K$ be a mixed monotone operator and*

$$T_1(tx, t^{-1}y) \geq t T_1(x, y), \ \forall t \in (0,1), \ x, y \in K. \tag{9}$$

*$T_2 : K \to K$ is an increasing $\gamma$-concave operator. Moreover,*

(i) *there exists $h_0 \in K_h$ such that $T_1(h_0, h_0), T_2 h_0 \in K_h$;*
(ii) *there exists $\sigma > 0$ such that $T_1(x, y) \leq \sigma T_2 x, \ x, y \in K$.*

*Then:*

(a) *$T_1 : K_h \times K_h \to K_h$ and $T_2 : K_h \to K_h$;*
(b) *there are $u_0, v_0 \in K_h$ and $\tau \in (0,1)$ satisfying*

$$\tau v_0 \leq u_0 < v_0, \ u_0 \leq T_1(u_0, v_0) + T_2 u_0 \leq T_1(v_0, u_0) + T_2 v_0 \leq v_0;$$

(c) *$T_1(x, x) + T_2 x = x$ exists a unique solution $x^*$ in $K_h$;*
(d) *for $x_0, y_0 \in K_h$, set*

$$x_n = T_1(x_{n-1}, y_{n-1}) + T_2 x_{n-1}, \ y_n = T_1(y_{n-1}, x_{n-1}) + T_2 y_{n-1}, \ n = 1, 2, \ldots,$$

*then $x_n \to x^*$, $y_n \to x^*$ as $n \to \infty$.*

**Remark 1.** *From Lemmas 3 and 4, we have two special cases:*

(i) *Let $T_2 = \theta$ in Lemma 3, we get the corresponding conclusion (see Corollary 2.2 in [27]);*
(ii) *Let $T_1 = \theta$ in Lemma 4, we have the corresponding conclusion (see Theorem 2.7 in [31]).*

## 3. Main Results

By using Lemmas 3 and 4, we will establish our main results for Label (1). Consider a Banach space $X = C[0,1]$, the norm is $\|u\| = \sup\{|u(t)| : t \in [0,1]\}$. Set $K = \{x \in C[0,1] | x(t) \geq 0, \ t \in [0,1]\}$, a normal cone.

**Lemma 5.** *(See [15]). Let $g \in C[0,1]$, $\beta\eta^{\alpha-2} \neq 1$ and $0 < \eta < 1$, then the unique solution of following three-point problem*

$$\begin{cases} D_q^{\alpha}x(t) + g(t) = 0,\ 0 < t < 1,\ 2 < \alpha < 3, \\ x(0) = D_q x(0) = 0,\ \ D_q x(1) = \beta D_q x(\eta) \end{cases} \tag{10}$$

*is*

$$x(t) = \int_0^1 G(t,qs)g(s)d_qs + \frac{\beta t^{\alpha-1}}{[\alpha-1]_q(1-\beta\eta^{\alpha-2})}\int_0^1 H(\eta,qs)g(s)d_qs, \tag{11}$$

*where*

$$G(t,s) = \frac{1}{\Gamma_q(\alpha)}\begin{cases} (1-s)^{(\alpha-2)}t^{\alpha-1} - (t-s)^{(\alpha-1)},\ 0 \le s \le t \le 1, \\ (1-s)^{(\alpha-2)}t^{\alpha-1},\ 0 \le t \le s \le 1, \end{cases} \tag{12}$$

$$\begin{aligned} H(t,s) &= {}_tD_qG(s,t) \\ &= \frac{[\alpha-1]_q}{\Gamma_q(\alpha)}\begin{cases} (1-s)^{(\alpha-2)}t^{\alpha-2} - (t-s)^{(\alpha-2)},\ 0 \le s \le t \le 1, \\ (1-s)^{(\alpha-2)}t^{\alpha-2},\ 0 \le t \le s \le 1. \end{cases} \end{aligned}$$

**Lemma 6.** *(See [15]). For $G(t,qs)$ in (11), we obtain*

*(1) $G(t,qs)$ is continuous and $G(t,qs) \geq 0$, $t,s \in [0,1] \times [0,1]$;*
*(2) $G(t,qs)$ is strictly increasing in $t \in [0,1]$.*

**Remark 2.** *For $G(t,qs)$ in (11), we can easily get*

$$G(t,qs) \leq \frac{1}{\Gamma_q(\alpha)}(1-qs)^{(\alpha-2)}t^{\alpha-1},\ t,s \in [0,1] \times [0,1].$$

*By (2) in Lemma 6, we have ${}_tD_qG(qs,t) \geq 0$, that is, $H(t,qs) \geq 0$. Obviously,*

$$H(t,qs) \leq \frac{[\alpha-1]_q}{\Gamma_q(\alpha)}(1-qs)^{(\alpha-2)}t^{\alpha-2} \leq \frac{[\alpha-1]_q}{\Gamma_q(\alpha)},\ t,s \in [0,1] \times [0,1].$$

*Next, four assumptions are listed:*

*($H_1$) $f : [0,1] \times [0,+\infty) \times [0,+\infty) \to [0,+\infty)$ and $g : [0,1] \times [0,+\infty) \to [0,+\infty)$ are continuous;*
*($H_2$) $f(t,u,v)$ is increasing relative to $u$ for fixed $t \in [0,1]$ and $v \in [0,+\infty)$, decreasing relative to $v$ for fixed $t \in [0,1]$ and $u \in [0,+\infty)$; $g(t,u)$ is increasing relative to $u$ for fixed $t \in [0,1]$;*
*($H_3$) for $\lambda \in (0,1), t \in [0,1], u \geq 0$, $g(t,\lambda u) \geq \lambda g(t,u)$ is satisfied, and there is $\gamma \in (0,1)$ such that $f(t,\lambda u,\lambda^{-1}v) \geq \lambda^{\gamma} f(t,u,v)$ for $u,v \geq 0$. In addition, $g(t,0) \not\equiv 0$;*
*($H_4$) there exists $\sigma > 0$ such that $f(t,u,v) \geq \sigma g(t,u)$, $\forall t \in [0,1], u,v \in [0,+\infty)$.*

**Theorem 1.** *Let $(H_1) - (H_4)$ be satisfied, then*

*(a) there are $u_0, v_0 \in K_h$ and $\tau \in (0,1)$ satisfying $\tau v_0 \leq u_0 < v_0$ and*

$$\begin{aligned} u_0(t) &\leq \frac{\beta t^{\alpha-1}}{[\alpha-1]_q(1-\beta\eta^{\alpha-2})}\int_0^1 H(\eta,qs)[f(s,u_0(s),v_0(s)) + g(s,u_0(s))]d_qs \\ &\quad + \int_0^1 G(t,qs)[f(s,u_0(s),v_0(s)) + g(s,u_0(s))]d_qs,\ t \in [0,1], \end{aligned}$$

$$\begin{aligned} v_0(t) &\geq \frac{\beta t^{\alpha-1}}{[\alpha-1]_q(1-\beta\eta^{\alpha-2})}\int_0^1 H(\eta,qs)[f(s,v_0(s),u_0(s)) + g(s,v_0(s))]d_qs \\ &\quad + \int_0^1 G(t,qs)[f(s,v_0(s),u_0(s)) + g(s,v_0(s))]d_qs,\ t \in [0,1], \end{aligned}$$

*where $h(t) = t^{\alpha-1}$ and $G(t,qs)$, $H(t,qs)$ are defined as in Lemma 5;*

*(b) BVP (1) has a unique positive solution $u^* \in K_h$;*
*(c) for $x_0, y_0 \in K_h$, set*

$$\begin{aligned} x_{n+1}(t) &= \frac{\beta t^{\alpha-1}}{[\alpha-1]_q(1-\beta\eta^{\alpha-2})}\int_0^1 H(\eta,qs)[f(s,x_n(s),y_n(s))+g(s,x_n(s))]d_qs \\ &\quad + \int_0^1 G(t,qs)[f(s,x_n(s),y_n(s))+g(s,x_n(s))]d_qs,\ n=1,2,\ldots, \end{aligned}$$

$$\begin{aligned} y_{n+1}(t) &= \frac{\beta t^{\alpha-1}}{[\alpha-1]_q(1-\beta\eta^{\alpha-2})}\int_0^1 H(\eta,qs)[f(s,y_n(s),x_n(s))+g(s,y_n(s))]d_qs \\ &\quad + \int_0^1 G(t,qs)[f(s,y_n(s),x_n(s))+g(s,y_n(s))]d_qs,\ n=1,2,\ldots, \end{aligned}$$

*then $\|x_n - u^*\| \to 0$, $\|y_n - u^*\| \to 0$ as $n \to \infty$.*

**Proof.** By Lemma 5, the solution $u$ of BVP (1) can be written by

$$\begin{aligned} u(t) &= \frac{\beta t^{\alpha-1}}{[\alpha-1]_q(1-\beta\eta^{\alpha-2})}\int_0^1 H(\eta,qs)[f(s,u(s),u(s))+g(s,u(s))]d_qs \\ &\quad + \int_0^1 G(t,qs)[f(s,u(s),u(s))+g(s,u(s))]d_qs. \end{aligned}$$

Now, we give two operators $T_1 : K \times K \to X$ and $T_2 : K \to X$ by

$$\begin{aligned} T_1(u,v)(t) &= \frac{\beta t^{\alpha-1}}{[\alpha-1]_q(1-\beta\eta^{\alpha-2})}\int_0^1 H(\eta,qs)f(s,u(s),v(s))d_qs \\ &\quad + \int_0^1 G(t,qs)f(s,u(s),v(s))d_qs, \end{aligned}$$

$$\begin{aligned} (T_2u)(t) &= \frac{\beta t^{\alpha-1}}{[\alpha-1]_q(1-\beta\eta^{\alpha-2})}\int_0^1 H(\eta,qs)g(s,u(s))d_qs \\ &\quad + \int_0^1 G(t,qs)g(s,u(s))d_qs. \end{aligned}$$

Obviously, $u$ is a solution of Label (1) if and only if $u = T_1(u,u) + T_2u$. By $(H_1)$, one has $T_1 : K \times K \to K$ and $T_2 : K \to K$. We will prove that $T_1$, $T_2$ satisfy all the assumptions of Lemma 3. The proof consists of three steps.

**Step 1.** The aim of this step is to prove that $T_1$ is a mixed monotone operator.

For $u_i, v_i \in K,\ i = 1,2$ with $u_1 \geq u_2,\ v_1 \leq v_2$, then $u_1(t) \geq u_2(t),\ v_1(t) \leq v_2(t)$ for $t \in [0,1]$. From $(H_2)$ and Lemma 6,

$$\begin{aligned} T_1(u_1,v_1)(t) &= \frac{\beta t^{\alpha-1}}{[\alpha-1]_q(1-\beta\eta^{\alpha-2})}\int_0^1 H(\eta,qs)f(s,u_1(s),v_1(s))d_qs \\ &\quad + \int_0^1 G(t,qs)f(s,u_1(s),v_1(s))d_qs \\ &\geq \frac{\beta t^{\alpha-1}}{[\alpha-1]_q(1-\beta\eta^{\alpha-2})}\int_0^1 H(\eta,qs)f(s,u_2(s),v_2(s))d_qs \\ &\quad + \int_0^1 G(t,qs)f(s,u_2(s),v_2(s))d_qs \\ &= T_1(u_2,v_2)(t). \end{aligned}$$

Thus, $T_1(u_1,v_1) \geq T_1(u_2,v_2)$, that is, $T_1$ is mixed monotone.

**Step 2.** Our aim of this step is to show that $T_1$ satisfies the condition (8) and the operator $T_2$ is sub-homogeneous.

From $(H_2)$ and Lemma 6, $T_2$ is increasing. Furthermore, for $\lambda \in (0,1)$ and $u, v \in P$, by $(H_3)$,

$$\begin{aligned} T_1(\lambda u,\lambda^{-1}v)(t) &= \frac{\beta t^{\alpha-1}}{[\alpha-1]_q(1-\beta\eta^{\alpha-2})}\int_0^1 H(\eta,qs)f(s,\lambda u(s),\lambda^{-1}v(s))d_qs \\ &\quad + \int_0^1 G(t,qs)f(s,\lambda u(s),\lambda^{-1}v(s))d_qs \\ &\geq \frac{\lambda^\gamma \beta t^{\alpha-1}}{[\alpha-1]_q(1-\beta\eta^{\alpha-2})}\int_0^1 H(\eta,qs)f(s,u(s),v(s))d_qs \\ &\quad + \lambda^\gamma \int_0^1 G(t,qs)f(s,u_2(s),v_2(s))d_qs \\ &= \lambda^\gamma T_1(u,v)(t), \end{aligned}$$

and thus $T_1(\lambda u,\lambda^{-1}v) \geq \lambda^\gamma T_1(u,v)$ for $\lambda \in (0,1),\ u,v \in K$. Hence, the operator $T_1$ satisfies (8). In addition, for any $\lambda \in (0,1),\ u \in K$, by $(H_3)$,

$$\begin{aligned} T_2(\lambda u)(t) &= \frac{\beta t^{\alpha-1}}{[\alpha-1]_q(1-\beta\eta^{\alpha-2})}\int_0^1 H(\eta,qs)g(s,\lambda u(s))d_qs + \int_0^1 G(t,qs)g(s,\lambda u(s))d_qs \\ &\geq \frac{\lambda\beta t^{\alpha-1}}{[\alpha-1]_q(1-\beta\eta^{\alpha-2})}\int_0^1 H(\eta,qs)g(s,u(s))d_qs + \lambda\int_0^1 G(t,qs)g(s,u(s))d_qs \\ &= \lambda T_2 u(t), \end{aligned}$$

that is, $T_2(\lambda u) \geq \lambda T_2 u,\ u \in P$. Thus, the operator $T_2$ is sub-homogeneous.

**Step 3.** The purpose of this step is to prove that $T_1(h,h),\ T_2 h \in K_h$. Furthermore, we also prove that there exists $\sigma > 0$ such that $T_1(x,y) \geq \sigma T_2 x,\ \forall x,y \in K$.

Firstly, in view of $(H_1), (H_2)$ and Lemma 6, for $t \in [0,1]$,

$$\begin{aligned} T_1(h,h)(t) &= \frac{\beta t^{\alpha-1}}{[\alpha-1]_q(1-\beta\eta^{\alpha-2})}\int_0^1 H(\eta,qs)f(s,h(s),h(s))d_qs + \int_0^1 G(t,qs)f(s,h(s),h(s))d_qs \\ &= \frac{\beta t^{\alpha-1}}{[\alpha-1]_q(1-\beta\eta^{\alpha-2})}\int_0^1 H(\eta,qs)f(s,s^{\alpha-1},s^{\alpha-1})d_qs + \int_0^1 G(t,qs)f(s,s^{\alpha-1},s^{\alpha-1})d_qs \\ &\leq \frac{\beta h(t)}{(1-\beta\eta^{\alpha-2})\Gamma_q(\alpha)}\int_0^1 (1-qs)^{(\alpha-2)}f(s,1,0)d_qs + \frac{h(t)}{\Gamma_q(\alpha)}\int_0^1 (1-qs)^{(\alpha-2)}f(s,1,0)d_qs. \end{aligned}$$

By the same arguments, for $t \in [0,1]$,

$$\begin{aligned} T_1(h,h)(t) &= \frac{\beta t^{\alpha-1}}{[\alpha-1]_q(1-\beta\eta^{\alpha-2})}\int_0^1 H(\eta,qs)f(s,s^{\alpha-1},s^{\alpha-1})d_qs + \int_0^1 G(t,qs)f(s,s^{\alpha-1},s^{\alpha-1})d_qs \\ &\geq \frac{h(t)}{\Gamma_q(\alpha)}\int_0^1 [(1-qs)^{(\alpha-2)} - (1-qs)^{(\alpha-1)}]f(s,0,1)d_qs. \end{aligned}$$

From $(H_2)$, $(H_4)$,

$$\int_0^1 f(s,1,0)d_qs \geq \int_0^1 f(s,0,1)d_qs \geq \sigma\int_0^1 g(s,0)d_qs > 0.$$

Set

$$l_1 = \left(\frac{1}{\Gamma_q(\alpha)} + \frac{\beta}{(1-\beta\eta^{\alpha-2})\Gamma_q(\alpha)}\right)\int_0^1 (1-qs)^{(\alpha-2)}f(s,1,0)d_qs,$$

$$l_2 = \frac{1}{\Gamma_q(\alpha)}\int_0^1 \left[(1-qs)^{(\alpha-2)} - (1-qs)^{(\alpha-1)}\right] f(s,0,1)d_qs.$$

Then, $l_2h(t) \leq T_1(h,h)(t) \leq l_1h(t), t \in [0,1]$. It follows that $T_1(h,h) \in K_h$. Similarly,

$$T_2h(t) \geq \frac{h(t)}{\Gamma_q(\alpha)}\int_0^1 \left[(1-qs)^{(\alpha-2)} - (1-qs)^{(\alpha-1)}\right] g(s,0)d_qs,$$

and

$$T_2h(t) \leq \left(\frac{1}{\Gamma_q(\alpha)} + \frac{\beta}{(1-\beta\eta^{\alpha-2})\Gamma_q(\alpha)}\right)h(t)\int_0^1 (1-qs)^{(\alpha-2)}g(s,1)d_qs.$$

Since $g(t,0) \not\equiv 0$, we also get $T_2h \in K_h$. Thus, the condition $(i)$ of Lemma 3 holds. Next, we will indicate that $(ii)$ of Lemma 3 is still satisfied. For $t \in [0,1]$, $u,v \in K$, from $(H_4)$,

$$\begin{aligned} T_1(u,v)(t) &= \frac{\beta t^{\alpha-1}}{[\alpha-1]_q(1-\beta\eta^{\alpha-2})}\int_0^1 H(\eta,qs)f(s,u(s),v(s))d_qs + \int_0^1 G(t,qs)f(s,u(s),v(s))d_qs \\ &\geq \frac{\sigma\beta t^{\alpha-1}}{[\alpha-1]_q(1-\beta\eta^{\alpha-2})}\int_0^1 H(\eta,qs)g(s,u(s))d_qs + \sigma\int_0^1 G(t,qs)g(s,u(s))d_qs \\ &= \sigma T_2u(t). \end{aligned}$$

Then, $T_1(u,v) \geq \sigma T_2u$ for $u,v \in K$. Therefore, by Lemma 3, we have: $u_0, v_0 \in K_h$ and $\tau \in (0,1)$ satisfying $\tau v_0 \leq u_0 < v_0$, $u_0 \leq T_1(u_0,v_0) + T_2u_0 \leq T_1(v_0,u_0) + T_2v_0 \leq v_0$; the equation $T_1(u,u) + T_2u = u$ has a unique solution $u^*$ in $K_h$; for $x_0, y_0 \in K_h$, set

$$x_n = T_1(x_{n-1},y_{n-1}) + T_2x_{n-1},\ y_n = T_1(y_{n-1},x_{n-1}) + T_2y_{n-1},\ n = 1,2,\ldots,$$

one obtains $x_n \to u^*$, $y_n \to u^*$ as $n \to \infty$. Namely,

$$\begin{aligned} u_0(t) &\leq \frac{\beta t^{\alpha-1}}{[\alpha-1]_q(1-\beta\eta^{\alpha-2})}\int_0^1 H(\eta,qs)[f(s,u_0(s),v_0(s)) + g(s,u_0(s))]d_qs \\ &\quad + \int_0^1 G(t,qs)[f(s,u_0(s),v_0(s)) + g(s,u_0(s))]d_qs,\ t \in [0,1], \end{aligned}$$

$$\begin{aligned} v_0(t) &\geq \frac{\beta t^{\alpha-1}}{[\alpha-1]_q(1-\beta\eta^{\alpha-2})}\int_0^1 H(\eta,qs)[f(s,v_0(s),u_0(s)) + g(s,v_0(s))]d_qs \\ &\quad + \int_0^1 G(t,qs)[f(s,v_0(s),u_0(s)) + g(s,v_0(s))]d_qs,\ t \in [0,1]; \end{aligned}$$

Label (1) has a unique positive solution $u^* \in K_h$; for $x_0, y_0 \in K_h$, the sequences

$$\begin{aligned} x_{n+1}(t) &= \frac{\beta t^{\alpha-1}}{[\alpha-1]_q(1-\beta\eta^{\alpha-2})}\int_0^1 H(\eta,qs)[f(s,x_n(s),y_n(s))+g(s,x_n(s))]d_qs \\ &+ \int_0^1 G(t,qs)[f(s,x_n(s),y_n(s))+g(s,x_n(s))]d_qs,\ n=1,2,\ldots, \end{aligned}$$

$$\begin{aligned} y_{n+1}(t) &= \frac{\beta t^{\alpha-1}}{[\alpha-1]_q(1-\beta\eta^{\alpha-2})}\int_0^1 H(\eta,qs)[f(s,y_n(s),x_n(s))+g(s,y_n(s))]d_qs \\ &+ \int_0^1 G(t,qs)[f(s,y_n(s),x_n(s))+g(s,y_n(s))]d_qs,\ n=1,2,\ldots \end{aligned}$$

satisfy $\|x_n - u^*\| \to 0$, $\|y_n - u^*\| \to 0$ as $n \to \infty$. □

**Theorem 2.** *Let $(H_1)$, $(H_2)$ and the following conditions be satisfied:*

$(H_5)$ *for $t \in [0,1], \lambda \in (0,1), u \geq 0$, there is $\gamma \in (0,1)$ such that $g(t,\lambda u) \geq \lambda^\gamma g(t,u)$ and $f(t,\lambda u,\lambda^{-1}v) \geq \lambda f(t,u,v)$ for $t \in [0,1], \lambda \in (0,1), u,v \geq 0$;*
$(H_6)$ *$f(t,0,1) \not\equiv 0$ for $t \in [0,1]$, and there is $\sigma > 0$ satisfying $f(t,u,v) \leq \sigma g(t,u)$, $\forall t \in [0,1], u,v \geq 0$.*

*Then:*

*(a) there is $u_0, v_0 \in P_h$ and $\tau \in (0,1)$ such that $\tau v_0 \leq u_0 < v_0$ and*

$$\begin{aligned} u_0(t) &\leq \frac{\beta t^{\alpha-1}}{[\alpha-1]_q(1-\beta\eta^{\alpha-2})}\int_0^1 H(\eta,qs)[f(s,u_0(s),v_0(s))+g(s,u_0(s))]d_qs \\ &+ \int_0^1 G(t,qs)[f(s,u_0(s),v_0(s))+g(s,u_0(s))]d_qs,\ t\in[0,1], \end{aligned}$$

$$\begin{aligned} v_0(t) &\geq \frac{\beta t^{\alpha-1}}{[\alpha-1]_q(1-\beta\eta^{\alpha-2})}\int_0^1 H(\eta,qs)[f(s,v_0(s),u_0(s))+g(s,v_0(s))]d_qs \\ &+ \int_0^1 G(t,qs)[f(s,v_0(s),u_0(s))+g(s,v_0(s))]d_qs,\ t\in[0,1], \end{aligned}$$

*where $h(t) = t^{\alpha-1}$ and $G(t,qs)$, $H(t,qs)$ are defined as in Lemma 5;*

*(b) BVP (1) has a unique positive solution $u^* \in K_h$;*
*(c) for any $x_0, y_0 \in K_h$, set*

$$\begin{aligned} x_{n+1}(t) &= \frac{\beta t^{\alpha-1}}{[\alpha-1]_q(1-\beta\eta^{\alpha-2})}\int_0^1 H(\eta,qs)[f(s,x_n(s),y_n(s))+g(s,x_n(s))]d_qs \\ &+ \int_0^1 G(t,qs)[f(s,x_n(s),y_n(s))+g(s,x_n(s))]d_qs,\ n=1,2,\ldots, \end{aligned}$$

$$\begin{aligned} y_{n+1}(t) &= \frac{\beta t^{\alpha-1}}{[\alpha-1]_q(1-\beta\eta^{\alpha-2})}\int_0^1 H(\eta,qs)[f(s,y_n(s),x_n(s))+g(s,y_n(s))]d_qs \\ &+ \int_0^1 G(t,qs)[f(s,y_n(s),x_n(s))+g(s,y_n(s))]d_qs,\ n=1,2,\ldots, \end{aligned}$$

*and we get $\|x_n - u^*\| \to 0$, $\|y_n - u^*\| \to 0$ as $n \to \infty$.*

**Proof.** We also consider two operators $T_1, T_2$. Given in the proof of Theorem 1, it has been shown that $T_1 : K \times K \to K$ is mixed monotone and $T_2 : K \to K$ is increasing. By $(H_5)$,

$$T_1(\lambda u, \lambda^{-1}v) \geq \lambda T_1(u,v),\ T_2(\lambda u) \geq \lambda^{\gamma} T_2 u,\ \lambda \in (0,1), u, v \in K.$$

From $(H_2), (H_6)$,

$$g(s,0) \geq \frac{1}{\sigma} f(s,0,1),\ f(s,1,0) \geq f(s,0,1),\ s \in [0,1].$$

Since $f(t,0,1) \not\equiv 0$, we obtain

$$\int_0^1 f(s,1,0)d_q s \geq \int_0^1 f(s,0,1)d_q s > 0,\ \int_0^1 g(s,1)d_q s \geq \int_0^1 g(s,0)d_q s \geq \frac{1}{\sigma}\int_0^1 f(s,0,1)d_q s > 0,$$

so

$$\begin{aligned} &\left(\tfrac{1}{\Gamma_q(\alpha)} + \tfrac{\beta}{(1-\beta\eta^{\alpha-2})\Gamma_q(\alpha)}\right) \int_0^1 (1-qs)^{(\alpha-2)} f(s,1,0)d_q s \\ \geq\ & \tfrac{1}{\Gamma_q(\alpha)} \int_0^1 \left[(1-qs)^{(\alpha-2)} - (1-qs)^{(\alpha-1)}\right] f(s,0,1)d_q s > 0, \end{aligned}$$

and

$$\begin{aligned} &\left(\tfrac{1}{\Gamma_q(\alpha)} + \tfrac{\beta}{(1-\beta\eta^{\alpha-2})\Gamma_q(\alpha)}\right) \int_0^1 (1-qs)^{(\alpha-2)} g(s,1)d_q s \\ \geq\ & \tfrac{1}{\Gamma_q(\alpha)} \int_0^1 \left[(1-qs)^{(\alpha-2)} - (1-qs)^{(\alpha-1)}\right] g(s,0)d_q s > 0. \end{aligned}$$

It can easily prove that $T_1(h,h),\ T_2 h \in K_h$. Furthermore, by $(H_6)$,

$$\begin{aligned} T_1(u,v)(t) &= \frac{\beta t^{\alpha-1}}{[\alpha-1]_q(1-\beta\eta^{\alpha-2})}\int_0^1 H(\eta,qs)f(s,u(s),v(s))d_q s + \int_0^1 G(t,qs)f(s,u(s),v(s))d_q s \\ &\leq \frac{\sigma\beta t^{\alpha-1}}{[\alpha-1]_q(1-\beta\eta^{\alpha-2})}\int_0^1 H(\eta,qs)g(s,u(s))d_q s + \sigma\int_0^1 G(t,qs)g(s,u(s))d_q s \\ &= \sigma T_2 u(t). \end{aligned}$$

Hence, $T_1(u,v) \leq T_2 u$, for $u, v \in K$. By Lemma 4, we can claim: there are $u_0, v_0 \in P_h$ and $\tau \in (0,1)$ satisfying $\tau v_0 \leq u_0 < v_0,\ u_0 \leq T_1(u_0,v_0) + T_2 u_0 \leq T_1(v_0,u_0) + T_2 v_0 \leq v_0$; the equation $T_1(u,u) + T_2 u = u$ has a unique solution $u^*$ in $K_h$; for $x_0, y_0 \in K_h$, set

$$x_n = T_1(x_{n-1}, y_{n-1}) + T_2 x_{n-1},\ y_n = T_1(y_{n-1}, x_{n-1}) + T_2 y_{n-1},\ n = 1,2,\ldots,$$

one has $x_n \to u^*,\ y_n \to u^*$ as $n \to \infty$. Namely,

$$\begin{aligned} u_0(t) \leq\ & \frac{\beta t^{\alpha-1}}{[\alpha-1]_q(1-\beta\eta^{\alpha-2})}\int_0^1 H(\eta,qs)[f(s,u_0(s),v_0(s)) + g(s,u_0(s))]d_q s \\ &+ \int_0^1 G(t,qs)[f(s,u_0(s),v_0(s)) + g(s,u_0(s))]d_q s,\ t \in [0,1], \end{aligned}$$

$$\begin{aligned} v_0(t) \geq\ & \frac{\beta t^{\alpha-1}}{[\alpha-1]_q(1-\beta\eta^{\alpha-2})}\int_0^1 H(\eta,qs)[f(s,v_0(s),u_0(s)) + g(s,v_0(s))]d_q s \\ &+ \int_0^1 G(t,qs)[f(s,v_0(s),u_0(s)) + g(s,v_0(s))]d_q s,\ t \in [0,1]; \end{aligned}$$

Label (1) has a unique positive solution $u^* \in K_h$; for $x_0, y_0 \in P_h$, the sequences

$$\begin{aligned} x_{n+1}(t) &= \frac{\beta t^{\alpha-1}}{[\alpha-1]_q(1-\beta\eta^{\alpha-2})}\int_0^1 H(\eta,qs)[f(s,x_n(s),y_n(s))+g(s,x_n(s))]d_qs \\ &+\int_0^1 G(t,qs)[f(s,x_n(s),y_n(s))+g(s,x_n(s))]d_qs,\ n=1,2,\ldots, \end{aligned}$$

$$\begin{aligned} y_{n+1}(t) &= \frac{\beta t^{\alpha-1}}{[\alpha-1]_q(1-\beta\eta^{\alpha-2})}\int_0^1 H(\eta,qs)[f(s,y_n(s),x_n(s))+g(s,y_n(s))]d_qs \\ &+\int_0^1 G(t,qs)[f(s,y_n(s),x_n(s))+g(s,y_n(s))]d_qs,\ n=1,2,\ldots \end{aligned}$$

satisfy $\|x_n-u^*\|\to 0$, $\|y_n-u^*\|\to 0$ as $n\to\infty$. □

In the sequel, we consider special cases of Label (1) with $g\equiv 0$ or $f\equiv 0$. Similar to the proofs of Theorems 1 and 2 and according to Remark 1, we can draw the following conclusions:

**Corollary 1.** *Assume $f$ satisfies $(H_1)-(H_4)$ and $f(t,0,1)\not\equiv 0$, for $t\in[0,1]$. Then: $(a)$ there are $u_0, v_0 \in K_h$ and $\tau\in(0,1)$ such that $\tau v_0 \le u_0 < v_0$ and*

$$\begin{aligned} u_0(t) &\le \frac{\beta t^{\alpha-1}}{[\alpha-1]_q(1-\beta\eta^{\alpha-2})}\int_0^1 H(\eta,qs)f(s,u_0(s),v_0(s))d_qs \\ &+\int_0^1 G(t,qs)f(s,u_0(s),v_0(s))d_qs,\ t\in[0,1], \end{aligned}$$

$$\begin{aligned} v_0(t) &\ge \frac{\beta t^{\alpha-1}}{[\alpha-1]_q(1-\beta\eta^{\alpha-2})}\int_0^1 H(\eta,qs)f(s,v_0(s),u_0(s))d_qs \\ &+\int_0^1 G(t,qs)f(s,v_0(s),u_0(s))d_qs,\ t\in[0,1], \end{aligned}$$

*where $h(t)=t^{\alpha-1}$ and $G(t,qs)$, $H(t,qs)$ are given as in Lemma 5; $(b)$ the following BVP*

$$\begin{cases} D_q^\alpha x(t)+f(t,x(t),x(t))=0, & 0<t<1,\ 2<\alpha<3, \\ x(0)=D_qx(0)=0,\ D_qx(1)=\beta D_qx(\eta), \end{cases} \tag{13}$$

*has a unique positive solution $u^*\in K_h$; $(c)$ for $x_0,y_0\in K_h$, set*

$$\begin{aligned} x_{n+1}(t) &= \frac{\beta t^{\alpha-1}}{[\alpha-1]_q(1-\beta\eta^{\alpha-2})}\int_0^1 H(\eta,qs)f(s,x_n(s),y_n(s))d_qs \\ &+\int_0^1 G(t,qs)f(s,x_n(s),y_n(s))d_qs,\ n=1,2,\ldots, \end{aligned}$$

$$\begin{aligned} y_{n+1}(t) &= \frac{\beta t^{\alpha-1}}{[\alpha-1]_q(1-\beta\eta^{\alpha-2})}\int_0^1 H(\eta,qs)f(s,y_n(s),x_n(s))d_qs \\ &+\int_0^1 G(t,qs)f(s,y_n(s),x_n(s))d_qs,\ n=1,2,\ldots, \end{aligned}$$

*and we get $\|x_n-u^*\|\to 0$, $\|y_n-u^*\|\to 0$ as $n\to\infty$.*

**Corollary 2.** *Assume $g$ satisfies $(H_1),(H_2)$ and $(H_5),(H_6)$, $g(t,0)\not\equiv 0$, for $t\in[0,1]$. Then:*
*(a) there are $u_0,v_0\in K_h$ and $\tau\in(0,1)$ such that $\tau v_0\le u_0<v_0$ and*

$$\begin{aligned} u_0(t) &\le \frac{\beta t^{\alpha-1}}{[\alpha-1]_q(1-\beta\eta^{\alpha-2})}\int_0^1 H(\eta,qs)g(s,u_0(s))d_qs \\ &\quad+\int_0^1 G(t,qs)g(s,u_0(s)),\ t\in[0,1], \end{aligned}$$

$$\begin{aligned} v_0(t) &\ge \frac{\beta t^{\alpha-1}}{[\alpha-1]_q(1-\beta\eta^{\alpha-2})}\int_0^1 H(\eta,qs)g(s,v_0(s))d_qs \\ &\quad+\int_0^1 G(t,qs)g(s,v_0(s))d_qs,\ t\in[0,1], \end{aligned}$$

*where $h(t)=t^{\alpha-1}$ and $G(t,qs)$, $H(t,qs)$ are given as in Lemma 5; (b) the following problem*

$$\begin{cases} D_q^{\alpha}x(t)+g(t,x(t))=0, & 0<t<1,\ 2<\alpha<3, \\ x(0)=D_qx(0)=0,\ D_qx(1)=\beta D_qx(\eta), \end{cases} \tag{14}$$

*has a unique positive solution $u^*\in K_h$; (c) for $x_0,y_0\in K_h$, set*

$$\begin{aligned} x_{n+1}(t) &= \frac{\beta t^{\alpha-1}}{[\alpha-1]_q(1-\beta\eta^{\alpha-2})}\int_0^1 H(\eta,qs)g(s,x_n(s))d_qs \\ &\quad+\int_0^1 G(t,qs)g(s,x_n(s))d_qs,\ n=1,2,\dots, \end{aligned}$$

$$\begin{aligned} y_{n+1}(t) &= \frac{\beta t^{\alpha-1}}{[\alpha-1]_q(1-\beta\eta^{\alpha-2})}\int_0^1 H(\eta,qs)g(s,y_n(s))d_qs \\ &\quad+\int_0^1 G(t,qs)g(s,y_n(s))d_qs,\ n=1,2,\dots, \end{aligned}$$

*and we obtain $\|x_n-u^*\|\to 0$, $\|y_n-u^*\|\to 0$ as $n\to\infty$.*

**Remark 3.** *In literature, we have not found such results as Theorems 1 and 2, and Corollaries 1 and 2 on fractional q-difference equation boundary value problems. The used methods in literature were not fixed point theorems for mixed monotone operators. Thus, our method is different from previous ones. We should point out that we can not only give the existence and uniqueness of solutions but also make an iteration to approximate the unique solution.*

## 4. Examples

**Example 1.** *We consider a problem:*

$$\begin{cases} D_q^{\frac{5}{2}}u(t)+u^{\frac{1}{5}}(t)+[u(t)+4]^{-\frac{1}{3}}+\frac{u(t)}{2+u(t)}t^3+3a=0, & t\in(0,1), \\ u(0)=D_qu(0)=0,\ D_qu(1)=\frac{1}{2}D_qu(\frac{1}{2}), \end{cases} \tag{15}$$

*where $q=\frac{1}{2}$, $\alpha=\frac{5}{2}$, $\beta=\eta=\frac{1}{2}$, $a>0$. Take $0<b<a$ and let*

$$f(t,u,v)=u^{\frac{1}{5}}+[v+4]^{-\frac{1}{3}}+b,\ g(t,u)=\frac{u}{2+u}t^3+3a-b,\ \gamma=\frac{1}{3}.$$

*Then, $f:[0,1]\times[0,+\infty)\times[0,+\infty)\to[0,+\infty)$ and $g:[0,1]\times[0,+\infty)\to[0,+\infty)$ are continuous, $g(t,0)=3a-b>0$. Furthermore, $f(t,u,v)$ is increasing relative to $u$ for fixed $t\in[0,1]$ and $v\in[0,+\infty)$,* *decreasing relative to v for fixed $t \in [0,1]$ and $u \in [0,+\infty)$, $g(t,u)$ is increasing relative to u for fixed $t \in [0,1]$. On the other hand, for $\lambda \in (0,1), t \in [0,1], u,v \geq 0$,*

$$g(t,\lambda u) = \frac{\lambda u(t)}{2+\lambda u(t)}t^3 + 3a - b \geq \frac{\lambda u(t)}{2+u(t)}t^3 + \lambda(3a-b) = \lambda g(t,u),$$

*and*

$$f(t,\lambda u,\lambda^{-1}v) = \lambda^{\frac{1}{5}}u^{\frac{1}{5}} + \lambda^{\frac{1}{3}}[v+4\lambda]^{-\frac{1}{3}} + b \geq \lambda^{\frac{1}{3}}\left\{u^{\frac{1}{5}} + [v+4]^{-\frac{1}{3}} + b\right\} = \lambda^{\gamma}f(t,u,v).$$

*Then, $(H_1)$–$(H_3)$ holds. Moreover, taking $\sigma \in (0, \frac{b}{3a-b}]$, one has*

$$f(t,u,v) = u^{\frac{1}{5}} + [v+4]^{-\frac{1}{3}} + b \geq b = \frac{b}{3a-b}\cdot(3a-b) \geq \sigma\left[\frac{u}{2+u}t^3 + 3a - b\right] = \sigma g(t,u),$$

*then $(H_4)$ holds. By means of Theorem 1, problem (15) has a unique positive solution $u^* \in K_h$, where $h(t) = t^{\frac{3}{2}}, t \in [0,1]$.*

**Example 2.** *In Example 4.1, we replace the nonlinear term $u^{\frac{1}{5}}(t) + [u(t)+4]^{-\frac{1}{3}} + \frac{u(t)}{2+u(t)}t^3 + 3a$ by*

$$\sin^2 t + u^{\frac{1}{3}}(t) + \frac{1}{2+u(t)} + \frac{u(t)}{1+u(t)} + 3.$$

*By Theorem 2, we can also show that problem (4.1) has a unique positive solution $u^* \in K_h$, where $h(t) = t^{\frac{3}{2}}, t \in [0,1]$. In fact, let*

$$f(t,u,v) = \sin^2 t + \frac{1}{2+v} + \frac{u}{1+u},\ g(t,u) = u^{\frac{1}{3}} + 3,\ \gamma = \frac{1}{3}.$$

*It is easy to check that $(H_1), (H_2)$ hold. We only show $(H_5), (H_6)$ are satisfied. For $\lambda \in (0,1), t \in [0,1], u,v \geq 0$,*

$$g(t,\lambda u) = \lambda^{\frac{1}{3}}u^{\frac{1}{3}} + 3 \geq \lambda^{\frac{1}{3}}[u^{\frac{1}{3}} + 3] = \lambda^{\gamma}g(t,u),$$

*and*

$$f(t,\lambda u,\lambda^{-1}v) = \sin^2 t + \frac{1}{2+\lambda^{-1}v} + \frac{\lambda u}{1+\lambda u} \geq \sin^2 t + \frac{\lambda}{2+v} + \frac{\lambda u}{1+u} \geq \lambda f(t,u,v).$$

*Furthermore, $f(t,0,1) = \sin^2 t + \frac{1}{3} \not\equiv 0$ and*

$$f(t,u,v) \leq 3 \leq u^{\frac{1}{3}} + 3 = g(t,u).$$

*Take $\sigma \in [1,\infty)$ and then $(H_5), (H_6)$ hold.*

**Remark 4.** *From Theorems 1 and 2 and Examples 1 and 2, we see that many boundary value problems can be studied by our methods under mixed monotone conditions. We can find that there are many functions that satisfy our conditions. In some works, the nonlinear terms required were super-linearity, sub-linearity or boundness, which guarantee existence of solutions, but the uniqueness has not been obtained.*

## 5. Conclusions

In this article, we investigate a fractional $q$-difference equation with three-point boundary conditions (1). We obtain the existence and uniqueness of positive solutions in a special $K_h$, where $h(t) = t^{\alpha-1}$. The used methods here are some theorems for operator equation $T_1(x,x) + T_2x = x$, where $T_1$ is a mixed monotone operator and $T_2$ is an increasing operator. Our methods are new to fractional $q$-difference equation boundary value problems. Thus, we can claim that we give an alternative answer to fractional problems and our results are very limited in the literature. Finally, two interesting examples are presented to illustrate the main results. We should note that, to get the uniqueness, we must need the conditions of mixed monotonicity and monotonicity for nonlinear terms.

**Funding:** This research received no external funding.

**Acknowledgments:** The research was supported by the Youth Science Foundation of China (11201272).

**Conflicts of Interest:** The author declares no conflict of interest.

## References

1. Al-Salam, W.A. Some fractional $q$-integrals and $q$-derivatives. *Proc. Edinb. Math. Soc.* **1966**, *15*, 135–140. [CrossRef]
2. Agarwal, R.P. Certain fractional $q$-integrals and $q$-derivatives. *Proc. Camb. Philos. Soc.* **1969**, *66*, 365–370. [CrossRef]
3. Annaby, M.H.; Mansour, Z.S. *$q$-Fractional Calculus and Equations, Lecture Notes in Mathematics*; Springer: Berlin, Germany, 2012; Volume 2056.
4. Ferreira, R.A.C. Nontrivial solutions for fractional $q$-difference boundary value problems. *Electron. J. Qual. Theory Differ. Equ.* **2010**, *70*, 1–10. [CrossRef]
5. Jackson, F.H. On $q$-functions and a certain difference operator. *Trans. R. Soc. Edinb.* **1908**, *46*, 253–281. [CrossRef]
6. Jackson, F.H. On $q$-definite integrals. *Quart. J. Pure Appl. Math.* **1910**, *41*, 193–203.
7. Purohit, S.D. A new class of multivalently analytic functions associated with fractional $q$-calculus operators. *Frac. Differ. Calc.* **2012**, *2*, 129–138. [CrossRef]
8. Rajković, P.M.; Marinković, S.D.; Stanković, M.S. Fractional integrals and derivatives in $q$-calculus. *Appl. Anal. Discret. Math.* **2007**, *1*, 311–323.
9. Ahmad, B.; Etemad, S.; Ettefagh, M.; Rezapour, S. On the existence of solutions for fractional $q$-difference inclusions with $q$-antiperiodic boundary conditions. *Bull. Math. Soc. Sci. Math. Roum.* **2016**, *59*, 119–134.
10. Ahmad, B.; Ntouyas, S.K. Existence of solutions for nonlinear fractional $q$-difference inclusions with nonlocal Robin (separated) conditions. *Mediterr. J. Math.* **2013**, *10*, 1333–1351. [CrossRef]
11. Ahmad, B.; Ntouyas, S.K.; Tariboon, J.; Alsaedi, A.; Alsulami, H.H. Impulsive fractional $q$-integro-difference equations with separated boundary conditions. *Appl. Math. Comput.* **2016**, *281*, 199–213. [CrossRef]
12. Almeida, R.; Martins, N. Existence results for fractional $q$-difference equations of order $\alpha \in ]2,3[$ with three-point boundary conditions. *Commun. Nonlinear Sci. Numer. Simul.* **2014**, *19*, 1675–1685. [CrossRef]
13. Li, X.; Han, Z.; Li, X. Boundary value problems of fractional $q$-difference Schröinger equations. *Appl. Math. Lett.* **2015**, *46*, 100–105. [CrossRef]
14. Li, X.; Han, Z.; Sun, S.; Zhao, P. Existence of solutions for fractional $q$-difference equation with mixed nonlinear boundary conditions. *Adv. Differ. Equ.* **2014**. [CrossRef]
15. Liang, S.; Zhang, J. Existence and uniqueness of positive solutions for three-point boundary value problem with fractional $q$-differences. *J. Appl. Math. Comput.* **2012**, *40*, 277–288. [CrossRef]
16. Marin, M. Weak solutions in elasticity of dipolar porous materials. *Math. Probl. Eng.* **2008**. [CrossRef]
17. Marin, M.; Agarwal, R.P.; Mahmoud, S.R. Modeling a microstretch thermoelastic body with two temperatures. *Abstr. Appl. Anal.* **2013**. [CrossRef]
18. Marin, M. An approach of a heat-flux dependent theory for micropolar porous media. *Meccanica* **2016**, *51*, 1127–1133. [CrossRef]
19. Sriphanomwan, N.; Patanarapeelert, N.; Tariboon, J.; Sitthiwirattham, T. Existence results of nonlocal boundary value problems for nonlinear fractional $q$-integrodifference equations. *J. Nonlinear Funct. Anal.* **2017**, *2017*, 28.
20. Tariboon, J.; Ntouyas, S.K. Three-point boundary value problems for nonlinear second-order impulsive $q$-difference equations. *Adv. Differ. Equ.* **2014**, *2004*, 31. [CrossRef]
21. Thiramanus, P.; Tariboon, J. Nonlinear second-order $q$-difference equations with three-point boundary conditions. *Comput. Appl. Math.* **2014**, *33*, 385–397. [CrossRef]
22. Zhai, C.; Ren, J. Positive and negative solutions of a boundary value problem for a fractional $q$-difference equation. *Adv. Differ. Equ.* **2017**, *2017*, 82. [CrossRef]

23. Ren, J.; Zhai, C. A fractional $q$-difference equation with integral boundary conditions and comparison theorem. *Int. J. Nonlinear Sci. Numer. Simul.* **2017**, *18*, 575–583. [CrossRef]
24. Zhai, C.; Ren, J. The unique solution for a fractional $q$-difference equation with three-point boundary conditions. *Indag. Math. New Ser* **2018**, *29*, 948–961. [CrossRef]
25. Yang, W. Positive solutions for three-point boundary value problem of nonlinear fractional $q$-difference equation. *Kyungpook Math. J.* **2016**, *56*, 419–430. [CrossRef]
26. Zhai, C.; Hao, M. Mixed monotone operator methods for the existence and uniqueness of positive solutions to Riemann–Liouville fractional differential equation boundary value problems. *Bound. Value Probl.* **2013**, *2013*, 85. [CrossRef]
27. Zhai, C.; Hao, M. Fixed point theorems for mixed monotone operators with perturbation and applications to fractional differential equation boundary value problems. *Nonlinear Anal.* **2012**, *75*, 2542–2551. [CrossRef]
28. Zhai, C.; Yan, W.; Yang, C. A sum operator method for the existence and uniqueness of positive solutions to Riemann–Liouville fractional differential equation boundary value problems. *Commun. Nonlinear Sci. Numer. Simul.* **2013**, *18*, 858–866. [CrossRef]
29. Zhai, C.; Xu, L. Properties of positive solutions to a class of four-point boundary value problem of Caputo fractional differential equations with a parameter. *Commun. Nonlinear Sci. Numer Simul.* **2014**, *19*, 2820–2827. [CrossRef]
30. Zhai, C.; Wang, L. $\varphi$-($h$, $e$)-concave operators and applications. *J. Math. Anal. Appl.* **2017**, *454*, 571–584. [CrossRef]
31. Zhai, C.; Yang, C.; Zhang, X. Positive solutions for nonlinear operator equations and several classes of applications. *Math. Z.* **2010**, *266*, 43–63. [CrossRef]

MDPI
St. Alban-Anlage 66
4052 Basel
Switzerland
Tel. +41 61 683 77 34
Fax +41 61 302 89 18
www.mdpi.com

*Symmetry* Editorial Office
E-mail: symmetry@mdpi.com
www.mdpi.com/journal/symmetry

www.ingramcontent.com/pod-product-compliance
Lightning Source LLC
Chambersburg PA
CBHW041215220326
41597CB00033BA/5977
* 9 7 8 3 0 3 9 2 1 7 3 2 8 *